# Molecular Biology and Genomics

# Molecular Biology and Genomics

### CORNEL MÜLHARDT

Translated by
E.W. Beese, M.D.
Munich, Germany

The Experimenter Series

AMSTERDAM • BOSTON • HEIDELBERG • LONDON
NEW YORK • OXFORD • PARIS • SAN DIEGO
SAN FRANCISCO • SINGAPORE • SYDNEY • TOKYO

Academic Press is an imprint of Elsevier

ELSEVIER

Academic Press is an imprint of Elsevier
30 Corporate Drive, Suite 400, Burlington, MA 01803, USA
525 B Street, Suite 1900, San Diego, California 92101-4495, USA
84 Theobald's Road, London WC1X 8RR, UK

This book is printed on acid-free paper.

**Library of Congress Cataloging-in-Publication Data**
Mülhardt, Cornel.
  [Experimentator. English]
  Molecular biology and genomics / Cornel Mulhardt ; translated by E.W. Beese.
    p. ; cm. – (The experimenter series)
  Translation of: Experimentator : Molekularbiologie. 4th ed.
  Includes bibliographical references and index.
  ISBN-13: 978-0-12-088546-6 (alk. paper)
  ISBN-10: 0-12-088546-8 (alk. paper)
  1.  Molecular biology—Research—Methodology.    2.  Genomics—Research—Methodology.
  I.  Title.    II.  Series.
  [DNLM: 1.  Genomics.    2.  Molecular Biology.    QU 450 M956m 2007a]
  QH506.M84713 2007
  572.8—dc22
                                                                    2006025989

**British Library Cataloguing-in-Publication Data**
A catalogue record for this book is available from the British Library.

ISBN 13: 978-0-12-088546-6
ISBN 10: 0-12-088546-8

For information on all Academic Press publications
visit our Web site at www.books.elsevier.com

Printed in the United States of America
06  07  08  09  10  11    9  8  7  6  5  4  3  2  1

# Contents

# Foreword

It is hard to believe that we are now already producing the fourth edition within such a brief time period of only five years. In this edition we have once again increased its dimensions; here, however, the basics have remained more or less unchanged, although the number of "specialties" covered continues to increase. This time, for instance, minisequencing, pyrosequencing, and RNA-interference have been included. In addition, the time has now come to write a short introduction concerning the planning of careers for younger researchers—after all, one does not live from the nobility of research alone. Numerous, minute corrections have had to be performed as well, since nothing is perfect. This book as well, in spite of the enormous amount of work which has already been carried out in its production, nevertheless continues to demonstrate room for improvement. One can do whatever is possible, but entropy always seems to win in the end! However, there are enough typographical errors still to be found here as well and we are therefore thankful for every piece of advice which we might receive!

# Foreword to the First Edition

Do you actually want to read a foreword?

Read the preface and concluding comments in Hubert Rehms' "Experimentator: Proteinbiochemie" ("Protein Biochemistry" in the "Experimentator" series from Spektrum Verlag) and, finally, after getting a taste for them, read those statements in Siegfried Bars' "Forschen auf Deutsch: Der Machiavelli für Forscher—und solche, die es werden wollen" ("How to Do Research in German: The Machiavellian Guideline for Researchers—and for Those Who Would Like to Become Such" from Verlag Harri Deutsch). Even if the beginner is not inclined to believe so, the wind actually blows as has been described here. The most successful researcher, namely, is not to be distinguished by any pioneering accomplishments, but rather through his/her clever actions in scenes involving publications and in his/her large number of friends. The goal is to keep the phase as short as possible between one's time as a student and that during which one is authorized to function as a professor. Once one has finally reached this goal, however, the research is then left to the others; in the same way in which a successful military strategist no longer reaches for his/her own saber, but instead only carries out his/her advances theoretically through the actions of ambitious military leaders. At this stage, one merely has an indirect relationship to research. Should you not have given up after completing this reading material, you may possibly demonstrate the necessary composure to dedicate yourself to research in molecular biology. This composure, namely, will prove to be quite essential in order to avoid being driven crazy from the endless series of failures which can be expected in the daily life which is to be experienced in the laboratory.

"Don't let yourself be discouraged [...]! The others are also working themselves to a frazzle without any success; it is quite normal that no results are to be observed in the beginning. Just keep it up!—Or take up a reasonable occupation."

With this in regard ...
Cornel Mülhardt

# Acknowledgments

I have always considered the effusive acknowledgments at the beginning of a book by the author directed to friends and relatives to represent the boring performance of one's duties. Today, I am finally aware that it truly is a matter near one's heart to express one's thanks for having survived a difficult phase of life. For their contribution toward the success of this book, I would like to thank Marc Bedoucha, Igor Bendik, Hans-Georg Breitinger, Dorothee Foernzler, Christophe Grundschober, Andreas Humeny, Frank Kirchhoff, Nina Meier, Nicoletta Milani, and Ruthild Weber. My special thanks go to Cord-Michael Becker, who has continuously supported me over the years and earned my warmest thanks.

Finally, there is also Spektrum Verlag and the project planner Ulrich Moltmann, who came up with the idea for such a book and thereby enabled me to fulfill a very old dream, Jutta Hoffmann, without whose occasional, friendly pressure I would certainly have given up before reaching the end, and Bettina Saglio, who must always struggle with my newest versions. If God does exist, may He/She bless them with more punctual authors.

It is known that nothing is so good that it cannot be improved upon. That is especially the case for such a book as this, a reason that would make me very happy to receive any constructive criticism.

# 1  What Is Molecular Biology?

*Gib nach dem löblichen
von vorn Schöpfung anzufangen!
Zu raschem Wirken sei bereit!
Da regst du dich nach ewigen Normen
durch tausend, abertausend Formen,
und bis zum Menschen hast du Zeit.*

*Yield to the noble inspiration
To try each process of creation,
And don't be scared if things move fast;
Thus growing by eternal norms,
You'll pass through many a thousand forms,
Emerging as a man at last.*

–Goethe, *Faust*[1]

Pursuits in the field of molecular biology involve genetic engineering and techniques such as cloning, and they may be interpreted as being quite adventuresome and almost divine. As a molecular biologist, declaring how you spend your days may garner boundless admiration from some people and disapproval from others. Consequently, you must consider carefully with whom you are speaking before describing your activities. You should not mention how many problems and how much frustration you must endure daily, because one group will be disillusioned, and the other will inevitably ask why you continue to work in this field.

I confess that I like the debates about whether we should clone humans, although the discussion often is not engaged at a scientific level. One individual makes a stupid suggestion, and one half of the media and the nation explain why the cloning of humans may not be justified by any means. It seems much ado about nothing, and I cannot comprehend what advantage it would be for me to clone myself. Why should I wish to invest $75,000 for a small crybaby, whose only common ground with me is that it looks like me as I did 30 years ago, if I could instead come to a similar result in a classic manner for the price of a bouquet of flowers for my wife and a television-free evening?

This example reveals the problems that are associated with molecular biology. Incited by spectacular reports in the media, everybody has his or her own, usually quite extreme, ideas on the benefits or disasters related to this topic. In reality, the picture is deeply distressing. Molecular biology, also known as the *molecular world*, generally deals with minute quantities of chiefly clear, colorless solutions. There are no signs of ecstatic scholars who have gone wild in the cinematic setting of flickering, steaming, gaudy-colored liquids. The molecular world deals with molecules, evidence for the existence of which has been attempted in many textbooks, although there is generally little more than a fluorescing spot to be seen on the agarose gel. Every procedure seems to take 3 days, and no Nobel Prize is associated with any of them.

---

[1] This and the following quotations are from Goethe's *Faust* (3rd ed. dtv-Gesamtausgabe, München 1966) and from Goethe's *Faust* (John Shawcross, translator. Allen Wingate Publishers, 1959).

Molecular biology is primarily a process of voodoo—sometimes everything works, but usually nothing works. Very unusual parameters seem to play a role in the result of an experiment, which should represent the subject of the research itself—the last, great taboo of modern science. Based on the empirical data, I have come to the realization that the results of an experiment can be calculated as the quotient of air pressure and the remaining number of scribbled note pages in the drawer raised to the power of grandmother's dog. Until I have been able to verify this experimentally, however, I will continue to limit myself to the classic explanations based on mathematics, which have admittedly enabled this profession to progress quite far.

## 1.1 The Substrate of Molecular Biology: The Molecular World for Beginners

*War es ein Gott, der diese Zeichen schrieb,*
*die mir das innre Toben stillen,*
*das arme Herz mit Freude füllen*
*und mit geheimnisvollem Trieb*
*die Kräfte der Natur rings um mich her enthüllen?*

*Was it a God, who traced this sign,*
*Which calms my raging breast anew,*
*Brings joy to this poor heart of mine,*
*And by some impulse secret and divine*
*Unveils great Nature's labors to my view!*

Molecular biology is by no means the biology of molecules, which is better characterized by biochemistry. Molecular biology is the study of life as reflected in DNA. It is a small, closely knit world that differentiates itself from all other related fields, including zoology, botany, and protein biochemistry. Only a few individuals in the field of molecular biology are biologists, and those who are biologists would be most happy to deny it. If you are not a biologist, be assured that you now find yourself in the best of company and that the small amount of material needed to master this work can quickly be understood.

Molecular biology deals with nucleic acids, which come in two forms: deoxyribonucleic acid (DNA) and ribonucleic acid (RNA). The chemical differences between the two substances are minimal. They are both polymers that are made of four building blocks each, deoxynucleotides in DNA and nucleotides in RNA.

The nucleotides of RNA are composed of a base constituent (adenine, cytosine, guanine, or uracil), a glucose component (ribose), and a phosphoryl residue, whereby two nucleotides are connected to each other through phosphoglucose bonds. In this way, one nucleotide can be connected to the next, forming a long chain known as a *polynucleotide*, which is designated as RNA. DNA is formed in almost the same manner. The deoxynucleotides of DNA are composed of a base constituent (adenine, cytosine, guanine, or thymine), 2′-deoxyribose (instead of ribose), and a phosphoryl residue.

More details can be found in textbooks of biochemistry, cell biology, or genetics. I limit my comments on these topics to a few aspects that are significant for laboratory practice.

*Deoxynucleotide* is a tongue twister, which is avoided whenever possible. Researchers instead speak of *nucleotides* when using these substances in the laboratory, although they generally refer to the deoxy variant.

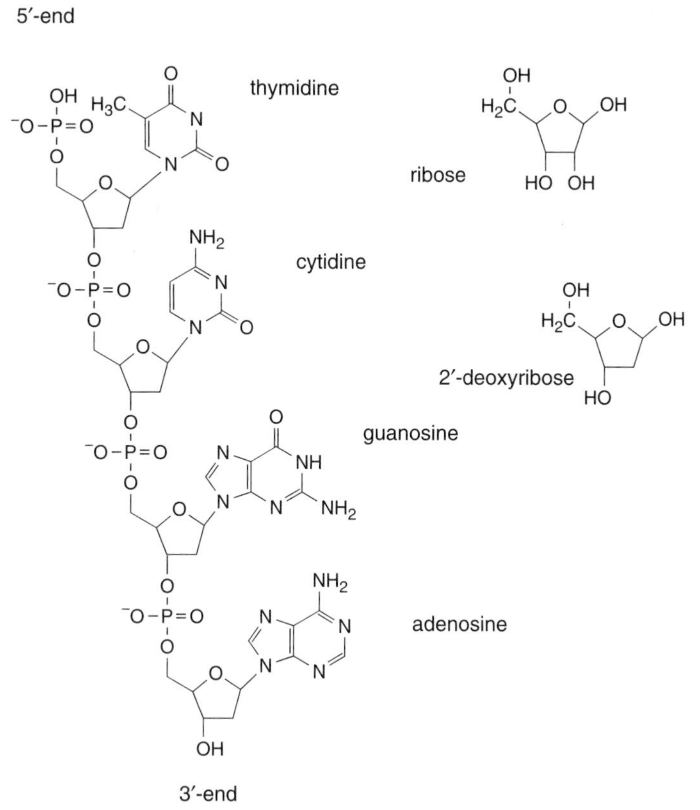

**Figure 1-1.** The four nucleotides and the way in which a DNA helix typically is portrayed.

The names of the nucleotides are based on the names of their bases: **adenosine** (A), **cytidine** (C), **guanosine** (G), **thymidine** (T), or **uridine** (U). Many confuse these bases with the nucleotides (Figure 1-1), although this rarely makes a difference in practice. The abbreviations A, C, G, T, and U have been used to immortalize all the DNA and RNA sequences known to the world.

The phosphoryl group of the nucleotides is connected to the 5′ carbon atom of the sugar component. The synthesis of nucleic acid begins with a nucleotide at whose 3′-OH group (i.e., the hydroxyl group at the 3′ carbon atom) a phosphodiester bond connects the phosphate group with the next nucleotide. Another nucleotide can be connected to the 3′-OH group of the nucleotide and so on. Synthesis proceeds in the 5′ to 3′ direction, because a 5′ end and a 3′ end exist at all times, and a new nucleotide is always added to the 3′ end. All known sequences are read from the 5′ end to the 3′ end, and attempts to alter this convention can only create chaos.

If no 3′-OH group exists, it is impossible to connect a new nucleotide. In nature, this situation is rarely observed, although it plays a large role in the sequencing performed in the laboratory. In addition to the four 2′-deoxynucleotides, 2′,3′-dideoxynucleotides are used in DNA synthesis. They can be connected to the 3′ end of a polynucleotide in the normal manner, but because a 3′-OH group is lacking, the DNA cannot be lengthened any further. This principle is essential for understanding sequencing, a subject that will be discussed in more detail later.

Each nucleotide or polynucleotide with a phosphoryl group at its 5′ end can be bonded (ligated) to the 3′ end of another polynucleotide with the aid of a DNA ligase. In this way, even larger DNA fragments can be bonded to one another. Without the phosphoryl group, nothing works, and elimination of a phosphoryl residue with a phosphatase can be used to help inhibit such ligations.

Two nucleotides bonded to one another are known as a *dinucleotide*, and three together are known as a *trinucleotide*. If more nucleotides are bonded together in a group, the structure is called an *oligonucleotide*, and if very many are bonded together, the entire structure is known as a *polynucleotide*. The boundary between oligonucleotides and polynucleotides is not defined precisely, but oligonucleotides usually are considered to contain fewer than 100 nucleotides.

Whether the structure is a mononucleotide, dinucleotide, oligonucleotide, or polynucleotide, the substance always represents a single molecule, because a covalent bond is formed between each of the initial molecules involved. The length of the chain formed in this way plays no role, because even a nucleic acid that is 3 million nucleotides long is a single molecule.

Polynucleotides have a remarkable and important characteristic: Their bases pair specifically with other bases. Cytosine always pairs with guanine, and adenine pairs with thymine (in DNA) or with uracil (in RNA); no other combinations are functional. The more bases that pair with one another, the more stable this combination becomes, because hydrogen bonds are formed between the bases of each pair, and the power of these bonds is cumulative. A sequence that pairs perfectly with another sequence is said to be *complementary*. The nucleic acid sequences of two complementary nucleic acids are completely different and pair in a mirror image form. The following example clarifies this arrangement:

5'-AGCTAAGACTTGTTC-3'

3'-TCGATTCTGAACAAG-5'

The orientation of both sequences proceeds in an opposite direction (reading from the 5' to 3' end), because the two chains must spatially be aligned in opposition for the pairing to function properly. In the normal 5' to 3' method of writing, the complementary sequences are read as follows:

5'-AGCTAAGACTTGTTC-3'

and

5'-GAACAAGTCTTAGCT-3'

The sequences appear to be so different that you must look very closely to see that they are complementary to one another, and you can see the value of the application of the 5' to 3' reading convention. A beginner can very quickly write a primer strand sequence of

3'-TCGATTCTGAACAAG-5'

as

5'-TCGATTCTGAACAAG-3'

so that the sequence, despite its normal appearance, looks completely different from that of the original sequence.

DNA typically is made up of two complementary chains, and RNA usually is made of a single chain. When pairs occur in RNA, they are found within the same chain and play a more significant role in the three-dimensional structure of the RNA, whereas double-stranded DNA must be considered a linear molecule.

Have you noticed the error? Double-stranded DNA does not consist of a single molecule; it is made up of a double-stranded structure with a chain and an opposing chain. The two chains can be separated from one another at any time if sufficient energy is supplied; this process is called *denaturation*. Ten seconds at 95°C (203°F) is sufficient for performing complete separation of both strands. If they join together again, a process known as *hybridization*, they are said to *anneal*. The entire procedure is almost arbitrarily reversible.

DNA is frequently called the *molecule of the life* by the yellow press. Why does such a very long, but quite boring molecule have such a lengthy name? Nature has optimally made use of the characteristics of nucleic acids to create a confusing diversity of life, and this is carried out in the following manner. Because the bases pair, a complementary strand can be synthesized from a single strand of DNA; another strand can be synthesized from this second strand, which is identical in structure to that of the initial strand. In this way, DNA can be duplicated as often as desired while preserving the sequence of its bases. The process, which is carried out by special enzymes called **DNA polymerases**, is known as **replication**.

Because RNA and DNA are similar chemically, a complementary RNA molecule can be synthesized from a DNA strand. This process, which is carried out with the aid of **RNA polymerases**, is known as **transcription**.

With an extremely complicated apparatus involving very many molecules (ribosomes), the sequencing information available in RNA can be used for the synthesis of a protein, a process known as **translation**. Proteins are long polymers, whose synthesis proceeds in much the same way as that of nucleic acids—an amino acid exists to which a second amino acid is added, then a third, and so on, until a polypeptide is finally formed, which is commonly known as a *protein*. Twenty different amino acids are known to exist, but with only four bases, how can the information for a protein be hidden in RNA? The trick is that three bases (codons) together contain the information to specify one amino acid. This code is interpreted in the same manner in almost all living cells on this planet: AAA signifies lysine, CAA represents glutamine, and so forth. Together, there are 64 triplet codons ($4^3$) that code for 20 different amino acids and 3 stop codons; the system is somewhat redundant, and the genetic code is therefore degenerate. For example, AAA and AAG each code for lysine. If you look through all of the 64 triplet codes, you can see that the first two bases are frequently decisive for the particular amino acid that is to be installed, whereas the third base often proves to be insignificant. This arrangement is very useful for molecular biologists, because they can cause mutations in DNA sequences without altering their coding characteristics.

Not every DNA sequence codes for a protein; in human chromosomes, more than 90% of the available sequences have no recognizable significance. The debate continues about whether these stretches should be considered as junk DNA or as elements with some unknown function, but let us instead concern ourselves with the miserable remnants.

The remnants are the **genes**, transcription units on the long DNA molecule that are made up of regulatory and transcribed regions (Figure 1-2). Their functions are indicated by their names. The regulatory region controls whether the other regions are transcribed or not. The transcribed area can be divided into two subregions, the **exons** and **introns**. This division reflects a peculiarity of transcription in higher organisms (in this case, everything that is not a bacterium): **splicing**. Directly after its synthesis, the transcribed RNA is immediately processed by a complex apparatus, the spliceosome; this process excises parts of the RNA sequence and reattaches the remnants, a complicated procedure that ultimately results in a final RNA with a somewhat shorter length than before. The sequences excised during the splicing are introns on the DNA level, and those that remain are exons. The remaining RNA can again be divided into a region that serves as an information source for the synthesis of a protein, also as a coding area or an **open reading frame** (ORF), and one region each at the 5′ and 3′ termini of the RNA, which are known as **untranslated regions** (UTR) and whose function is still being investigated. The 5′ end of the RNA has a methylated G nucleotide (**5′ cap**), and a **poly A tail** (sequence of 100 to 250 adenosines) is located at the 3′ end. The A residues are not encoded on the DNA but are added by the action of poly A polymerase using ATP as a substrate.

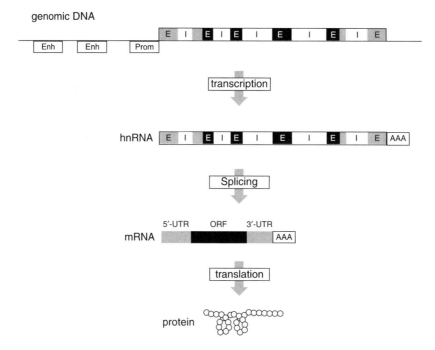

genomic DNA

transcription

hnRNA

Splicing

5′-UTR   ORF   3′-UTR

mRNA

translation

protein

**Figure 1-2. The eukaryotic gene and its designations.** The gene contains a regulatory region with enhancers (Enh) and promoters (Prom), and a transcribed area where RNA is translated. The heterogeneous nuclear RNA (hnRNA) is a one-to-one copy of the DNA with a poly A tail (AAA) that contains exons (E) and introns (I). The introns are excised while still in the cell nucleus through the action of a spliceosome complex to eventually form messenger RNA (mRNA). Only the open reading frames (ORF) of the mRNA are translated into a protein, whereas the untranslated regions (UTR) at the 5′ and the 3′ termini are in part responsible for the stability of the RNA, in part for their localization, and in part for nothing at all.

The structure signals the existence of mRNA[2] and protects it from attack by exonucleases, thereby increasing the half-life of the transcript, which codes for a protein.

All the regions described are arbitrary. It is currently impossible to discern whether a DNA sequence originates from the exon of a gene or from one of its many junk sequences. Many clever people are working to rectify this situation, and their eventual success will save us all a lot of work.

In molecular biology, experimenters also work with a handful of proteins, including polymerases, restriction enzymes, kinases, and phosphatases, in processing nucleic acids, although the functions of these proteins are not fully understood, which is not absolutely necessary as long as they do function. If this is not the case, the experimenter must usually only order new ones, and this part of the work therefore needs no further explanation.

---

[2] Messenger RNA (mRNA) is one of the three primary classes of RNA. Although it represents only approximately 2% of the RNA in a cell, it does play an important role because mRNA contains the sequencing information for the synthesis of proteins. Ribosomal RNA (rRNA) represents the most important component of the ribosomes, and transfer RNA (tRNA) occurs as small RNA segments to which the amino acids are coupled before they are installed in the protein sequence. Normal cells contain several other types of RNA, which are rarely mentioned, with the exception of small interfering RNA (siRNA) and microRNA (miRNA), which target corresponding mRNAs and thereby participate in post-transcriptional gene silencing.

# 1.2 What Is Required for This Work?

For a job in a laboratory with the authorization to carry out experiments in genetic engineering, you need three pipettes with which you can transfer volumes between almost 0 and 1000 µL; everything else seems to be a luxury. Most teams have high demands and little equipment. Even a place to write is frequently not available, or it must be shared with somebody else. The experimenter should be happy with what she or he gets and should try to make the best of it.

Laboratory requirements include a large quantity of bottles for buffers, graduated cylinders, glass pipettes, centrifuges, chemicals, refrigerators, freezers, deep-freezers, polymerase chain reaction (PCR) machines, and scales. Anyone who is not yet a professional in this field should refrain from trying to set up a laboratory, at least if he or she does not want to lose a year of life.

What should the workplace look like? Everybody is happy to see a well-ordered workplace—tidiness is half the battle, as the old adage goes. An alternate approach says that a chaotic workplace, where no staff member is to be found, at least looks as if work were being performed, whereas a spotless, totally empty surface cannot convince a boss that anyone has much enthusiasm for the work.

The people in a laboratory can be differentiated into anarchists and nitpickers. Find out to which group you belong. Because both are determined genetically and can be influenced only to a small degree, do not waste your time trying to change your colleagues. Learn to live with the situation, although you do not have to be satisfied with everything. Some colleagues, for instance, tend to extensively use premixed solutions from others because they are too lazy to prepare their own. Experiments are compromised when the experimenter is faced with an empty bottle of buffer. A popular remedy is to change the labels on the bottles according to the motto "when the label says Tris, there is guaranteed to be no Tris inside"; this is effective as long as the colleague is not aware of the system of re-labeling employed. This method is at best an emergency brake for extremely incomprehensive contemporaries, and it does not help to improve relationships with other colleagues. Worse, experimenters can forget their own systems after the next vacation.

How should a molecular biologist work? Many are in the habit, with great conviction, of upholding the principle of creative chaos. This may work well until the experimenter finds five racks in the refrigerator filled with test tubes that have accumulated over the past 3 weeks, and each of them is numbered consecutively from 1 to 24. You should get used to following a certain system in your work from the start so that you can understand what you have done, especially during some particularly "fervent" phases.

You should get used to a reasonably **clean mode of working**. Nucleic acids are quite stable molecules that fear only one thing: nucleases. Unfortunately, a little bit of nuclease is to be found almost everywhere, beginning with our own fingers. A slovenly worker will soon find a plasmid-DNA preparation that is made up of many small DNA pieces that are no longer useful. Consequently, standard practice is to autoclave everything with which the DNA might have come in contact: pipette tips, reaction vessels, glass bottles, and solutions. Some of this labor can be avoided if you learn to work in a clean manner (i.e., free from nucleases). The plastic reaction vessels, for instance, are practically always free of nucleases on delivery, even if it is not so indicated on the packaging, and they do not necessarily need to be autoclaved if you are capable of tipping them out of the bag without touching them, which may take some practice. The acid test is working with bacterial media. Anyone who is capable of using the same bottle of LB medium for an entire week without any bacteria or yeast growing profusely inside has proved that he or she has learned the tricks of the trade. However, even the most extensive autoclaving is useless if you subsequently rummage in the glass containing the sterile Eppendorf tubes using your dirty little fingers!

Experimenters should become accustomed to using a **laboratory log book**. This is required to maintain an overview, and it is a regulation. Never wait any longer than a week to make your entries because you can quickly forget many of the details. The best thing to do is complete this work at the end of a long, successful workday, perhaps while enjoying a cup of coffee. At the same time, you can consider what you intend to do the next day. I am well aware that this suggestion is about as realistic as the wish that all people will be good tomorrow.

# 1.3 Safety in the Laboratory

A molecular biologist is confronted with a whole series of **safety regulations** in the laboratory. Many do not have much to do with biology, and the same guidelines are valid in a chemical laboratory, such as those that primarily deal with the question of safe working methods and with the prevention of accidents. For instance, eating, drinking, and smoking are forbidden in the laboratory, and safety codes demand that laboratory coats; firm, closed, surefooted shoes (i.e., no sandals or no high heels); and a pair of goggles be worn at all times. Normal eyeglasses were allowed in the past, but they are no longer considered adequate. Employers are obligated to make safety goggles with corrective lenses available to their employees, but it is no surprise that such demands from people who spend their days pipetting miniscule amounts of saline and protein solutions from one plastic vessel into another are chiefly ignored. Nevertheless, everyone is urged to conform to these rules.

Experimenters encountering dangerous substances are obliged to wear **protective gloves** for their own protection and for the protection of their fellow workers. For instance, it is not enough to put on gloves when working with ethidium bromide; you must also take them off when you are no longer working with the substance to avoid leaving a wafer-thin film of poison behind on door handles, telephone receivers, and water faucets.

Gloves are essential, but life with them is not easy. The most unsuitable gloves in the laboratory are vinyl gloves because they are difficult to put on, fit badly, and are easily punctured. Latex gloves, which can be found in different sizes and thicknesses, are substantially better. They fit like a second skin, but they are unfortunately allergenic, especially the powdered varieties, and over the course of months and years, most people develop skin problems. If you ultimately develop a severe allergy to latex, even the powder on the gloves of your neighbor may cause you to have problems. The newest design for gloves is the nitrile glove, which is not as elastic as latex and is a little more expensive, but it is not associated with any allergenic risk.

A second group of safety regulations involves **interaction with radioactive substances**. Handling these substances is forbidden unless the particular institution is authorized to handle them. Even with institutional authorization, the experimenter must seek further permission and instruction. Authorities responsible for radiation protection will explain in detail which radionuclides and what quantities are allowed and in which rooms and under the observance of which safety regulations you may work.

Biologic safety is further regulated by the **genetic engineering safety regulations**. These regulations establish that the responsibility lies fully with the project manager, including the administrative aspects and the instruction and supervision of employees. Project managers must complete a scientific or medical university education and must have at least 3 years of experience in the field of genetic engineering. They must have completed a program of continuing education dealing with the topic of genetic engineering, safety precautions, and legal regulations.

Safety regulations for genetic engineering are naturally interesting for the young researcher. They assign organisms that have been altered through genetic engineering to specific **risk groups**. Group 1 generally includes all microorganisms and cell cultures representing no risk, those in group 2 demonstrate a slight risk, those in group 3 pose a moderate risk, and those in group 4 have a high risk for causing harm if they are released into the environment. These designations have been established by scientific and governmental organization. In Germany, classification has been directed by the Zentrale Kommission für die biologische Sicherheit (ZKBS; the Central Committee for Biological Safety), which is equivalent to the American Biological Safety Association (ABSA). *Escherichia coli*

K-12, the forbearer of most strains used in the laboratory, is classified in risk group 1, in the same way as, for instance, the laboratory mouse, *Drosophila melanogaster*, or the CHO, PCL2, or HeLa cell lines. In the higher-risk groups are primarily viruses such as the adenovirus (group 2), hepatitis B virus (group 2), hepatitis C virus (group 3), and human immunodeficiency virus (group 3) and cell lines that can release these viruses.

The risk groups correlate with **stages of safety**, which designate in which type of rooms the respective experiments may be carried out. A laboratory for biosafety level 1 must have four walls, a ceiling, windows, and at least one door, whereby the room can be kept closed. It cannot be a storeroom (because "the work should be performed in sufficiently large rooms and/or regions"), and it must be designated as being a working area for genetic engineering. The list of these regulations is considerably longer. The laboratory, for instance, must have a sink, at which investigators can wash their hands at the end of the workday, after cleaning the room and fighting all the "bugs." Other regulations duplicate previously established ordinances. For instance, the experimenter must wear a laboratory coat and may not perform any oral pipetting. Although eating, drinking, smoking, and applying makeup are forbidden, the storage of food, personal items, and cosmetics is allowed as long as these things do not come into contact with "organisms that have been changed by way of genetic engineering."

Biosafety level 2 represents a more earnest situation in laboratories. There is a warning sign signifying *biohazards*, a disinfectant must stand on the sink, and the doors and windows must be kept closed. No aerosols may be released, work surfaces must be disinfected, and everything that comes into contact with genetically modified organisms (GMOs) must be autoclaved and disinfected. The responsible authorities must also provide protective clothing. Such laboratories may be entered only by those who are explicitly authorized.

In level 3 biosafety laboratories, the windows do not need to be kept closed, because the ordinance anticipates that they cannot be opened. Investigators can enter such a laboratory only by way of a sluice, in which suitable protective clothing is donned. An autoclave must be available, the laboratory must be maintained at a negative pressure, and an individual may not work there alone.

A level 4 biosafety laboratory is more or less a shelter into which those who are weary of life must enter the region disguised as astronauts. The two-page list of official safety precautions is twice as long as that for the other safety stages, and it almost relieves experimenters of the desire to work. There are no level 4 laboratories existing in Germany, but there are four in the United States, with another three under construction. Still, they are rare elsewhere, so a broader description here is superfluous.

**Disposal** of biologic wastes and sewage is regulated through the safety regulations for genetic engineering. The wastes from biosafety level 1 facilities may be eliminated without any preliminary treatment, as long as no increase in or danger of infection is expected. Beginning with biosafety level 2, such waste must be inactivated or autoclaved. You should make it a habit to autoclave cultures, plates with cell cultures, pipette tips that have been contaminated by DNA, bacteria, and similar by-products of your productive actions before disposing of them to prevent being accused by somebody of having worked sloppily. This approach also corresponds with the rules of **Good Laboratory Practice** (GLP) established by the Organization for Economic Cooperation and Development (OECD). These guidelines also regulate other areas of laboratory life fully (e.g., the method by which experiments are documented), and they are used for well-managed laboratory work in Germany. Anyone who is interested in the GLP can find information in the Internet (http://www.oecd.org/ehs/glp.htm; http://www.glpguru.com; http://www.uic.edu/~magyar/Lab_Help/Ighome.html).

You can learn more about genetic engineering and safety through Internet sources. For example, the Robert-Koch-Institute, Germany's highest authority in this field, publishes useful information on the Internet, such as a list of organisms and vectors, general opinions concerning topics related to genetic engineering, and other information rated by the ZKBS (http://www.rki.de, under the keyword genetic engineering). You also can consult the local commissions for biologic safety.

**Literature**

Ashbrook PC, Renfrew MM. (1990) Safe Laboratories: Principles and Practices for Design and Remodelling. Chelsea, Michigan: Lewis Publishers.

Rayburn SR. (1990) The Foundations of Laboratory Safety. New York: Springer-Verlag.

Block S. (1991) Disinfection, Sterilization and Preservation. Philadelphia: Lea & Febiger.

Flemming DO, Hunt DL (eds). (1995) Laboratory Safety. Principles and Practices. Washington, DC, American Society for Microbiology.

# 2 Fundamental Methods

*Heute, seh ich, will mir nichts gelingen.*

*Today, I see, I'm not in clover.*

In this chapter, some of the methods described are required almost daily in the practice of molecular biology. For this reason, records must be kept in a more detailed fashion, so that a mixture can be remade in the event of an emergency.

The material in this chapter allows you to see if you have what it takes to be a molecular biologist, because it is not the larger concepts that lead to failure in practice; it is instead the smaller things in the daily routine. There are more than enough ingenious techniques to be found in the literature, but what do you do if the DNA does not precipitate as expected? A researcher with more than one method available will ultimately be the winner.

## 2.1 Differences in Nucleic Acids

As described in Chapter 1, all nucleic acids are principally constructed in the same manner. Nevertheless, they do differ in some points according to their specific origin, which is of considerable significance for the practice of molecular biology. A DNA chain made up of 500 base pairs (bp) is substantially easier to handle than one 500,000 bp long, and genomic DNA is obtained in a totally different manner from plasmid DNA.

**Genomic DNA** constitutes the genome, the complete gene complement of an organism. In higher organisms, the genomic DNA is found within the cell nucleus. It is composed of a duplicate set of chromosomes, the numbers of which vary according to the particular organism.

Each chromosome is an individual DNA double strand, which is millions of nucleotides long and must therefore be maintained by many proteins that are bonded to the DNA. For the purification of genomic DNA, the trick is to dispose of these proteins without causing the DNA to crumble too extensively.

The total number of base pairs in a genome varies in different organisms. In humans, it is about 3.2 billion; in the fruit fly, *Drosophila*, it is approximately 180 million base pairs. The size of the genome is not proportional to the size of the organism. The number of base pairs in different mammals, for instance, is more or less equal, whereas some plants can reach considerably higher numbers. Most of the genome is made up of *junk* DNA. The share of important sequences is represented by a maximum of 10% of the DNA, and the remainder appears to be nonsense. Much of it consists of *parasitic* DNA, repetitive sequences and transposons (so-called jumping DNA elements) that increase in our genome over time. Ll elements, the most frequent transposons in the mouse, are estimated to make up 5% to 10% of the murine genome. A first analysis of the human genome showed that the different repetitive elements that we carry make up 40% to 50% of the entire amount of DNA (Li et al., 2001)—that would be 1.6 billion bp that most probably will turn out to be useless (Table 2-1).

**Bacteria** also contain genomic DNA, although it is constructed differently. It takes the form of an individual, ring-shaped chromosome, which is not localized in the nucleus of the cell but is instead

**Table 2-1.** Examples of Genome Size and Chromosome Number

| Organism | Number of Chromosomes | Genome Size (bp) |
| --- | --- | --- |
| Humans | $2 \times 23$ | $3.2 \times 10^9$ |
| Mouse | $2 \times 20$ | $2.7 \times 10^9$ |
| Fruit fly | $2 \times 4$ | $1.8 \times 10^8$ |
| Tobacco | $2 \times 24$ | $4.8 \times 10^9$ |
| Corn | $2 \times 10$ | $3.9 \times 10^9$ |
| *Saccharomyces cerevisiae* (yeast) | $\sim 17$ | $1.5 \times 10^7$ |
| *Escherichia coli* | 1 | $4.64 \times 10^6$ |
| Phage $\lambda$ | 1 | 48,502 |

fixated at a site on the cellular membrane. The size of the bacterial chromosome is substantially smaller than that seen in higher organisms; in *Escherichia coli*, it is made up of only 4.64 million bp. The typical experimenter recognizes the bacterial genome only as something that contaminates his or her plasmid DNA preparations.

**Plasmids** are ring-shaped DNA molecules that exist in bacteria in addition to the bacterial chromosome, but they are substantially smaller. Plasmids commonly occur in nature and permit the transmission of useful genes from bacterium to bacterium. They can reproduce autonomously (i.e., independent of the bacterial chromosome) because they have their own *origin of replication*. They may also contain an arbitrary number of other sequences, a characteristic in molecular biology that is made use of so frequently and with such excessive delight that these plasmids are discussed somewhat more comprehensively in Chapter 6 (Section 6.2.1).

Bacteriophages, generally known as **phages**, are bacterium-specific viruses that replicate at the expense of the bacteria. To do so, the phage swims in the culture medium until it encounters a bacterium it can infect. It brings the normal reproduction of the bacterium to a standstill and instead causes it to produce many new phages. At the end of the process, the bacterium is lysed (i.e., the bacterium explodes), and it releases the new phages into the culture medium, where the process begins anew.

Phages, unlike the plasmids, are not elements of the bacterium, but independent organisms. The phage "chromosome" appears to be quite different from that of a plasmid. In bacteriophage $\lambda$, for instance, the chromosome consists of a linear DNA double strand, whereas phage M13 has ring-shaped, single-stranded DNA. Other phages have a genome that is coded on an RNA molecule. The size of the phage genome is relatively small: 48 kb in phage $\lambda$ and only 6.4 kb in phage M13. Often, the phage does not require its entire genome to survive. In phage $\lambda$, for instance, a researcher can insert fragments of foreign DNA up to 13 kb long and replicate these together with the phage DNA. The development of this technology was a decisive step in the cloning of genes, because whole genomes could be packed into phages in this manner. Since then, phages have become less important, because more favorable vectors are available for most of these purposes (see Chapter 6, Section 6.2.3).

**RNA** develops through the transcription of DNA by an RNA polymerase. An organism contains different types of RNA, which fulfill different tasks. Only the *messenger RNA* (mRNA) is really of interest to the experimenter, because it represents the transcription product of the genes that he or she wants to investigate. On the 5' end, it usually has a 7-methylguanosine cap (5' cap), and a polyadenosine sequence is found at the tail of the 3' end, which is used primarily for the specific purification of mRNA.

Other RNA types include ribosomal RNA (rRNA) and transfer RNA (tRNA). Except for specific research concerned with these substances, they are irritating contaminants that make up approximately

98% of the total RNA of a cell. Fortunately, they can usually be ignored. If not, the work becomes somewhat more laborious.

**Literature**
Li WH, Gu Z, Wang H, Nekrutenko A. (2001) Evolutionary analyses of the human genome. Nature 409:847–849.

# 2.2 Precipitation and Concentration of Nucleic Acids

A characteristic of DNA is that it is mostly found in the wrong concentration. This is not a concern if the concentration is too high, but it is a serious problem if the opposite occurs, and that is usually the case when an experimenter purifies DNA. Another typical problem is that the DNA solution may contain high concentrations of saline or of a particular salt, which may cause some later applications to be disturbed. The most popular solution in both cases is alcohol precipitation, which is found in almost every protocol concerned with the preparation of DNA, but there are an astonishing number of different methods to choose from.

Alternatively, the researcher can desalinate these nucleic acids with commercial concentrators or concentrate them by way of precipitation or with the aid of a Savant Speed Vac (i.e., a vacuum centrifuge [see Section 2.2.3]). DNA can also be precipitated with polyethylene glycol (PEG) (see Section 2.3.2), although this proves to be interesting only for the purification of larger DNA quantities.

## 2.2.1 Alcohol Precipitation

The principle of alcohol precipitation is simple. Dilute a nucleic acid solution with monovalent (i.e., singly charged) saline and add alcohol to it (how this functions is difficult to explain and does not help much in practice). The nucleic acid precipitates practically spontaneously if the solution has been mixed thoroughly, and it then is pelleted through centrifugation and purified of salts and alcohol remnants by washing it with 70% ethanol. A pleasant side effect of this method is that many small, water-soluble substances (other than the salts) remain in the supernatant, effectively achieving purification. Many varieties of this simple method exist.

## Salts

**Sodium acetate** is the standard saline solution used by experimenters. The concentration of the DNA solution (i.e., final concentration) before the addition of any alcohol should be 0.3 M. To accomplish this, the experimenter uses a 3 M sodium acetate solution at a pH of 4.8 to 5.2.

**Sodium chloride** (with a final concentration of 0.2 M) is used if the DNA solution contains sodium dodecyl sulfate (SDS), because this remains soluble in 70% ethanol and disappears with the supernatant.

**Ammonium acetate** (with a final concentration of 2 to 2.5 M) is used if the solution contains nucleotides (dNTPs) or oligonucleotides up to 30 bp long that should not be co-precipitated. It should not be used if the DNA will be phosphorylated, because the T4 polynucleotide kinase is inhibited by the ammonium ions.

**Lithium chloride** (with a final concentration of 0.8 M) is readily soluble in ethanol, and the precipitated nucleic acid contains less saline. It is not to be used for RNA, which is employed for cell-free translation or reverse transcription, because chloride ions inhibit the initiation of protein synthesis in most cell-free systems and inhibit the activity of RNA-dependent DNA polymerases. LiCl (without ethanol) can also be used for the precipitation of large RNA molecules (discussed later).

**Potassium chloride** functions as well, although it precipitates in large quantities in the presence of isopropanol. **Potassium acetate** is frequently used in the preparation of plasmid DNA through the aid of alkali lysis, because potassium forms an almost insoluble precipitate together with the SDS used, and you can almost eliminate the SDS quantitatively in an elegant manner.

## Alcohols

**Ethanol** (EtOH) is used as the standard alcohol for precipitations. It can be easily eliminated because it evaporates quickly. Precipitation requires a high concentration of approximately 70 volume percent (which corresponds to a 2.5-fold increase in volume of the DNA solution to be precipitated). That level frequently causes problems with centrifugation, because a substantially large centrifugation beaker is required. Another problem is the high cost, because pure ethanol is subject to the tax on spirits. There are some denatured ethanols suitable for precipitation, and they cost only about one fifth of the price of pure ethanol. Roth, for instance, offers an acetone-denatured ethanol that is suitable for precipitating nucleic acids. Many institutes have their own patrons who cover the cost of denatured ethanol at least to a small degree. At the current price of approximately $80 per liter of ethanol, this may prove to be quite a savings over the course of a year, and it may be worth looking into. Money is too scarce to waste it on DNA precipitations.

**Isopropanol** is far more effective than ethanol in the precipitation of nucleic acids. Consequently, 0.6 to 0.8 volume percent of isopropanol at room temperature is sufficient for a quantitative precipitation. This is interesting, especially for larger volumes. Remnants of isopropanol can be removed less easily than ethanol, and DNA pellets should be rewashed with 70% (v/v) ethanol.

**Butanol** is also suitable for the precipitation of DNA, although it functions completely differently. Butanol is barely miscible with water, and two phases to develop. The butanol phase can, in slight quantities of water (up to 19 volume percent), dissolve small amounts of saline. This can be used to desalinate smaller volumes. A volume of DNA solution dissolves in a 10-fold volume of butanol (1- and 2-butanol are equally effective), and the preparation is mixed for about 1 to 5 minutes, until the aqueous phase has disappeared. Next, centrifuge, rinse with 70% ethanol, and take up the dried (frequently not visible) pellet into $H_2O$.

## Concentration of Nucleic Acids

Whereas nucleic acids at high concentrations ($>250\,ng/\mu L$) precipitate very rapidly, the process is somewhat more difficult when working with substantially lower concentrations. At concentrations between 1 and $250\,ng/\mu L$, you can increase the yield by placing the preparation in the refrigerator (discussed later). In an attempt to increase the concentration of nucleic acids, such as to $100\,ng/\mu L$, another trick is to add **tRNA**, as long as the presence of RNA in the solution does not cause any disturbances later. An alternative is **glycogen**, which functions in the same manner, and it can be purchased from Roche or Sigma.

Small pellets are usually difficult to see and can be recognized with the naked eye only at concentrations higher than $2\,\mu g$. Relatively new on the market is another carrier (Pellet Paint Co-Precipitant) from Novagen, whose particularity is that it is coupled with a dye that **gives the pellet a distinct color**. For some, this product may also prove to be interesting for normal DNA precipitation. However, some caution is appropriate. The original version contains a fluorescent dye, which leads to intensive backgrounds or to error signals in the event that methods using ultraviolet (UV) light are employed (e.g., in automatic sequencers). For this purpose, a second variety of nonfluorescing Pellet Paint NF is more suitable.

## Temperature

Precipitation is a rapid process at nucleic acid concentrations greater than 250 ng/μL. In contrast to popular belief, it is not necessary to perform these precipitations at −20°C (−4°F). Instead, the precipitation proceeds more quickly and more effectively at higher temperatures. It is sufficient to let the mixture stand at room temperature for 5 minutes and then centrifuge it. At concentrations between 1 and 250 ng/μL, it is advisable to refrigerate the DNA-saline-alcohol mixture at −20°C (−4°F) for 30 minutes to 24 hours before centrifugation (the lower the concentration, the longer the period needed) to achieve yields of 80% to 100%.

## Centrifugation

It has become customary to carry out centrifugation at 4°C (39°F), although the source of this habit remains unknown. The nuclease activity should perhaps be reduced in very unclean DNA preparations—considering the most dreadfully exaggerated centrifugation times of 30 minutes or more, which would make some sense. The fact is that centrifugation can be accomplished just as well at room temperature, and the acquisition of an expensive, refrigerated centrifuge can be avoided.

Small volumes (up to 1.5 mL) are centrifuged for 5 to 15 minutes at 12,000 g in a table-top centrifuge (Figure 2-1). At higher DNA concentrations (starting at 250 ng/μL), as are expected in maxi-preparations of plasmid, the DNA precipitates spontaneously and frequently aggregates to form clouds. In this case, centrifugation at 3000 to 5000 g, which can be attained in a medium-sized centrifuge, is sufficient. At low DNA concentrations (10 to 250 ng/μL), it is better to increase the time and the speed, such as 30 minutes of centrifugation at 10,000 g. When working with very low concentrations, it is frequently helpful to pray, or if you own an ultracentrifuge, follow the guidelines of Shapiro (1981), who has been able to obtain yields of 80% for concentrations of up to 10 pg of DNA/μL after 24 hours of precipitation at −20°C (−4°F) through ultracentrifugation in the range of 100,000 g.

**Literature**
Shapiro DJ. (1981) Quantitative ethanol precipitation of nanogram quantities of DNA and RNA. Anal Biochem 110:229–231.

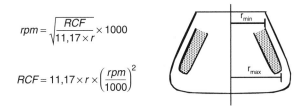

$$rpm = \sqrt{\frac{RCF}{11{,}17 \times r}} \times 1000$$

$$RCF = 11{,}17 \times r \times \left(\frac{rpm}{1000}\right)^2$$

**Figure 2-1. Relation between the number of revolutions per minute and centrifugal force.** The radius varies depending on the position of a particle in the tube, mostly orienting itself to the maximum radius ($r_{max}$), which is found in the rotor handbook. r, radius (in cm); RCF, relative centrifugal force in acceleration due to gravity ($g$); rpm, revolutions per minute.

## 2.2.2 Concentrators

Concentrators are primarily employed in the field of protein biochemistry, although they are increasingly found in the world of nucleic acids (Table 2-2). Most common are the products from the Millipore (previously Amicon) company, although comparable products are offered by other companies (e.g., Pall Filtron Corporation).

The concentrator has small units containing a filter, through which the solution to be concentrated by means of centrifugation is pressed. The DNA is held back by the filter and can be recovered in a second, very short stage of centrifugation. Manufacturers offer different systems with a loading capacity of 0.5 to 15 mL of the initial solution and with filters that have molecular weight cutoffs of 1 to 100 kDa.

With concentrators, you can concentrate DNA according to the *molecular weight cutoff size* of the filters, and you can carry out completely different tasks. Small molecules such as salines or unbonded nucleotides can be eliminated from a solution through the use of a labeling reaction, because these substances are not held back by the filters. The solution is diluted using $H_2O$ or buffers until the concentration of the undesirable substance has dropped to an acceptable level and can be centrifuged to concentrate the DNA sufficiently. If this procedure is done repeatedly, the undesirable substance can be eliminated almost entirely. Polymerase chain reaction (CR) assays can be freed from primers, nucleotides, and polymerases.

It is irritating that the molecular weight cutoff of the filters is oriented to proteins, which are mostly smaller and of a different form from that of nucleic acids. When in doubt, question the manufacturer about which filters are suitable for which fragment lengths. Fortunately, the companies have recognized the molecular biologists as a target group and increasingly offer tailor-made products for their purposes, such as for PCR purifications. The experimenter frequently lacks essential information about the molecular weight cutoff and the binding behavior of the membranes, which must be obtained from the manufacturer.

The *molecular weight cutoff* (MWCO) of the filter membrane (NMWL = *nominal molecular weight limit*) determines the size of the protein (in kDa) that can be retained at a rate of 80% to 90%. The index permits conversion to nucleotides (according to Millipore).

**Advantages:** The application of concentrators requires little work, and on a small scale, it can be performed quite quickly; centrifugation, depending on the unit and filter, requires only 10 to 60 minutes. The saline concentration does not increase during this procedure.

**Disadvantages:** The concentrator is less suitable for solutions with higher concentrations of nucleic acid, because the filter pores clog very quickly, after which even hours of centrifugation no longer proves to be helpful. They are quite expensive.

**Suggestions:** The units can be used repeatedly as long as the filter has not been torn or clogged. An example is provided by units with omega membranes from Pall Filtron.

**Table 2-2.** Molecular Weight Cutoff of Millipore Concentrators

| Microcon Centrifugal Filter Unit | Molecular Weight Cutoff (kDa) | ssDNA (nt) | dsDNA (bp) |
| --- | --- | --- | --- |
| 3 | 3 | 10 | 10 |
| 10 | 10 | 30 | 20 |
| 30 | 30 | 60 | 50 |
| 50 | 50 | 125 | 100 |
| 100 | 100 | 300 | 125 |

dsDNA, double-stranded DNA; ssDNA, single-stranded DNA.

## 2.2.3 Savant Speed Vac

Smaller volumes can be dried in a Savant Speed Vac. This method deals with a table-top centrifuge to which a vacuum can be applied. The vacuum results in a reduction in the boiling point of the solvent (usually water or ethanol) and in more rapid evaporation of the fluid. To prevent gas bubbles from being released from the DNA solution and being distributed throughout the entire device, the experimenter centrifuges simultaneously to prevent the development of foam. Because the liquid loses warmth as a result of evaporation, a Speed Vac system should be heatable; otherwise, the preparation in the tube will freeze rather than dry.

Small remnants of fluids can be eliminated very quickly in this manner. However, the larger the amount of liquid, the more time is required to carry out this procedure. It is therefore recommended only for smaller volumes ($< 500\,\mu L$ or, better still, $< 50\,\mu L$).

The Savant Speed Vac is frequently used after precipitation to remove ethanol remnants from the pellets. Nevertheless, care must be taken, especially because the larger pellets may become overdried, and they then can be dissolved only with difficulty.

**Disadvantage:** Because only the solvent evaporates, there is enrichment of salts in the solution, which may be disturbing during the course of further handling.

## 2.2.4 Salting Out

Precipitation with the aid of high concentrations of saline (i.e., salting out) is used for RNA. High molecular-weight RNA (mRNA, rRNA), in contrast to smaller RNA strands (tRNA, 5S RNA), is insoluble in solutions of high ionic strength and can therefore be precipitated specifically through the addition of **lithium chloride** and then separated from the smaller RNA or from the DNA strands.

Plasmid DNA mini-preparations can be purified using RNA by adding $300\,\mu L$ of ice-cold, 4 M LiCl to $100\,\mu L$ of DNA solution, leaving the mixture on ice for 30 minutes and then centrifuging for 10 minutes at $12,000\,g$. From the supernatant, the DNA is precipitated through the addition of $600\,\mu L$ of isopropanol, and the pellet is then washed with 70% ethanol.

# 2.3 Purification of Nucleic Acids

In addition to precipitation, the purification of nucleic acids represents the other standard procedure of the industrious and successful experimenter. This process is encountered while working with DNA from material from tissues, bacteria, and gels. A discussion of the most common purification methods follows.

## 2.3.1 Phenol-Chloroform Extraction

Nucleic acid solutions commonly contain undesirable contaminants that are chiefly made up of proteins. A classic method of purifying is phenol-chloroform extraction, by which the nucleic acid solution is extracted by successively washing with a volume of phenol (pH 8.0); a volume of a phenol, chloroform, and isoamyl alcohol mixture (25:24:1 [v/v]); and a volume of chloroform (Figure 2-2). Centrifugation is performed intermittently, and the upper, aqueous phase is conveyed into a new vessel while avoiding the interphase. In this way, the contaminants are denatured and accumulate in the organic phase or in the marginal layer between both phases, and the nucleic acids are preserved in

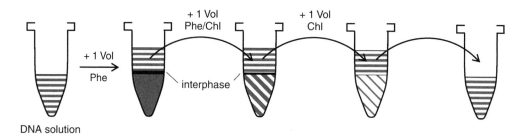

DNA solution

**Figure 2-2. Phenol-chloroform extraction.**

the aqueous phase. Through a final alcohol precipitation, the remaining phenol remnants are removed from the solution.

Instead of filling with alcohol, remnants of phenol, chloroform, or other organic solvents are removed using **ether**. To do so, mix the DNA solution thoroughly with an equal volume of diethyl ether and then centrifuge for a few seconds to achieve a full separation of the two phases. The (upper) ether phase is removed with the pipette, the aqueous phase is mixed a second time with ether, and the ether residues are allowed to evaporate for 15 minutes in the air or, in the event of larger volumes, for 15 minutes using vacuum evaporation. This method, however, is less commonly used because ether has a numbing effect and is highly flammable. Because it is heavier than air, ether can accumulate at ankle level while fresh air may still be encountered at the level of the nose. You should therefore work with ether only when using a well-functioning fume hood.

Not every vessel used in centrifugation is resistant to organic solvents such as phenol and chloroform. Polystyrene and polycarbonate, two substances that are very popular because of their transparency, dissolve (unfortunately very poorly, so that this occurs relatively slowly) and then break down during centrifugation. The cleaning of this stinking, corrosive mess from the rotor and the centrifuge represents one of the most lasting impressions in the life of an experimenter.

The protocol can be simplified by eliminating the first step: cleaning with phenol. The isoamyl alcohol can be eliminated as well, because a 1:1 mixture of phenol and chloroform is equally effective.

Although the contaminants that collect in the interphase are frequently easy to see, it is generally difficult to pipette out the aqueous phase without taking some of them along. Frequently, the losses brought about by this are relatively extensive. Eppendorf offers an interesting product called Phase Lock Gel for this purpose. It is a thick gel already pipetted into tubes of different sizes that does not react with the phenol or the aqueous phase, and after centrifugation, it forms a thick, stable layer between both phases, making it easy to pipette off the supernatant. If this is too expensive, the silicon vacuum grease available for centrifuges can be used, but you must then determine the proper quantity by trial and error.

Caution is warranted because phenol is corrosive and poisonous. The skin can be burned on contact, and phenol is absorbed into the system, where it may lead to damage of the nervous system, liver, and kidneys and, in the event of larger quantities, to death. Phenol vapors cause irritation of the eyes and the respiratory tract, and can cause nausea, vomiting, and even worse. Take great care with phenol, and observe all safety guidelines (e.g., gloves, goggles, ventilation). In the event of skin contact, wash well with soap and water. When phenol is splashed in the eye, rinse intensively with water, and consult a physician afterward. The practice employed previously of producing phenol by equilibrating solid phenol over the course of many hours with Tris(base) solution must be considered absurd because of the potential for danger, criminal because of the improper disposal of waste material, and uneconomical because of the time demanded by this process. Do yourself and your environment a favor, and buy a prepared phenol solution, even if it is somewhat more expensive.

**Advantages:** It is cheap and well tested, and it functions reliably.
**Disadvantages:** It stinks, is corrosive, and may prove dangerous to health.

## 2.3.2 Precipitation Using Polyethylene Glycol

A less poisonous but more lengthy method for the purification of DNA precipitates involves the use of PEG. The DNA solution is brought to a final concentration of 0.5 M NaCl and 10% polyethylene glycol 8000 (w/v), which is mixed thoroughly and incubated for 1 to 15 hours at 0°C to 4°C (32°F to 39°F). The mixture is then centrifuged, and the pellets are washed twice with 70% ethanol.

The key point is the yield, because incubation is required for at least 12 hours at 4°C (39°F) to perform a complete precipitation; 1 hour of incubation on ice results in a yield of about 50%. This method is therefore not very suitable for the precipitation of small quantities of DNA.

## 2.3.3 Protein-Binding Filter Membranes

For anyone who has treated DNA with any enzyme and would like to remove this enzyme from the reaction preparation after a successful incubation, there is now an interesting alternative to the usual phenol-chloroform extraction. This involves a by-product from the world of filter membranes, a membrane that demonstrates high affinity to proteins that bind DNA only to a small degree (Micropure-EZ, Millipore). Heat-stabilized enzymes that can be inactivated only with difficulty (e.g., restriction enzymes, polymerases, phosphatases) can be eliminated very quickly and with little effort in this manner.

## 2.3.4 Anion-Exchange Columns

Because of the hazardous wastes that result from phenol- and chloroform-containing compounds, these substances must be detoxified according to the regulations and with great difficulty. Phenol-chloroform purification has therefore been increasingly replaced by faster, cleaner, and for the industry, more profitable purification methods. In principle, there are two different methods: anion-exchange columns and glass milk (discussed later).

Anion-exchange columns are composed of a matrix, such as cellulose or dextran, which contains positively charged groups (Figure 2-3). The spine of the nucleic acids contains phosphate groups that are charged negatively above a pH of 2. For this reason, nucleic acids bind excellently in anion-exchange columns. Because the strength of bonds in DNA, RNA, and proteins is different and depends on the pH and on the ionic strength of the buffer, this method is suitable for the purification even of very crude DNA preparations.

In principle, the DNA preparations are brought to a certain saline concentration and then purified in anion-exchange columns. The highly charged nucleic acids bind to the material of the columns, whereas the proteins, which are far less negatively charged, run through these columns. With a buffer of a higher ionic strength, the RNA can be washed with the aid of the column, and the DNA is ultimately eluted using a buffer of an even higher ionic strength or a lower pH. To dispose of the salt, the DNA is precipitated with alcohol (i.e., isopropanol or ethanol). The composition of the buffers differs somewhat depending on the material in the columns, and the experimenter must follow the recommendations of the column manufacturers. Mostly, the buffers must be purchased, because the manufacturers do not give away their secrets concerning the composition of their products. Macherey-Nagel is an exception, and Qiagen has been less secretive over the past few years.

Unfortunately, DNA purification with the aid of columns occasionally performs somewhat mystically. It usually functions, although not always. Do not despair if your neighbor in the laboratory

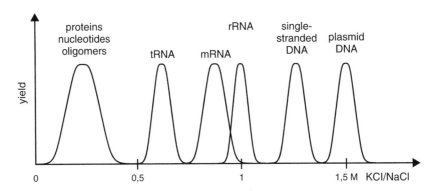

**Figure 2-3. Typical elution patterns of a commercial anion-exchange column at pH 7.0.** The purification of DNA over anion-exchange columns is relatively simple because of the high bonding strength of the DNA to the column matrix. Other nucleic acid structures can also be separated with the aid of such columns if the washing buffers and elution buffers have a suitable ionic strength. The elution pattern depends on the pH, and the pH of the elution buffers usually must be in the range of 8.5 to keep the saline concentration as low as possible (according to Qiagen).

claims that he or she has never had any problems with it. The most probable cause is that no DNA exists in the initial solution, which is chiefly a bacterial lysate. This can best be examined by precipitating an aliquot of the initial solution with ethanol and, after digestion with DNase-free RNase (otherwise you will see only a large RNA spot later), applying the remnants of this pellet on an agarose gel. The second possibility is to carry out the classic error of "a lot helps a lot", because anion-exchange columns have only a certain binding capacity. If the quantities of bacterial lysate added to the column are too large, non-DNA components (mostly bacterial RNA, which makes up approximately 90% of the nucleic acids in the lysate) will saturate the binding sites so that the DNA yield decreases. A third possibility is that the buffers are not correct. If the ion concentration of the wash buffer is too high, for instance, the DNA is eluted during the elution stage, and the yield drops. This can happen even with the manufacturer's own buffers; therefore, do not have too much respect for such kits.

Even if the manufacturers are not willing to betray their secrets, ion-exchange columns can be used repeatedly. Make use of columns from manufacturers who betray the composition of their buffers, so that these substances do not become the limiting factor. Consider the fact that the elution of DNA is never complete, and contaminants from earlier purifications remain. Consequently, it is best to use one single column only for the purification of a specific DNA. Autoclaving the columns is not recommended, because the material of the column and the plastic components suffer from this process, and DNA, although it breaks apart during autoclaving, is not always fully destroyed.

The capacity of the columns has unfortunately been reduced in recent years. It was previously possible to obtain more than $800\,\mu g$ from a single maxi-preparation column, but now, it is frequently difficult to obtain a nominal $500\,\mu g$. It can be helpful to collect the bacterial lysate after a single passage through the column and then perform a second elution, which can increase the yield to about 60% to 70% with little effort.

## 2.3.5 Glass Milk

Glass milk (originally a suspension of small glass beads) is a completely different kind of anion exchanger. Silicates (glass is nothing else) bind specifically to DNA in the presence of high concentrations of chaotropic salts that, after washing with $H_2O$ or Tris-EDTA buffers (see Appendix 1, Standard Solutions), can easily be eluted to be used in further procedures.

Although this method is primarily used for the purification of small quantities of DNA from agarose gels and is therefore described more comprehensively there (see Chapter 3, Section 3.2.2), it is also suitable for the separation of DNA from solutions. Many kits for the preparation of plasmid DNA on a small scale, for cleaning of PCR preparations, or for the elimination of enzymes from reaction mixtures function according to this principle. Because of their easier handling, silicate membranes are now used rather than glass beads, which are only rarely used.

## 2.3.6 Cesium Chloride Density Gradient

The classic but now unfashionable method for the purification of larger quantities of DNA is centrifugation of a cesium chloride density gradient (Figure 2-4). Cesium chloride (CsCl) has a remarkable characteristic. If a CsCl solution is centrifuged long enough at a sufficiently high speed, a higher concentration of the heavy salt collects at the bottom of the centrifuge tube, and the concentration continuously decreases toward the surface so that a density gradient occurs. Because proteins, RNA, and DNA can be distinguished by means of their densities, this gradient can be used for DNA purification.

It can also be used to separate plasmid DNA from bacterial DNA by adding **ethidium bromide**. Ethidium bromide becomes intercalated between the bases of the DNA and transforms them into a DNA-ethidium bromide complex, which has a lower density than DNA. Plasmid DNA, as a circular form of DNA with additional helices (*supercoiled DNA*), is under a great deal of tension and consequently binds less intensively to ethidium bromide. The density of plasmid DNA is therefore higher than that of linearized DNA, and both types of DNA can be separated from each other in a CsCl density gradient; under UV light, the band with the bacterial DNA is located a short distance above that of the plasmid DNA band. CsCl density gradients are also used for the separation of genomic DNA derived from bacteria or plants, as well as for that of proteins, RNA, and other contaminants, although only a single DNA band can be seen in that case.

Centrifuge at **room temperature**. Cesium chloride precipitates easily at low temperatures that, considering the high centrifugation rates, can quickly lead to maximum credible accidents (MCAs) due to ultracentrifugation. Those responsible for safety in your institute will be able to show you some interesting results of this.

**Advantage:** At least you can see something.

**Disadvantages:** The centrifugation takes a long time (more than 14 hours at 350,000 g) and requires much work. Ethidium bromide is a mutagenic agent, which means that gloves and safety clothing are required and that the hazardous wastes must be taken into consideration. You can avoid messing up

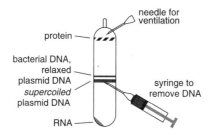

**Figure 2-4. Continuous cesium chloride density gradient.**

the laboratory by acquiring some experience and by being very careful. Use ion-exchange columns if possible.

### 2.3.7 Dialysis

You have the possibility of dialyzing DNA. This method is suitable for carefully disposing of small molecules, such as salts and phenol remnants, in the preparation of genomic DNA.

Dialysis tubing is delivered by the running meter in a dry form and must first be allowed to swell and then be washed. The fastest method is to cook a piece of tubing in $H_2O$ for 10 minutes. The tubing is then clamped at the lower end, the DNA solution is pipetted into the tubing, and the upper end is clamped. The filled tube hangs in 1 L of TE buffer, and dialysis is carried out at 4°C (39°F) for 2 to 16 hours, depending on the volume of the DNA solution. The buffer must be stirred during dialysis. In the event that larger volumes of DNA are being used, the buffer should be changed every few hours. Use gloves during all steps of the procedure to avoid any contamination with nucleases.

**Advantage:** This is the most gentle method for eliminating any contamination caused by smaller molecules.

**Disadvantage:** The process requires an extremely long period. It is not suited for smaller volumes, because a large amount of the material is lost in the tubing.

Microdialysis is suitable for smaller volumes. Many manufacturers offer a variety of systems with different volumes of yield for volumes of 10 to 5000 µL, and the loss of material is relatively small. One can also help oneself: An interesting, small protocol advises the experimenter to fill a Petri plate with 25 mL of $H_2O$ or a buffer and place a dialysis filter membrane with a pore diameter of 0.025 µm on the surface of this fluid (available from Millipore in a number of different diameters). He or she then pipettes 5 to 100 µL of DNA solution onto the filter and removes it after approximately 30 minutes.

## 2.4 Determining the Concentration of Nucleic Acid Solutions

The third in the league of techniques is the determination of concentration. Here again, the wider the choice, the greater the trouble.

### 2.4.1 Optical Density Measurements with the Aid of Absorption Spectrometry

The determination of the concentration by measuring the **optical density** (OD) at a wavelength of 260 nm is fast and simple and therefore very popular. It requires a quartz cuvette and a photometer with a UV lamp. The concentration of nucleic acids can be calculated from the absorbance or OD at 260 nm, the dilution, and a multiplication factor that is specific for DNA, RNA, or oligonucleotides. From the relationship of $OD_{260\,nm}$ and $OD_{280\,nm}$, information can be obtained about the protein contaminants in a solution.

The problem is that a fair-sized quantity of DNA or RNA must be available, because the photometer provides reliable measurements only in a range between 0.1 and 1 OD, which represents a concentration of nucleic acids between 5 and 50 µg/mL. With semi-micro and ultra-micro cuvettes, the measurement volumes can be decreased along with the quantity of DNA or RNA; you cannot go below 100 ng. The photometer may not play along, the measurement error may be too large, or the necessary small change for these quite expensive and rarely used cuvettes may be lacking. The cuvettes with these smallest of measurement volumes are extremely difficult to handle.

OD measurements make sense if you are sure that the preparation contains only DNA or RNA. Because these nucleic acids cannot be distinguished from each other, it is useless to attempt to determine the DNA concentration photometrically in a crude plasmid DNA preparation, because the proportion of RNA is about 90%.

Another problem with OD measurements is that there is something wrong with this method. All of the books on the possible methods are in agreement that the concentration is calculated from the $OD_{260}$ and that the purity of the preparation is calculated using the ratio of $OD_{260}/OD_{280}$. Using these methods, the nucleic acid solution should demonstrate a ratio of 1.8 to 2.0, although experience has shown that even highly purified preparations occasionally do not have ratios of more than 1.5. A publication (Wilfinger et al., 1997) may have the key to this problem. The researchers showed, independent of the pH and salt concentration of the water used in the measurement, that the $OD_{260}/OD_{280}$ ratio of a SINGLE nucleic acid preparation ranged from 1.5 to 2.2, and a calculated concentration of $0.55\,\mu g/\mu L$ to $0.7\,\mu g/\mu L$ was found, a difference of 25%. Evidence of protein contaminations using the OD ratio may demonstrate completely different results as well, ranging from clearly measurable to not at all demonstrable. To obtain reproducible results, the study authors recommend making $OD_{260/280}$ measurements in a 1 to 3 mM $Na_2HPO_4$ buffer at a pH of 8.5. Examples are provided for these calculations in Figure 2-5.

All nucleic acids, including fully developed DNA, oligonucleotides, and monomeric nucleotides, have absorbance at a wavelength of 260 nm. In a mixed solution such as for PCR, for instance, measuring the amount of DNA that has been synthesized during amplification will prove unsuccessful.

**Literature**

Willinger WW, Mackey K, Chomczynski P. (1997) Effect of pH and ionic strength on the spectrophotometric assessment of nucleic acid purity. Biotechniques 22:474–481.

concentration:

$$c\left[\frac{\mu g}{ml}\right] = OD_{260} \times V \times F$$

molarity of dsDNA:

$$c\left[\frac{\mu mol}{ml}\right] = \frac{OD_{260} \times V \times F}{M_w \times L} = \frac{OD_{260} \times V}{13,2 \times L}$$

molarity of oligonucleotides:

$$c\left[\frac{\mu mol}{ml}\right] = \frac{OD_{260} \times V \times F}{M_w \times L} = \frac{OD_{260} \times V}{10 \times L}$$

or more precisely:

$$c\left[\frac{\mu mol}{ml}\right] = \frac{OD_{260} \times V \times 1000}{15 \cdot A + 7,1 \cdot C + 12 \cdot G + 8,4 \cdot T}$$

c = concentration of undiluted DNA solution
OD260 = absorption at 260 nm
V = dilution factor
F = multiplication factor:
  50 for dsDNA,
  40 for RNA,
  33 for ssDNA,
  20 for oligonucleotides
L = length in kb (!)
Mw = molecular weight per base (330 g/mol) or base pair (660 g/mol)

**Figure 2-5. Determination of the concentration and the molarity of nucleic acid solutions.** For example, if the concentration needs to be determined for $30\,\mu L$ of a DNA solution (plasmid with a length of 4.3 kb), $1\,\mu L$ is diluted 1:200 and measured, and the absorption is 0.43 OD. The concentration of the initial solution is $c = 0.43 \times 200 \times 50\,\mu g/mL = 4.3\,\mu g/\mu L$, and the molarity is $c = 0.43 \times 200/(13.2 \times 4.3)\,\mu mol/mL = 1.5\,\mu mol/mL$. If it was a primer with the sequence AAGGAATTCCTT, the concentration would be $c = 0.43 \times 200 \times 20\,\mu g/mL = 1.7\,\mu g/\mu L$, with a molarity of $c = 0.43 \times 200 \times 1000/(15 \times 4 + 7.1 \times 2 + 12 \times 2 + 8.4 \times 4) = 86,000/131.8\,\mu mol/mL = 0.65\,\mu mol/\mu L$. dsDNA, double-stranded DNA; ssDNA, single-stranded DNA.

## 2.4.2 Determination of Concentration by Means of Agarose Gels

Agarose gels can be used for the measurement of small quantities of nucleic acids. The detection limit of this method is about 5 ng of DNA per band. If you apply an aliquot of this solution together with a dilution series of DNA of a known concentration to the gel, you can estimate the DNA concentration based on a comparison of the band intensities (Figure 2-6). A common variant of the dilution series is the DNA molecular-weight marker, which is also applied directly to the agarose gel. Because the fragment length and the quantity that has been applied are usually known, this band can be used as a marker for quantity.

The work involved in this method is more substantial than that for photometric determination, and the results are associated with some uncertainty. Nevertheless, for small quantities of DNA, it is frequently the only method available to get an idea of the dimensions without having to sacrifice the entire DNA fragment for such a measurement. Even for larger quantities of DNA, it can be useful after determining the concentration by means of a photometer to investigate the actual quantities by means of an agarose gel.

## 2.4.3 Dot Quantification

Dot quantification represents an abbreviated version of determining the concentration by means of agarose gel. It requires a series of DNA solutions of different concentrations (e.g., 0 to $20\,\mu g/mL$); $4\,\mu L$ each of a DNA solution and an ethidium-bromide solution ($1\,\mu g/mL$) are mixed together, and the same thing is done with $4\,\mu L$ of the DNA solution, whose concentration is being determined. Put plastic foil on the bank of UV lights, and then pipette the solutions side by side and drop by drop. This work of art is photographed, and the concentrations are determined by comparison with the standards. A disadvantage is that this is a very inaccurate method.

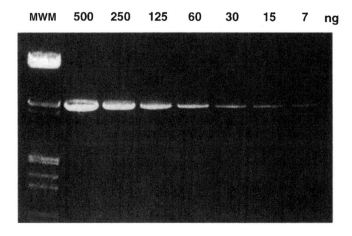

**Figure 2-6. Determination of concentration by means of agarose gel.** The quantification by means of agarose gel is not particularly exact (especially with images from a CCD camera), and it is easy to make a mistake by up to a factor of 2. If possible, it is better to determine the concentration with the photometer and confirm the value with the help of an agarose gel to avoid bad surprises, which are also a threat in photometric determinations. MWM, molecular-weight marker.

## 2.4.4 Fluorometric Determination

A less well-known alternative is the determination of the concentration with the aid of fluorescing dyes that absorb light of a certain wavelength and can emit light of a different wavelength. This method is easier and substantially more sensitive than spectrophotometric determinations. The DNA is mixed with **Hoechst 33258** (e.g., from Sigma), a DNA-specific fluorescent dye that has only a slight affinity to RNA and proteins and that is measured by a fluorometer. The excitation occurs at a wavelength of 365 nm, and the emitted light is measured at 460 nm. An instrument especially designed for this purpose can be obtained from Hoefer. The price, at about $3250, is on the order of that required for a photometer. With this instrument comes a set of special fluorometer cuvettes, which have four clear lateral surfaces.

For the measurement of DNA solutions with concentrations between 5 and 500 ng/mL, it is sufficient to use a dye concentration of 0.1 μg/mL. With 1 μg/mL, you can even determine concentrations of up to 15 μg/mL, although higher concentrations of dye decrease the sensitivity for the measurement of smaller DNA concentrations. With a special adapter, you can detect 1 ng of DNA in volumes of 3 μL. The dye, however, has a higher affinity for AT-rich rather than to GC-rich sequences. Labarca and Paigen (1980) have described the method.

A stock solution of Hoechst 33258 (1 mg/mL [w/v] in $H_2O$) can be stored for about 6 months at a temperature of 4°C (39 °F). For the working solution, the stock solution is diluted 1:1000 or 1:10000 with TNE buffer (10 mM Tris HCl, pH 7.4/200 mM NaCl/1 mM EDTA). Le Pecq (1971) also describes the use of ethidium bromide as a fluorescent dye, although it has a sensitivity that is worse than that of Hoechst 33258 by a factor of about 20.

**Literature**
Labarca C, Paigen K. (1980) A simple, rapid, and sensitive DNA assay procedure. Anal Biochem 102:344–352.
Le Pecq JB. (1971) Use of ethidium bromide for separation and determination of nucleic acids of various conformational forms and measurement of their associated enzymes. In: Glick D (ed): Methods of Biochemical Analysis, vol 20, pp 41–86. New York, John Wiley & Sons.

## 2.4.5 Nucleic Acid Dipsticks

Invitrogen offers a DNA DipStick Kit with which the concentration of DNA, RNA, and oligonucleotide solutions can be determined using only 1 μL of solution. The nucleic acid solution is pipetted onto a test strip, the strip develops a color, and the intensity of the resulting spot is compared with a standard. The greatest advantage of this method is that it can be performed with 1 μL of solution and is nevertheless quite rapid. The sensitivity, however, is only mediocre, with a lower limit of 100 ng/mL.

## 2.4.6 Enzymatic Evidence

Another method for quantifying even the smallest quantities of DNA is available (DNAQuant DNA Quantitation System, Promega). The evidence is derived from a classic, coupled, enzymatic reaction, which apparently runs in an inverted manner. In a first step, the DNA is degraded in the presence of T4 DNA polymerase and pyrophosphate (i.e., pyrophosphorylation). The dNTPs that develop are converted in a second step into ATP and dNDPs using nucleoside diphosphate kinase and ADP (i.e., transphosphorylation). The emerging ATP emits light in a third reaction with the help of luciferase and a luminogenous substrate, which can be measured with the help of a commercially available luminometer. Because the light production is proportional to the quantity of DNA transformed, the quantity of DNA applied in the solution can be determined easily with the help of a DNA quantitation standard. In this way, DNA quantities between 20 pg and 1 ng or DNA concentrations between 10 and 500 pg/μL can be measured. To achieve reliable results, the DNA must be smaller than 6000 bp and must have free ends so that the polymerase has access to the DNA. Plasmid and chromosomal DNA must therefore be linearized or shortened before being degraded by restriction digestion. Because of the specificity of the T4 DNA polymerase, neither single-stranded DNA nor RNA is measurable with this method (unless it comes to the formation of a dimer or of a hairpin). Compared with other methods of detection, this method is suitable when the necessity of a precise quantification of small quantities of DNA justifies the great amount of laboratory work.

## 2.5 Methods of DNA Preparation

There are many different types of DNA and many methods of preparation. Because the preparation of DNA is one of the chief occupations in this trade, many people have concerned themselves with the topic, and the number of methods and variants is almost endless. A couple of the most useful are presented here, although it is always beneficial to contact a colleague, because he or she might have already discovered something even better.

### 2.5.1 Preparation of Plasmid DNA on a Small Scale

To carry out 48 or 96 mini-preparations is a very nerve-wracking affair. Consequently, the method must be fast and simple. Here are three methods that differ foremost in the type of bacterial lysis. The efforts demanded and the time required are almost equivalent in every case. The decision to use one or another method is therefore a question of personal preference, although it occasionally depends on the subsequent application of the DNA. Important for the yield and quality of the DNA are the bacterial strain, plasmid (*low-copy* or *high-copy* plasmid), and culture medium (i.e., minimal nutrient, nutrient rich, or very nutrient rich) used. If one method delivers unsatisfactory results, another should be attempted. The yield for 1.5 mL of culture amounts to about 2 to 10 μg of plasmid DNA if it is a high-copy plasmid.

### Alkaline Lysis

Alkaline lysis is my favorite method, because it is possible to do almost anything with DNA that has been prepared in this manner. In a mini-scale preparation, it is very fast and can yield large quantities of DNA with an unusual purity. Performed properly, it can be used effectively for any preparations carried out with the use of measurement columns. It has also become the basis for almost all commercial plasmid-DNA purification kits.

The pellet derived from 1.5 mL of a bacterial culture is resuspended in 150 μL of solution I (50 mM glucose/25 mM Tris/10 mM EDTA), and 150 μL of solution II (0.2 N NaOH/1% SDS [w/v]) is added. This preparation is then mixed thoroughly and incubated for 30 seconds until the bacteria are lysed, which is recognizable by the slightly slimy consistency of the solution. You then add 150 μL of solution III (3 M potassium acetate, pH 5.2) and mix it well again. The fluid becomes clear, although it contains flakes of potassium-SDS, which are only slightly soluble and precipitate out. Another two drops of chloroform are added, in which potassium-SDS is pelletized better. This solution is mixed and centrifuged for 2 to 5 minutes at room temperature. The clear supernatant (approximately 400 μL) is poured into a new vessel without the SDS flakes, 1 mL of ethanol is added, and the solution is mixed and centrifuged (15,000 g) for 10 to 15 minutes at 4°C. The pellet is dried and dissolved in 50 to 100 μL of TE plus 1 to 2 μL of DNase-free RNase A (10 mg/mL).

If a quick and dirty preparation is all that is required, you can use the solution at this point for restriction digestion. If it is mandatory that the DNA be clean, incubate for 30 minutes at room temperature to give the RNase time to work and then perform a phenol-chloroform extraction, including precipitation with alcohol.

The DNA pellet is easily visible after the first precipitation, but you should not be deceived because more than 90% of this precipitate is made of RNA, proteins, and salts. Nevertheless, the size of the pellet is generally proportional to the quantity of DNA and provides information concerning the yield of plasmid. After degradation with RNase and phenol-chloroform extraction, it is considerably smaller because the concentration of RNA has been reduced substantially.

**Suggestions:** Instead of solution I, you can also use TE solution or, in an extreme case, simply water. Solution II ages very quickly and should ideally be prepared in a fresh form, although it can be kept for up to 4 weeks. Solution III is a 3 M potassium acetate solution with a pH of 5.2, which contains far more acetate or acetic acid than potassium at this pH. The easiest approach is to mix 60 mL of 5 M potassium acetate, 11.5 mL of reagent-grade acetic acid, and 28.5 mL of $H_2O$. Instead of solution III, you can use 3 M sodium acetate at a pH of 4.8.

After the first precipitation, the DNA is surprisingly pure and poor in nucleases, and it can easily be stored overnight or even over the weekend at 4°C (39°F). You can add the RNase to solution I at the beginning instead of at the end (5 to 10 μL of RNase A [10 mg/mL]), and you should first make use of a 10-minute incubation period before the first precipitation with ethanol; in this case, however, the pellets are smaller.

A nice alternative is demonstrated in the method of Good and Nazar (1997). The bacteria are scratched off of the agar plate with a toothpick, which has been broken halfway, about 1 cm from the end and has been formed into a sort of golf club with the aid of tweezers, a tool that is also well suited to resuspend the bacteria in solution I later. The alkaline lysis is filled with plasmid DNA and then with polyethylene glycol 8000 (see Section 2.3.2).

**Literature**

Birnboim HC, Doly J. (1979) A rapid alkaline extraction procedure for screening recombinant plasmid DNA. Nucleic Acids Res 7:1513–1523.

Good L, Nazar RN. (1997) Plasmid mini-preparations from culture streaks. Biotechniques 22:404–406.

## Boiling Method

The boiling method is simple and fast, and it delivers very dirty DNA with a large quantity of bacterial DNA and proteins. The pellets from 1.5 mL of bacterial culture are resuspended in 300 μL of STET (8% [w/v] sucrose/5% (w/v) Triton X-100/50 mM EDTA/50 mM Tris HCl at pH 8.0; sterilized with a filter and stored at 4°C [39°F]), 200 μg of lysozyme is added, and this is then mixed and incubated on ice for 5 minutes. The tube is placed in boiling water for 1 to 2 minutes and then centrifuged for 15 minutes at 15,000 g. The supernatant is transferred to a new tube, 200 μL of isopropanol is added, and it is centrifuged (15,000 g) at 4°C (39°F) for 15 minutes. The pellets are washed with 70% ethanol, dried, and dissolved in 50 μL TE plus 1 μL of RNase A (10 mg/mL).

The method becomes attractive if you refrain from transferring the supernatant to another vessel and instead remove the pellets with a sterile toothpick so that the entire preparation can be processed within a single tube.

## Lithium Mini-Preparation

A very appealing feature of the lithium mini-preparation method is that it is a little faster than the others. The pellet from a 1.5-mL bacterial culture is resuspended in 100 μL of TELT (50 mM Tris HCl, pH 8.0/62.5 mM of NaEDTA/2.5 M of LiCl, 4% Triton X-100 [w/v]); it is stored at $-20°C$ [$-4°F$]), and 100 μL of phenol-chloroform (1:1) is added. It is vortexed briefly and centrifuged for 1 minute at 15,000 $g$. Then, 75 μL of the aqueous supernatant is transferred into a new tube, 150 μL of cold ethanol is added, and it is mixed and centrifuged for 10 minutes at 4°C (39°F) at 15,000 $g$. The pellets are washed briefly with 70% ethanol, dried, and dissolved in 50 μL of TE plus 1 μL of RNase A (10 mg/mL).

## 2.5.2 Preparation of Plasmid DNA on a Large Scale

In principle, a maxi-preparation does not differ extensively from a mini-preparation, at least if the specific dimensions are not considered. In the first step, you manufacture a crude bacterial lysate, thereby eliminating a large proportion of the bacterial proteins and membranes, and the bacterial genome is removed simultaneously because the chromosomal DNA (because of its larger parts, structure, and anchoring) is centrifuged off while the smaller and free plasmid molecules are preserved in the supernatant. This necessitates a somewhat careful procedure, because the share of bacterial DNA will otherwise increase extensively. In a second step, in which you eliminate the RNA and the residual proteins, the final purification of the DNA occurs.

True maxi-preparations have become rare because experimenters can perform most actions with the quantity of DNA that is obtained with a mini-preparation. We limit ourselves in the laboratory chiefly to 100 mL cultures in TB medium, which corresponds with an LB culture of 500 mL and delivers a yield of far more than 500 μg of plasmid DNA.

All records begin with inoculating the bacterial medium with 1 mL of the freshest possible overnight culture (a single bacterial colony is also adequate if it is not too old) and incubating it overnight at 37°C (99°F) so that the bacteria can attain a maximum density. The culture should then be transferred to an Erlenmeyer flask with a flow spoiler (i.e., with a spout that extends from inside) and diluted with a sufficient volume (guideline: five times the volume of the medium), which is then mixed on a flatbed agitator at 200 to 400 rpm so that the bacteria are well aerated. It is a good sign when it forms a foam. The bacteria produce higher yields.

There are three well-known methods each for bacterial lysis and subsequent DNA purification. The methods can be combined arbitrarily with one another. With all three methods for lysis, the volume of fluids can be reduced through precipitation by using a 0.8-fold volume of isopropanol instead of a 2.5-fold volume of ethanol.

## Bacterial Lysis by Means of Alkaline Lysis

With alkaline lysis, the bacterial DNA is removed effectively because DNA is denatured at a highly alkaline pH. After neutralization with solution III, both strands of the plasmid DNA are quickly hybridized again, while the far larger chromosomal DNA remains single stranded and precipitates out.

The bacterial pellet from 500 mL of LB culture is resuspended in 20 mL of solution I (see Section 2.5.1) plus 0.5 mL of RNase A (10 mg/mL). After the addition of 40 mL of solution II, mix well but not too furiously, and then let the cloudy solution stand for 10 minutes. With increasing lysis, the "soup" is converted into a slimy broth to which 30 mL of solution III and 1 mL of chloroform are added. This is mixed thoroughly but not too furiously, and it is then centrifuged (5000 $g$) at 4°C (39°F) for 15 to 30 minutes. Chloroform is an organic solvent that has been known to dissolve many plastic credit cards. Therefore, use only centrifuge tubes made of polyethylene, polypropylene, or other chloroform-resistant materials. The supernatant is incubated for another 30 minutes at 4°C (39°F) to guarantee the most complete digestion by RNase.

To increase the yield, you can add 1 mL of lysozyme (25 mg/mL) and then incubate the solution for 10 minutes at room temperature after resuspending the bacteria. If no chloroform-resistant centrifuge tubes are available, you can also dispense with the chloroform. However, pelletizing of the contaminants is clearly worse, and it is best to centrifuge the supernatant a second time.

The cleaner the supernatant is after centrifugation, the easier it is to perform the next stage of purification. We filter it through a large-pored (pore size of 0.8 μm), disposable filter, which must be clean but not necessarily sterile.

## Bacterial Lysis by Means of Boiling

The bacterial pellet from 500 mL of LB culture is resuspended in 20 mL of STET solution (see Section 2.5.1) plus 2 mL of lysozyme (10 mg/mL). It is transferred into a fire-resistant glass vessel and heated to boiling with a Bunsen burner (beware of the boiling retardation), and it is then heated for another minute in boiling water. The solution is cooled on ice, and the cold, slimy liquid is centrifuged ($\sim$25,000 $g$, 20 minutes), preferably using a swing-out rotor. The supernatant is transferred into a new vessel, 0.5 mL RNase A (10 mg/mL) is added, and the solution is incubated at 4°C (39°F) for 30 minutes.

If the lysozyme does not function properly, the yield tends to be almost nonexistent. The duration of the heating is critical and varies for different strains of bacteria. It is best to first test the conditions for the particular strain used.

## Bacterial Lysis by Means of Triton

The bacterial pellet from 500 mL of LB culture is resuspended in 5 mL of STET solution (25% [w/v] sucrose/50 mM Tris HCl, pH 8.0/100 mM EDTA; stored at 4°C [39°F]) plus 1.5 mL of lysozyme (10 mg/mL) plus 2 mL of 0.5 M EDTA plus 25 μL of RNase A (10 mg/mL), and it is incubated for 15 minutes on ice. Then add 2.5 mL of lysis solution (3% [v/v] Triton X-100/200 mM Na EDTA, pH 8.0/150 mM Tris HCl, pH 8.0; stored at 4°C [39°F]), mix thoroughly but carefully, and incubate for another 20 minutes at 4°C (39°F). Centrifuge the viscous solution for 60 minutes at 40,000 $g$ and 4°C (39°F), and decant the supernatant from the gelatinous pellets.

The method is somewhat critical, and the lysozyme solution should be fresh. If the pellet is not sufficiently firm after centrifugation and cannot be separated from the remainder, repeat the centrifugation process using a higher speed and, in case of doubt, even ultracentrifugation.

## Lysate Purification by Means of Cesium Chloride Gradient Centrifugation

Although lysate purification by means of cesium chloride gradient centrifugation was formerly the method of plasmid DNA purification, it is now used unwillingly because of the amount of work required. This method (see Section 2.3.6) does have some advantages that should not be underestimated. It is the only method that shows how much DNA is available, it is very clean, and very large quantities of DNA can be purified. You can attain *supercoiled* plasmid DNA, which is more suitable for the transfection of eukaryotic cells than relaxed plasmid DNA.

## Lysate Purification by Means of Anion-Exchange Chromatography

The use of anion-exchange chromatography has exceeded the use of cesium chloride gradients by far, primarily because of easier operation, but also because this procedure functions more rapidly. Most manufacturers (e.g., Qiagen, Macherey-Nagel, Promega) offer completed columns in different sizes and for different types of DNA, as well as for buffers and for different procedures. These columns are good, but you should not rely on statements from the manufacturers blindly. If the catalogs are to be believed, the DNA purified with the respective product is pure, much purer than the result found using a product from some competitor, and it can be used for everything. Practice, however, shows that preparations only sometimes function with commercial kits and that the purity of the DNA is not guaranteed. Opinions differ about which manufacturer produces the best column, and "religious wars" sometimes occur. In most cases, however, convictions are based on two unsuccessful preparations and much superstition. Kits that specify the composition of their buffers should be preferred because the accompanying buffers are frequently used up more rapidly than the columns. Further remarks concerning anion-exchange chromatography are found in Section 2.3.4.

## Lysate Purification by Means of Polyethylene Glycol Precipitation

PEG precipitation, which is a standard procedure for the precipitation of phages from phage DNA preparations, is relatively useless for the purification of DNA, because it is not as clean as the two methods mentioned previously and because the overnight incubation is quite lengthy. The amount of work required, however, is quite minimal, so that this method is suitable for any DNA that does not have to be so clean (Figure 2-7). It is described in Section 2.3.2.

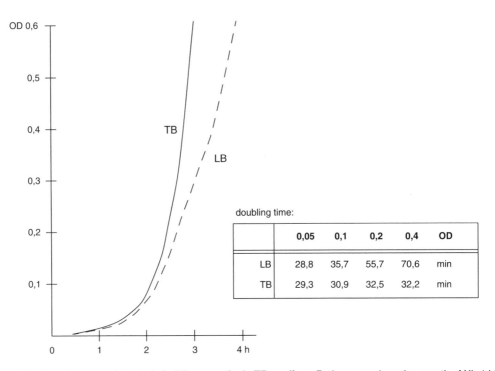

**Figure 2-7. Growth curve of bacteria in LB respectively TB medium.** Both curves show the growth of XL-1 blue MRF bacteria (Stratagene) in LB or TB medium, or both media. Although LB is considered to be a nutrient-rich medium, much richer ones are available that can deliver three to four times higher bacterial yields from overnight cultures, and the bacteria seem to fare well during the growth phase, as demonstrated by the doubling times.

## 2.5.3 Bacterial Media

Bacteria are frequently considered to be relatively easy to care for, although they nevertheless require their feed. The bacterial media can be classified as minimal media or nutrient-rich media, which are used in a liquid form or, through the addition of 1.5% (w/v) agar-agar, changed to a solid medium. Normally, the experimenter uses a nutrient-rich media, because the bacteria multiply more rapidly. Most frequently, Luria broth (LB) medium is employed (10 g of tryptone or peptone/5 g of yeast extract/5 to 10 g of NaCl/1 mL of 1 N NaOH per liter), which is reflected in all protocols. It is not always the most suitable medium. To obtain large yields of bacteria, such as for plasmid-DNA preparations, it is better to use terrific broth (TB) medium (12 g of tryptone, 24 g of yeast extract and 4 mL of glycerin per liter, to which is added a 0.1-fold volume of 1 M $KHPO_4$ solution at pH 7.5), because the bacteria then grow to about four times as densely. Such a rich medium can also be interesting for the production of competent cells, because the bacteria remain in the phase of optimal growth for a longer period than in the LB medium, in which the bacteria go into a sort of hibernation at an $OD_{595\,nm}$ of 0.1.

The selection of a bacterial medium also depends on the particular behavior in the laboratory, the requirements of the specific bacteria, and the limitations imposed by their application. It is best to follow the recommendations of colleagues or to refer to the standard laboratory literature.

Bacterial media can be used repeatedly as long as they are nutrient rich enough and sufficiently buffered. Although LB is less suitable for such eccentric attempts, I have been able to successfully cultivate bacteria and to isolate DNA three times from a single TB culture by tipping the supernatant from the culture back into the culture tube after centrifugation; only during the fourth attempt did the bacteria finally stop replicating. The bacteria multiply so quickly in this process that you can continue with the second DNA preparation immediately after having finished with the first.

## 2.5.4 Preparation of Phage DNA

Although the handling of phages used to be a classic because all DNA banks were cloned into phage vectors, they have gone somewhat out of fashion because there are more practical vectors available that can take larger DNA fragments into consideration. The real weakness of phages is DNA preparation, for which there is really no ingenious method. The preparation demands too much work and delivers yields that are too small or produces relatively unclean DNA.

The phage particles can be obtained from plate lysates or from liquid cultures. For a **plate lysate**, a sufficient amount of phage is plated to obtain a confluent (i.e., sufficiently lysed) plate. To this is added the phage solution (approximately 1 to 100 mL according to the phage titer) with 150 mL of a bacterial overnight culture (be sure to use the bacterial strain fitting to the phages used). It is incubated for 10 minutes at 37°C (97°F); approximately 3 to 6 mL of lukewarm, still liquid and soft agar (LB with 0.7% agar) are added. It is mixed briefly, and the fluid is distributed evenly on an LB agar plate, which is incubated overnight at 37°C (97°F). Phages are eluted with SM solution (see Appendix 1), using 5 mL for a 10-cm disk and 10 mL for a 15-cm plate, and it is pipetted on the plate and incubated overnight at 4°C (39°F), preferably while shaking gently. The eluate is suitable as a phage stock solution or can be used for the preparation of small quantities of phage DNA. Nevertheless, the agar is likely to contain contaminants that must also be eluted and generally do not disappear during preparation. They very efficiently inhibit restriction enzymes, and you must frequently employ a 10-fold quantity of enzyme to obtain a satisfactory yield, if it functions at all. You can avoid the problem by replacing the agar with *molecular biology grade* agarose, which is definitely more expensive.

The problem with **liquid cultures** is that they must contain exactly the right balance of bacteria and phages when preparing the culture to obtain an optimal yield. If too many phages are used, all of

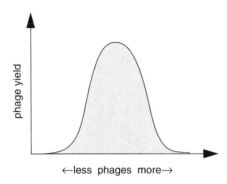

**Figure 2-8. The yield of phages depends on the relationship of the bacteria and the phages.**

the bacteria will be lysed before their concentration has increased sufficiently. If too many bacteria are used, they will reach their stationary phase and discontinue growth before the phages have had a chance to replicate sufficiently. Because the optimal relationship depends on the exact conditions employed (phages, bacteria, and media), it is worthwhile to perform a test series at the beginning of the work (Figure 2-8).

A protocol provides guidance: $50\,\mu L$ of bacteria from an overnight culture (corresponds with approximately $5 \times 10^7$ bacteria) and $10^5$ phages (most plans make use of 50 to 500 times as many bacteria as phages) are incubated for 15 minutes at 37°C (97°F). Then, 150 mL of LB medium is added, and the mixture is incubated at 37°C (97°F) while shaking vigorously. After about 4 to 5 hours, the culture becomes increasingly cloudy, and it then becomes almost clear within another hour. This is the point of the maximum concentration of phages; if you let the culture continue to grow, it will again become cloudy, because resistant bacteria will get out of hand. If the culture does not become cloudy, too few bacteria have been used in the solution, and if it does not become clear, there were too few phages.

Before use, the lysate must be centrifuged ($5000\,g$, 4°C [39°F], 15 minutes) to dispose of the remaining living bacteria and the remainder of the lysed bacteria. You can then proceed to DNA extraction.

## $\lambda$ Phage DNA Mini-preparation I by Means of Polyethylene Glycol Precipitation

In the standard method, 3 mL of phage lysate is incubated with $10\,\mu L$ of RNase A solution (see Appendix 1) and $1\,\mu L$ of DNase I ($10\,mg/mL$) for 1 hour at 37°C (97°F). Afterward, add 3 mL of PEG solution (20% [w/v] PEG 6000/2 M NaCl in SM solution) (see Appendix 1), incubate for 1 hour on ice, centrifuge ($10,000\,g$, 4°C [39°F], 20 minutes), and resuspend the pellets formed in 0.5 mL of SM solution. Crude contaminants are centrifuged out (10,000 to $15,000\,g$, room temperature, 2 minutes), the supernatant is then incubated with $5\,\mu L$ of 10% (w/v) SDS and $5\,\mu L$ of 0.5 M EDTA (pH 8.0) for 15 minutes at 68°C (154°F). You then perform a phenol-chloroform purification (see Section 2.3.1), precipitate the DNA with ethanol, and place the dried pellets in 20 to $40\,\mu L$ of TE (see Appendix 1).

## $\lambda$ Phage DNA Mini-preparation II by Means of DEAE Cellulose

$\lambda$ Phage DNA mini-preparation II by means of DEAE cellulose is more rapid than PEG precipitation but likely to be even dirtier. The protocol is based on the publication from Manfioletti and Schneider

(1988), although they use cetyl trimethyl ammonium bromide (CTAB, Sigma) for the final step in extraction, with which nucleic acids and acidic polysaccharides can be precipitated specifically. The publication is worth taking a look at.

The DEAE cellulose is prepared by suspending 10 g of DEAE cellulose (DE 52 ion-exchange cellulose according to Whatman) in 200 mL of 0.05 N HCl, which is neutralized with 400 μL of 10 N NaOH, from which the supernatant is decanted. The cellulose is then washed three or four times with a fivefold volume of SM solution. The cellulose then is suspended in a one-third-fold volume of SM solution. If the cellulose is to be stored longer than 1 week at 4°C (39°F), add 0.01% of sodium azide (which is extremely poisonous) to protect the medium from the malicious microbes, and wash once more with SM before use.

Mix 0.6 mL of phage lysate and 0.6 mL of suspended DEAE cellulose thoroughly. The mixture is centrifuged (10,000 g at room temperature for 5 minutes), mixed with supernatant containing 120 μL of denaturing solution (10 mM Tris HCl, pH 8.0/2.5% [w/v] SDS/0.25 M EDTA, pH 8.0), and incubated for 1 minute at 70°C (158°F). The reaction material is allowed to cool, and 75 μL of 5 M potassium acetate is added. It is incubated for 15 minutes on ice and centrifuged (10,000 g at room temperature for 5 minutes). The DNA is precipitated from the supernatant using a 0.8-fold volume of isopropanol. The washed and dried pellets are dissolved in 10–50 μL of TE.

## λ Phage DNA Mini-preparation III by Means of Zinc Chloride

The λ phage DNA mini-preparation III by means of zinc chloride is described further with the maxi-preparation II. Is was conceived by Santos (1991) as being a preparation for only a few milliliters of lysate, and the original version is described here.

Add 20 μL of 2 M ZnCl to the lysate (already degraded by DNase 1 and RNase A), incubate for 5 minutes at 37°C (97°F), and centrifuge (10,000 g, 1 minute). The pellets are then resuspended in 500 μL of TES buffer (0.1 M Tris HCl, pH 8.0/0.1 M EDTA/0.3% SDS) and incubated for 15 minutes at 60°C (140°F), and 60 μL of 3 M potassium acetate at pH 5.2 is then added. It is placed on ice for 10 to 15 minutes and centrifuged (10,000 g, 4°C [39°F], 1 minute), and the DNA is precipitated from the supernatant using isopropanol.

## λ Phage DNA Maxi-preparation I Using a Cesium Chloride Density Gradient

Phage DNA maxi-preparation I using a cesium chloride density gradient is by far the cleanest method for attaining λ DNA, although this procedure takes a long time and is quite laborious. It does not function with small quantities of phages.

Dissolve 30 g of NaCl in 500 mL of phage lysate, centrifuge (4000 g, 4°C [39°F], 10 minutes), dissolve 50 g of polyethylene glycol 6000 in the supernatant, and precipitate the phages overnight at 4°C (39°F). After centrifugation (4000 g, 4°C [39°F]), 10 minutes), the pellet is dissolved in 5 mL of SM solution (see Appendix 1, Standard Solutions), the preparation is mixed with 5 mL of chloroform and centrifuged (6000 g, 4°C [39°F], 10 minutes), and the aqueous phase is transferred into a new vessel. The chloroform step is repeated until PEG is almost eliminated (i.e., you can no longer see a white interphase after centrifugation). The aqueous phase is applied to a CsCl density gradient (three levels: 1.3, 1.5, and 1.7 g of CsCl/mL SM solution; see Fig. 2-9) and ultracentrifuged (80,000 g, 20°C [68°F], 1.5 hours). The phages collect as a pale bluish band at the phase boundary between the 1.5 and 1.7 g/mL solutions. The phage band is pipetted off and dialyzed for 30 minutes in SM solution (see Chapter 2, Section 2.3.7); subsequently, 100 μL of each of the phage solutions, 5 μL of 0.5 M EDTA, and 1 μL of 10% (w/v) SDS are added and incubated for 10 minutes at 65°C (149°F). This step decomposes the phage particles into DNA and proteins; the latter are

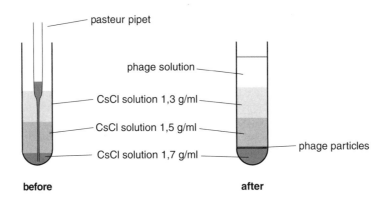

| density | 1,7 g/ml | 1,5 g/ml | 1,3 g/ml |
|---|---|---|---|
| quantity of CsCl | 56,24 g | 45,41 g | 31,24 g |
| volume of solution SM | 43,76 ml | 54,59 ml | 68,76 ml |

**Figure 2-9. Cesium chloride staging gradient.** Density gradients are practical because much smaller centrifugation times are required than with the use of a continuous gradient. The sharpest possible boundaries between the individual CsCl solutions are decisive. It is easiest to begin with the solution of the lowest density and, with the help of a Pasteur pipette, carefully add this to form a lower layer under the solutions of higher density. The gradient is then covered with a layer of the phage solution. Manufacture of the CsCl solutions requires precise work. The table indicates which quantities of CsCl must be dissolved in which volumes of SM solution to obtain a solution of the specified density. The proportion of CsCl of the total weight for a CsCl solution of a desired density can be calculated using the following equation: % (w/w) CsCl = 137.48 − 138.11/desired density.

removed by phenol-chloroform extraction (see Section 2.3.1), and the DNA solution is dialyzed in TE for at least 60 minutes.

The CsCl density gradient I is somewhat difficult to produce (Figure 2-9). It is easiest if you first pipette the solution with the lowest density from the centrifuge cup and carefully, with the aid of a Pasteur pipette, add it to form a lower layer below the solutions with higher densities.

## λ Phage DNA Maxi-preparation II by Means of Zinc Chloride Precipitation

The quality of the DNA is not outstanding with λ phage DNA maxi-preparation II by means of zinc chloride precipitation, but the amount of laboratory work is substantially reduced, and there is a substantial gain in time. The method has been derived from Santos (1991).

A 100-mL liquid culture must be prepared. One hour before the end of the lysis, 25 μL of DNase I (10 mg/mL) and 50 μL of RNase A are added (see Chapter 13, Section 13.2) and incubated for 1 hour at 37°C (99°F) under continuous agitation (in this way, the degradation step can be skipped). The lysate is centrifuged (5000 g, 5 minutes), and the supernatant is mixed with 1 mL of 1 M $ZnCl_2$, incubated for 5 minutes at 37°C (99°F), and again centrifuged (5000 g, 5 minutes). The pellet is carefully resuspended in 3 mL of $H_2O$, and two phenol-chloroform purification procedures are carried out (see Section 2.3.1). To displace the zinc from the DNA, add a 0.1-fold volume of 0.5 M $MgCl_2$, incubate for 10 minutes, precipitate with a 0.1-fold volume of 3 M sodium acetate (pH 4.8) and a

0.8-fold volume of isopropanol, and then wash with 70% (v/v) ethanol. If the DNA yield is large, a DNA cloud develops during precipitation.

The $MgCl_2$ step is important in this procedure, because $Zn^{2+}$ restriction enzymes and other enzymes are inhibited. If you have time, you should dialyze the DNA against TE (see Chapter 13, Section 13.2).

**Literature**
Manfioletti G, Schneider C. (1988) A new and fast method for preparing high quality lambda DNA suitable for sequencing. Nucleic Acids Res 16:2873–2884.
Santos MA. (1991) An improved method for the small scale preparation of bacteriophage DNA based on phage precipitation by zinc chloride. Nucleic Acids Res 19:5442.

## 2.5.5 Preparation of Single-Stranded DNA with the Aid of Helper Phages

From plasmids with an M13 replication origin (so-called phagemids), you can obtain single-stranded DNA through transfection of the bacteria with the aid of helper phages. This characteristic is sometimes very useful, because single-stranded DNA can be much more easily sequenced, for instance, than double-stranded DNA. Helper phages (e.g., M13K07, R408) can be obtained from Promega, Stratagene, or Clontech. They contain all the information to reproduce themselves and to also initiate single-strand reproduction and the packaging of plasmid DNA.

The manufacture of single-stranded DNA is simple. Cultivate a plasmid-containing bacteria in a suitable medium until it has reached an $OD_{600}$ of 0.1, and then add helper phages with a *multiplicity of infection* (MOI) equivalent to 20 (i.e., 20 times more phages than bacteria, for which 1 $OD_{600}$ corresponds to about $3 \times 10^8$ bacteria/mL), and incubate the culture for another 4 hours or overnight. You will obtain a phage lysate that can be processed similar to a $\lambda$ lysate, such as with PEG precipitation and phenol-chloroform purification.

## 2.5.6 Preparation of Genomic DNA

Genomic DNA is extremely long. The 3 billion base pairs of a mammal's genome have distributed themselves into about 20 chromosomes, making an average length of about 150 million base pairs per chromosome (a chromosome is one DNA double strand that is composed of two complete molecules). The difficulty lies in purifying such gigantic molecules without hacking them into pieces that are far too small. With the standard methods, this is not simple, and you can consider yourself to be successful if you eventually have pieces about 200,000 bases long.

Genomic DNA is usually to be obtained from tissues, blood cells, or cultivated cells. The most laborious method is the use of tissues. Liver is the most beneficial source because many liver cells are polyploid, and the DNA yield is therefore very high. However, be sure to remove the gallbladder during preparation, because this structure is rich in nucleases.

The tissues are first weighed and then deep-frozen in liquid nitrogen (do not forget to wear goggles). They are then ground to a fine powder in a mortar pre-cooled with liquid nitrogen. It is much easier if the fresh tissues are ground into small pieces that are separately deep frozen, because even the most flabby liver becomes as hard as stone at −70°C (−94°F) to −196°C (−321°F), and larger pieces can frequently be ground only with the help of a hammer. Do not forget to pre-cool the pestle. It is helpful to construct a cover for the mortar out of thick aluminum foil, because the tissue pieces may otherwise splash out when using vigorous force.

The powder is dissolved in proteinase K buffer free of any clumps (100 mM NaCl/10 mM Tris HCl, pH 8.0/50 mM EDTA, pH 8.0/0.5% SDS/20 μL of RNase A/0.1 mg/mL of proteinase K), using

1.2 mL of buffer per 100 mg of tissue. This produces a rather unsavory, slimy solution. The tissue broth is mixed or shaken overnight at 50 °C (122 °F) so that the consistency becomes somewhat more liquid. Add an equal volume of phenol-chloroform solution (see Appendix 1, Standard Solutions), mix thoroughly, and centrifuge for 15 minutes at 2000 to 5000 g. The supernatant is transferred to a new vessel, and the interfacial phase should be avoided. It is again extracted by shaking with a phenol-chloroform solution, centrifuged, and transferred to a new vessel. Add a 0.1-fold volume of 5 M LiCl and a twofold volume of ethanol, and then mix very carefully. If a sufficient quantity of DNA is available in the preparation, a Medusa-like, cloudy object develops from the precipitated, long-chain DNA, which becomes increasingly compact during the course of mixing. The transparent, sticky DNA cloud is fished out with a Pasteur pipette, which has been melted until it has the form of a small crochet hook, and the DNA is then carefully washed in 70% ethanol to eliminate the remnants of phenol and salts. The cloud turns white and becomes substantially smaller. All of the fluid is removed, and the DNA is dried in the air. It is then dissolved in TE, a slow process that is carried out over 1 to 2 days. With gentle movement of this solution at room temperature or at 65°C (149°F), the time required for this procedure can be accelerated.

With cultivated cells, this method can be performed much more easily. After they have been washed with PBS and pelleted, they are resuspended in a suitable volume of proteinase K buffer. For smaller quantities, 300 µL is sufficient; for more than $3 \times 10^7$ cells, calculate 1 mL of buffer per $10^8$ cells. Then proceed as described previously.

The preparation of genomic DNA from **blood samples** is somewhat more difficult. The DNA quantities acquired are not as large, but contamination with heme may occur, which can inhibit PCR reactions. Care must be taken in the choice of the anticoagulant. EDTA and citrate are very suitable, but heparin may cause difficulties.

For the purification of DNA from blood, kits are available from many manufacturers (e.g., Pharmacia, Qiagen, Promega, AGS). They can make life easier, especially considering the great need for clinical diagnoses.

For larger preparations, previous isolation of the lymphocytes is recommended. A twofold volume of Ficoll-Paque (Pharmacia; Ficoll-Paque is an aqueous solution with a density of 1.077 g/mL that is made up of 5.7 g of Ficoll 400 and 9 g of sodium diatrizoate with calcium EDTA per 100 mL of solution) is carefully covered with a layer of 1 volume of blood in a centrifuge tube and centrifuged for 30 minutes at 250 g. The lymphocytes accumulate at the boundary layer between the blood plasma and the Ficoll, and they can be removed with a Pasteur pipette. The erythrocytes form a sediment on the floor of the vessel.

Performed carefully, the method can produce DNA fragments of approximately 50 to 100 kb. If you require DNA fragments that are as long as possible, you should carry out a precipitation with ethanol instead of dialysis before performing a phenol-chloroform purification (see Section 2.3.7). Dialysis is performed at least twice for a total of 24 hours in a 100-fold volume of TE solution.

Frozen tissues can be preserved indefinitely at −70°C (−94°F). However, you should consider the fact that the tissue must be removed from the vessel in which it has been stored. Because the tissue may not be allowed to thaw before being ground with a mortar, it is generally helpful to use the handle of a hammer to make use of the pieces of liver that have become solidified in the vessel. It is therefore best to freeze tissue pieces individually in liquid nitrogen and to then place them in a (precooled) storage vessel, or you can freeze the material immediately within small, plastic sacks.

# 3 The Tools

*Laß du die große Welt nur sausen,*
*wir wollen hier im Stillen hausen.*

*Leave the world to its carousing*
*Let us here in peace be housing!*

## 3.1 Restriction Enzymes

Restriction enzymes are more correctly called restriction endonucleases. Without these structures, modern molecular biology is barely conceivable, and this book might not have come into existence.

The term *restriction endonuclease* cannot be considered to be very sensible in English or German. The origin of this term is an observation reported by Luria in the beginning of the 1950s (Luria, 1953). It was noticed that the bacteriophages examined in bacterial strain A replicated very well, whereas they replicated very poorly in strain B. They were *restricted* to a specific strain of bacteria; the term *restricted* was first used in this context by Arber and Dussoix in 1962. The few phages that are able to develop in strain B continue to grow well in strain B, but no longer in strain A. This phenomenon was later explained by the discovery of endonucleases that, because of specific methylation patterns, were able to differentiate the bacterium's own DNA from foreign DNA. If the phages have first proliferated successfully in a new bacterial strain, their DNA subsequently carries the specific methylation pattern of the new bacterial strain, and they no longer recognize this DNA as being foreign.

Restriction enzymes are truly miracles of biology. Depending on their specificity, they recognize four to eight base pairs in a DNA strand, in which only a small pile of phosphates and residual glucose can be seen. Nevertheless, they are exceptionally precise, because they slice out only their specific target sequence and nothing else as long as the proper buffering conditions are maintained. How they attain this specificity is one of the puzzles of molecular biology, a situation with which the experimenter fortunately does not have to concern himself or herself.

### 3.1.1 Nomenclature

The names of restriction enzymes appear to be rather cryptic at the beginning, although they are in no way to be compared with the first attempts at writing made by bonabos (i.e., pygmy chimpanzees). The first three letters, written in italics, are composed of the initial letter of the genus and the first two letters of the species name of the bacterium from which the enzyme was isolated (e.g., *Escherichia coli* = *Eco*). After it, in roman typeface, is the designation of the stem or type (e.g., *Eco*R), which is followed by the number of the isolated restriction enzyme in roman numerals (e.g., *Eco*RI).

Using this method of construction, the restriction enzymes are divided into three categories. **Type I** restriction enzymes are composed of three subunits; the S subunit recognizes the DNA sequence, the M subunit methylates it, and the R subunit cleaves it. They recognize a specific DNA sequence, but

**Table 3-1.** Examples of Isoschizomers and Neoschizomers

| C↓CCGGG GGGCC↑C | CC↓CGGG GGGC↑CC | CCC↓GGG GGG↑CCC | CCCG↓GG GG↑GCCC | CCCGG↓G G↑GGCCC |
|---|---|---|---|---|
| Cfr9I PspAI XmaI XmaCI | | PspALI SmaI | | |

they cut by chance so that their application in molecular biology is limited. Of exceptional interest is their specific behavior. If the enzyme comes in contact with unmethylated DNA, it is (primarily) cleaved; hemimethylated DNA, as is found at the end of the replication, is methylated.

Isoschizomers are interchangeable, and neoschizomers are not, because they cut differently even if they recognize the same sequence. In Table 3-1, *Xma*I and *Psp*AI are isoschizomers, and *Xma*I and *Sma*I are neoschizomers relative to one another.

**Type III** restriction enzymes have several subunits, recognize specific sequences, and sever the DNA 20 to 25 nucleotides removed from these sites. Only five type III restriction enzymes have been described, and their significance in daily laboratory life is relatively small.

Truly interesting for the experimenter are the **type II** restriction enzymes. They have been the key to molecular biology over the past few decades, because they sever DNA at exactly definable sites. Almost all commercial restriction enzymes belong to this type. In each of the bacterial strains explored over the past 25 years, only one or two type II restriction enzymes were found, and more than two were found in approximately 25% of the cases; the record has been six. A total of 2750 of such restriction enzymes has emerged from these screenings, demonstrating 211 different specificities (Roberts and Macelis, 1996). Type I restriction enzymes, with only 29 individual cases, play almost no role.

Type II restriction systems are composed of two independent proteins, restriction enzymes and methylase, which both recognize the same target sequence. The restriction enzymes sever the DNA, and the methylase preserves hemimethylated DNA completely from methylation and protects the DNA from being degraded by its own restriction enzymes. This mechanism, by which the bacteria protect themselves against the degradation of their own DNA by these restriction enzymes, occasionally presents a substantial problem for the experimenters and others, because other organisms sometimes also methylate their DNA, unfortunately exactly where the experimenter would like to cut. Fortunately, the redundancy is rather large, and various restriction enzymes from different bacteria recognize the same sequence, although they do differ in their sensitivity to methylation because of their various origins.

Enzymes that recognize the same sequences and cleave them in a similar manner are known as **isoschizomers**. In addition to their sensitivity to methylation, they may differ in their buffer relationships, which may be of interest, especially in the event of double digestion, although the difference is occasionally seen in the price alone. Isoschizomers should not be confused with **neoschizomers**, which recognize the same sequences but sever the DNA differently (see Table 3-1). There is little consideration of this behavior even in catalogs, and soon, everything will be combined under the heading of *isoschizomer*.

Most of the restriction enzymes used in the laboratory recognize a sequence of four, six, or eight bases (i.e., 4-, 6-, or 8-cutters), although 5-cutters are used occasionally. The recognized sequences are normally palindromes (i.e., symmetrical) and consequently identical in both strands. Considered statistically, 4-cutters sever every 256 bp, 6-cutters every 4096 bp, and 8-cutters every 65536 bp. In practice, the frequency varies somewhat, depending on the specific organism and DNA, because the frequency and distribution of the bases is not random. Most enzymes cleave within their recognized sequence, although there are some exceptions that can be recognized based on their asymmetrical recognition sequence.

Almost all enzymes generate fragments with a 5′-phosphate and a 3′-OH end, which is important for ligation. The ends are smooth (i.e., *blunt ends*), with both strands cut at the same site, or they overlap (i.e., *sticky ends*). If the 5′ end is longer, it is referred to as a 5′ overhang; otherwise, it is a 3′ overhang. The overhang usually is two or four bases long. Because longer overhangs can be ligated more easily, experimenters prefer to use neoschizomers as much as possible for cloning, because they produce four-base overhangs.

Enzymes that cut at a defined distance from the recognized sequence are sometimes very useful for cloning when the experimenter wishes to retain no remnants of the detection sequence. Highly interesting in this respect are the restriction enzymes *Bbs*I/*Bpi*I, *Bpu*AI (GAAGAC), *Bsa*I/*Eco*31I (GGTCTC), *Bsm*BI/*Esp*3I (CGTCTC), and *Bsp*MI (ACCTGC), because they produce an arbitrary, 4-base, 5′ overhang. By installing the recognition sequences in a primer and amplifying them in a cloned fragment by means of a polymerase chain reaction (PCR), vectors can be cloned (e.g., fragments in a *Bam*HI spliced vector that contain one or several internal *Bam*HI cleavage sites) that normally can be cloned only with much effort.

## 3.1.2 The Activity Test

A unit is the quantity of an enzyme that is necessary for fully degrading 1 μg of the indicated substrate DNA within 60 minutes at the correct temperature and in the correct buffer from 50 μL of a preparation of reaction material. This definition, however, says little about the quantity of enzymes that should be purchased. *Apa*I, for example, is cut 12 times in the adenovirus-2 genome but only once in the almost 50% larger λ genome. According to the DNA substrate, which is used for the *Apa*I activity test, the enzyme quantity per unit can differ by a factor of about 17!

This definition allows comparison of the different enzymes to an equally small degree. λ-DNA demonstrates a single cutting site for *Xho*I but two sites for *Sal*I and even eight for *Bel*I. The enzyme quantity in one unit of *Bel*I consequently demonstrates an eightfold higher activity than that found in one unit of *Sal*I, as long as λ DNA is used as a substrate.

The buffer can enable some important tricks. In the presence of bovine serum albumin (BSA), many enzymes reveal a higher activity (discussed later), and fewer enzymes are required to perform an activity test. In the normal restriction buffer, the BSA is left out for reasons of economy. Only few manufacturers provide exact information in their catalogs concerning the specific test conditions so that it is impossible to make an accurate price comparison. If the enzyme-specific information (e.g., *Sal*I slices only *supercoiled* plasmid DNA very poorly) and sequence-specific problems (e.g., *slow sites* that can be cut only slowly) are considered in addition to this confusion in definitions, you can find yourself to be near despair.

There are other characteristics of the delivered enzyme solution that are of interest and can be measured with the aid of two tests. In the **overdigestion assay**, the substrate DNA is digested for 16 hours with different quantities of the enzyme. The test provides information concerning the maximum enzyme quantity at which well-defined DNA bands can be seen. This quantity also provides information on the contamination with nonspecific endonucleases and exonucleases. In the **ligation assay**, the substrate DNA is digested completely, ligated again, and digested again with the same restriction enzyme. The quantity of non-religated fragments provides information on the contamination with phosphatases and exonucleases, whereas the quantity of ligation products that cannot be cleaved provides additional information on the exonucleases. Normally, the manufacturer tests the quality of the batches with both of these assays to guarantee a certain standard of quality. However, this information can occasionally be of interest to the experimenter, because some enzymes perform poorly. If the segments cut by *Nci*I or *Bst*NI, for example, are poorly religated or cannot be religated at all, it is wise to dispense with them for cloning.

Some manufacturers (e.g., New England Biolabs, Promega) test some enzymes for their exonuclease content (important for blue or white selection after cloning) (see Chapter 6, Section 6.2.1), and some (e.g., Promega) test for their in-gel digestion (important in pulse-field gel electrophoresis) (see Section 3.2.5).

Restriction enzymes are stored at $-20°C$ $(-4°F)$ like most enzymes. If they are taken from the freezer, they should be put on ice or, better still, in a cooler (e.g., StrataCooler from Stratagene, Labtop Cooler from Nalgene), which keeps the enzymes at $-20°C$ $(-4°F)$ for a longer period and is not as moist. There are also coolers designed especially for refrigerator temperatures (0 to $4°C$ [32 to $39°F$]).

Many restriction enzymes are sold at far higher concentrations than is useful in daily laboratory practice. This presumably involves a conscious marketing strategy of the manufacturers, because it is true for the most commonly used enzymes, such as *Eco*RI or *Hind*III, which are comparatively inexpensive. Because the average, old laboratory fogey always employs microliter enzyme solutions for his or her digestions, regardless of how much enzyme the source contains, the companies still make their share. Money can be saved by diluting these enzymes. Many manufacturers make reference to this practice, and some even market suitable dilution buffers. Because the stability decreases over time, you should dilute only such quantities as can be used within 1 week.

## 3.1.3 Making a Restriction Digestion

From the enzyme activity list of the manufacturers, the experimenter selects the fitting buffer, pipettes, DNA, water, buffers, and enzymes. The appropriate items are mixed thoroughly and incubated for an hour at $37°C$ $(97°F)$. Usually, this approach works.

In the **standard digestion** method for analytical restriction digestion, 0.2 to 1 μg of DNA is used. Under certain circumstances, using slightly smaller quantities of DNA, one can create difficulties with the identification, because the detection limit for ethidium bromide in agarose gel is approximately 10 to 20 ng of DNA per band. With larger quantities, the problem may be an overload of the agarose gel, so that the band becomes a cloudy spot with branches extending upward that progresses more quickly than corresponds with the fragment size, invariably making an exact estimation of the size incorrect. This phenomenon occurs at a DNA concentration of 0.3 to 1 μg per band, depending on the size of the DNA fragment. The volume of the reaction material should be approximately 20 μL, because this amount usually fits in the sample well of an agarose gel.

The **digestion of PCR fragments**, according to Turbett and Sellner (1996), functions very well with unpurified PCR fragments. The enzyme is added to the commercial PCR solution, and digestion begins as usual. This method works well with many common enzymes, although some require somewhat more time or an extra shot of magnesium. For analysis of large quantities of PCR solution, perform the digestion overnight to avoid many additional manipulations.

If you want to reuse the generated fragments for cloning, purify the PCR reaction material beforehand, because the Taq polymerase demonstrates sufficient activity at $37°C$ $(97°F)$ to fill up overlapping ends or to add nonspecific adenosine (Table 3-2). Even if the cleavage sites are at the ends of a fragment, such as in the region of the primer, you should fully purify the fragment. In addition to the incorrect buffer conditions and the residual activity of the Taq polymerase, there may be another problem. Many restriction enzymes cut better in the middle of a fragment than at its ends, because they require a certain number of bases to the left and to the right of the cleavage site for full activity. Many enzymes are satisfied with one to four bases, whereas others (e.g., *Sal*I) require up to 20. This peculiarity is troublesome, because the slight difference in size between digested and undigested DNA makes it practically impossible to examine whether the digestion has been

**Table 3-2.** Activity of Restriction Enzymes in Taq Polymerase Buffer

| Activity | Restriction Enzymes | Comments |
|---|---|---|
| Full activity | *Aat*II, *Alu*I, *Ava*I, *Bam*HI, *Bst*OI, *Dra*I, *Hae*III, *Hha*I, *Hinc*II, *Hind*III, *Hpa*II, *Kpn*I, *Msp*I, *Pvu*II, *Rsa*I, *Sac*II, *Sma*I, *Stu*I, *Taq*I, *Xba*I, *Xma*I | |
| Slightly reduced activity | *Ava*II, *Bgl*II, *Bst*XI, *Hinf*I, *Mbo*I, *Mbo*II, *Pst*I, *Ssp*I | Prolong incubation time or increase the quantity of enzymes. |
| Reduced activity | *Bsp*1286I, *Sal* I, *Sau*3AI | Restriction enzyme buffer to be added (to a final, 0.5-fold concentration) |
| Star activity | *Eco*RI | Restriction enzyme buffer to be added (to a final, 0.5-fold concentration) or $Mg^{2+}$ (to a final 10 mM concentration) |

completed. The solution to the problem can be found in the catalog from New England Biolabs, in the appendix under "Cleavage Close to the End of DNA Fragments." The list provides information with which to gauge the probable success of a digestion. The article from Moreira and Noren (1995) also contains a very useful survey concerning the propensity of fragment ends to be digested by the most common restriction enzymes.

Frequently, a digestion can be performed by simultaneously using two enzymes (i.e., **double digestion**), as long as both enzymes demonstrate sufficiently high activity in the buffer ($\geq 75\%$). If no such buffer is to be found, digestion is first carried out with the enzyme that requires the lower salt concentration, and then the solution is stocked with buffers until the optimal concentration is reached for the second enzyme. After the first digestion, purification is done with phenol-chloroform using ethanol precipitation or by cleaning with glass milk, which is the cleanest, but also the most laborious alternative and the method associated with the highest quantities being lost.

Double digestions can become problematic if two restriction cleavage sites are located next to one another (as is occasionally to be observed with the *multiple cloning site* of vectors), because some enzymes no longer recognize their cleavage sites if they are situated at the end of the DNA fragment (discussed earlier). If a cut is made at one cleavage site, the cut also will occur at the other site. There is not much an experimenter can do about this. Sometimes, very long digestion times can help, other cleavage sites may be tried.

Some restriction enzymes can digest **single-stranded DNA**, although with reduced efficiency (1% to 50%). Primarily, *Hha*I, *Hin*PII, and *Mnl*I are suitable, and *Hae*III, *Bst*NI, *Dde*I, *Hga*I, *Hinf*I, and *Taq*I are somewhat less effective.

After the digestion, the enzyme can be disposed of by **heat inactivation**. The activity is disposed of rather than the enzyme itself, which is frequently just as good. To do so, the reaction materials are heated to 65°C (149°F) for 20 minutes. Unfortunately, this does not work with all restriction enzymes, and in that case, routine cleaning is the only solution. A list of the enzymes that can be heat inactivated can be found in every catalog in which restriction enzymes are listed.

## 3.1.4 Difficulties Associated with Restriction Digestion

Although the restriction digestion represents one of the simplest procedures in molecular biology, things can go wrong. Some of the possible errors and problems are discussed in this section.

**Instead of bands, only a smear is recognized in the agarose gel**. The solution has been contaminated by nonspecific nucleases. The most probable source is your own fingers. Be careful in the use of reaction vessels; many individuals routinely grasp the inside of the lid to the vessel and are subsequently surprised to see that their solution has been contaminated with nucleases. Another possibility is that you have digested **genomic DNA**. Because distribution of the cleavage sites in the DNA occurs more or less by chance, a static mixture of fragments between almost 0 and 50 kb long is obtained from the digestion of genomic DNA, which appears as a uniform smear in the gel. Genomic DNA also contains many repetitive elements in which the distribution of the cleavage sites is no longer random so that some of the many restriction enzymes present in this smear produce clearly defined, typical bands. With some practice, you can use these bands to see whether the digestion has functioned or the DNA has been degraded by nonspecific nucleases. Under certain conditions, you may even save yourself several days of work.

**Instead of bands, a gigantic, bright spot is seen in the agarose gel at the lower end of the tracing**. A classic error with the use of plasmid DNA mini-preparations is forgetting the RNase digestion in the DNA preparation. The fat spot is the bacterial RNA.

**The DNA is digested incompletely or not at all, although you are sure that it contains a corresponding cleavage site**. There are several possible causes.

- **The solution was not mixed properly**. Restriction enzymes, like most enzymes, are stored in a buffer that contains 50% glycerin. If you do not mix this properly, the enzyme will drop to the floor of the tube and will remain there.

- **The restriction enzyme has given up the ghost**. This occurs rarely, primarily when the enzyme is old, although most enzymes are active well beyond their expiration dates (unless a colleague has repeatedly dispensed with putting the enzymes on ice while pipetting). A few restriction enzymes remain stable for only a few months, even when stored correctly.

- **The enzyme has low stability**. In this rare variant, the restriction enzyme has given up the ghost. Some enzymes can be stored normally, but after they have been diluted, they become very unstable, even at the right incubation temperature, and they have a half-life of less than 1 hour. It helps to add more enzyme to the reaction.

- **The buffers may be incorrect**. Even the best experimenter makes mistakes. In the best case, the enzyme still cleaves, and the reaction may occur very slowly. However, the enzyme reaction may sour, change its specificity, and cleave at sites where this should not occur. This behavior is known as **star activity**, and it can be induced in all restriction enzyme experiments. Typically, this occurs in only a handful of experiments in practice, which include sensitivity to low salt concentrations ($<25$ mM), high pH ($>8.0$), high glycerin concentrations ($>5\%$), organic solvents (e.g., ethanol, DMSO, dimethyl formamide), or large quantities of enzymes. Life can be made easier by the use of a universal buffer, with which most common enzymes have an activity of more than 75%. A buffer with 33 mM Tris acetate, pH 7.9, 10 mM Mg acetate, 66 mM K acetate, 0.1 mg/mL of BSA, and 0.5 mM DTT, for example, is suitable. However, you should be thoroughly knowledgeable about the enzymes that you routinely use and ensure that they do not demonstrate cleavage activity in the universal buffer.

- **The temperature may be incorrect**. Most enzymes function at 37°C (97°F), with a few exceptions (Table 3-3). In case of doubt, you should take a look at the catalog or package circular.

- **The BSA was forgotten**. Many restriction enzymes demonstrate a higher activity in the presence of BSA, although you generally can get along without it. If you so desire, you can routinely add BSA to digestions. Even if it does not always help, it will do no harm, which is the reason why many manufacturers routinely add it to their restriction buffers. Acetylated BSA is used at a final concentration of 0.1 µg/µL. Because the modification inactivates available traces of nucleases, never use the BSA from the protein chemists next door, even if they have ample quantities of it. Commercially available (modified) BSA is offered in concentrations that are much too high. You can manufacture a 10× solution and freeze it. You can also add BSA directly to your 10× buffer

**Table 3-3.** Unusual and Unstable Restriction Enzymes

| Properties | Enzymes | Incubation Temperature |
|---|---|---|
| Unusual incubation temperatures | *Sma*I | 25°C (77°F) |
| | *Csp*I | 30°C (86°F) |
| | *Ban*I, *Bcl*I, *Bsa*OI, *Bss*HII, *Bst*71I, *Bst*XI, *Bst*ZI, *Sfi*I | 50°C (122°F) |
| | *Bst*EII, *Bst*OI | 60°C (140°F) |
| | *Bsa*MI, *Bsr*BRI, *Bsr*SI, *Taq*I, *Tru*9 I, *Tth*111I | 65°C (149°F) |
| Unstable at −20°C (−4°F) | *Msp*A1I, *Nla*III, *Pml*I | |
| Unstable in digestion solution | *Nde*I, *Sfc*I, *Sma*I | |
| Star activity with the false buffer | *Apo*I, *Ase*I, *Bam*HI, *Bss*HII, *Dde*I, *Eco*RI, *Eco*RV, *Hind*III, *Hin*fI, *Kpn*I, *Mam*I, *Pst*I, *Pvu*II, *Sal*I, *Sau*3AI, *Sca*I, *Sgr*AI, *Taq*I, *Xmn*I | |

for restriction digestion, but be sure to freeze the buffer again, or you will soon find some fungi and bacteria laughing at you from within the vessel.

- **There are contaminants in the DNA preparation**. Occasionally, the DNA can be polluted with some mysterious substances that inhibit the activity of the restriction enzymes. Frequently, this phenomenon is observed with λ DNA that has been won from plate lysates, but the purity of genomic DNA also occasionally leaves much to be desired. It is often helpful to add more enzyme or increase the volume of the solution substantially, such as up to 50 to 200 μL per 10 μg of genomic DNA.
- **There may be unfavorable sequences**. Not all restriction cleavage sites are severed equally well. The cleavage site of *Eco*RI at the right end in λ DNA, for instance, is cleaved 10 times more quickly than that in the middle of the molecule (Thomas and Davis, 1975). The three *Sac*II cleavage sites in the middle of the λ DNA are cut 50 times more rapidly than the fourth recognition site at the right end. This evidently results from the adjacent sequences. Differences in the cleavage velocity, if they appear at all, are rarely more than a factor of 10 and are therefore not very relevant, because the situation usually is one of *overdigestion* (i.e., too much enzyme has been employed). In some restriction enzymes such as *Nar*I, *Nae*I, and *Sac*II, potential cleavage sites are not severed at all or only very poorly. They belong to a group of restriction enzymes that have two recognition sites that must both be occupied for the enzyme to cut the fragment. Other examples are *Bsp*MI, *Eco*RII, and *Hpa*II (Oller et al., 1991). They can undergo transactivation (i.e., with the aid of another molecule) by adding oligonucleotides, and the addition of spermidine occasionally helps.
- **Methylation may be a factor**. The protection that restriction enzymes offer bacteria against the intrusion by foreign DNA is based on the distinction of their own DNA from the foreign DNA because of the specific methylation patterns. This has an unpleasant side effect, because methylated cleavage sites are not digested with all restriction enzymes. Methylated bases, for example, are found in large quantities in the DNA of mammalian tissues (e.g., genomic DNA) or of bacteria (e.g., plasmids, phage DNA), although not in DNA produced in vitro, such as cDNAs or PCR products. A list concerning the sensitivity of restriction enzymes and their sensitivity to methylation can be found in the article by McClelland and colleagues (1994), or you can obtain information at the web site of New England Biolabs (http://rebase.neb.com).
- The *E. coli* strains used in the laboratory contain three methylases: Dam methylase, which recognizes and methylates G$^m$ATC; Dcm methylase, which recognizes C$^m$CAGG and C$^m$CTGG); and *Eco*KI methylase, which recognizes A$^m$AC($N_6$)GTGC and GC$^m$AC($N_6$)GTT. Most bacterial strains are Dam$^+$, Dcm$^+$, and M$^+$*Eco*KI. If the recognition sequences of these methylases overlap with the recognition sequences of restriction enzymes, frequently nothing more will occur as long as the restriction enzyme is sensitive to methylation. This problem can be avoided by using a bacterial strain for plasmid DNA replication that is Dam$^-$ and Dcm$^-$.

- Sometimes, methylation can be used to advantage. There is a pleasant method on mutagenesis based on the fact that *Dpn*I cleaves only methylated DNA; the methylated template DNA is then destroyed through restriction digestion, whereas the mutated (and not methylated), newly generated DNA survives with the aid of PCR (see Chapter 9, Section 9.2).

## 3.1.5 Works of Reference for Restriction Digestion

Because New England Biolabs has set a standard with its exemplary catalog, other manufacturers have increasingly expanded their catalogs with a few useful pages dealing with the topic of restriction enzymes. A regular expedition through these areas is highly recommended. You will be astounded by what you find there. The NEB catalog continues to be the front-runner in terms of usefulness.

The Restriction Enzyme Database (**REBASE**), which can be found on the Internet (http://rebase.neb.com), is recommended enthusiastically. REBASE provides a current review of restriction enzymes, whether and where they can be obtained, a list of publications concerning the enzymes, and much more.

Another service available from a similar source (http://tools.neb.com/NEBcutter/index.php3) may also prove useful. At this site, you can search DNA sequences for their open reading frames and cleavage sites. Even the band pattern to be expected on an agarose gel can be viewed there—a dream come true for molecular biologists.

**Literature**

Arber W, Dussoix D. (1962) Host specificity of DNA produced by *Escherichia coli*. I. Host controlled modification of bacteriophage lambda. J Mol Biol 5:18.

Chirikjian JG (ed). (1981) Gene Amplification and Analysis, vol 1. Restriction Endonucleases. Philadelphia, Elsevier.

Chirikjian JG (ed). (1987) Gene Amplification and Analysis, vol 5. Restriction Endonucleases and Methylases. Philadelphia, Elsevier.

Dussoix D, Arber W. (1962) Host specificity of DNA produced by *Escherichia coli*. II. Control over acceptance of DNA from infecting phage lambda. J Mol Biol 5:37.

Luria SE. (1953) Host-induced modifications of viruses. Cold Spring Harb Symp Quant Biol 18:237.

McClelland M, Nelson M, Raschke E. (1994) Effect of site-specific modification on restriction endonucleases and DNA modification methyltransferases. Nucleic Acids Res 22:3640–3659.

Moreira R, Noren C. (1995) Minimum duplex requirements for restriction enzyme cleavage near the termini of linear DNA fragments. Biotechniques 19:56–59.

Oller AR, Vanden Bfroek W, Conrad M, Topal MD. (1991) Ability of DNA and spermidine to affect the activity of restriction endonucleases from several bacterial species. Biochemistry 30:2543–2549.

Roberts RJ, Macelis D. (1996) REBASE–restriction enzymes and methylases. Nucleic Acids Res 24:223–235.

Thomas M, Davis RW. (1975) Studies on the cleavage of bacteriophage lambda DNA with *Eco*RI restriction endonuclease. J Mol Biol 91:315–328.

Turbett GR, Sellner LN. (1996) Digestion of PCR and RT-PCR products with restriction endonucleases without prior purification or precipitation. Promega Notes Magazine 60:23.

# 3.2 Gels

*Sie gehen ihren stillen Schritt*
*Und nehmen uns doch auch am Ende mit.*

*For all they seem so still and shy,*
*You'll see they'll let us join them by and by.*

To an experimenter, casting gels often represents the struggle for one's daily bread. This work usually involves agarose gels, and it is not possible to imagine molecular biology without these gels. Over time, experimenters develop substantial expertise with gels and the many things that can go wrong. The DNA may not remain in the sample wells, or it runs in the wrong direction. Sometimes, unusual results are found, probably because the gel was cast erroneously using tap water. It is fascinating to see everything that can go wrong in the process of such a simple, everyday technique!

## 3.2.1 Agarose Gels

The simplest and most effective method for separating and identifying 0.5- to 25-kb-long fragments of DNA is agarose gel electrophoresis. Agarose is heated in the electrophoresis buffer until it is dissolved. With the help of a gel tank and a comb, the gel is cast in sample wells. As soon as the agarose has hardened, the gel is placed in a flow-migration chamber, electrophoresis buffer is added until the gel is covered by 1 mm of this solution, the DNA solutions are pipetted into the sample

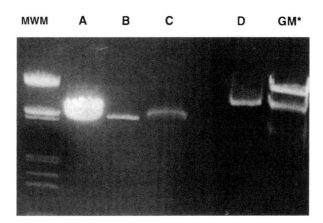

**Figure 3-1. Problems can occur with agarose gels. MWM** refers to the molecular-weight marker. **A,** There is 3 µg of DNA. Bands or zones of this kind permit only a rough estimate of the dimensions. **B,** There is 100 ng of DNA, electrophoresed under optimal conditions. **C,** There is 100 ng of DNA, dissolved in a high-salt solution and then electrophoresed. The fragment appears to be larger than it is, because it migrates slowly out of the high-salt solution. This effect is made use of with a salt trap (see Section 3.2.2). **MWM\*** is a molecular-weight marker. **D,** There is 200 ng of DNA. Agarose gels should be cast with an electrophoresis buffer. If the experimenter grasps the water bottle by mistake, however, the result will be shocking. The bands appear to be quite good, but strange clouds appear in such cases.

wells, and a current (usually between 50 and 150 volts) is applied. After the DNA has migrated sufficiently, the gel is stained with a dye and observed under ultraviolet (UV) light. Usually, the results are also recorded photographically, unless you want to have performed such work for 1.5 days and have nothing to show for it.

A Bunsen burner is no longer used to boil agarose. The microwave oven used is similar to the familiar model at home. When purchasing a new unit for the laboratory, you should select an instrument with the fewest possible number of buttons and one without a grill to avoid a situation in which some blind trainee grills his or her agarose by mistake and thereby eliminate the chance that more extensive damage may occur to the laboratory. Besides, the interior of a modern microwave unit is not very high, and an additional grill may cause the loss of a few valuable centimeters. As a rule of thumb, 1 minute of heating at full power is used per 100 mL of solution. The microwave requires watching during its use, because agarose solutions have a great tendency to boil over, necessitating twice as much time to clean the microwave as for casting the gel. The vessel used for bringing it to a boil should only be filled only halfway. Viscous solutions (e.g., liquid agarose) tend to demonstrate a delay in boiling, which means that the solution occasionally boils after it is removed from the oven, a situation that can cause terrible burns. Even if the solution does not boil over, a sudden release of steam can scald your fingers. Gloves are therefore mandatory when using the microwave. The best are padded gloves made of leather, as are found in specialty shops for the construction industry.

## Electrophoresis Buffer

The two most frequently used buffers are Tris-acetate-EDTA buffer (**TAE**) and Tris-borate-EDTA buffer (**TBE**) (see Appendix 1, Solutions). TAE is presumably the frontrunner, because it can be stored as a 50-fold concentrated stock solution, which saves preparation time and is suitable for later purification of DNA from the gel. A clear disadvantage, however, is the minimal buffer capacity. The gel migration must be run at lower currents (0.5 to 5 V/cm interelectrode distance), because the gel would otherwise melt. TBE has a much better load-carrying ability (>10 V/cm) and permits a much faster electrophoresis. TBE has two major disadvantages. Borate causes difficulties in preparing DNA from the agarose gel, and it precipitates in a malicious manner when using a 5- to 10-fold stock solution. An interesting alternative is **TTE** (90 mM Tris base/30 mM taurine/1 mM EDTA), which demonstrates a similar buffer capacity to that of TBE, but it does not precipitate in a 20-fold concentration of the stock solution and proves to be suitable for DNA preparations.

## The Agaroses

Although experimenters usually do not have to consider the agarose gel they use in the same way as for NaCl or yeast extracts, there are some differences in the agaroses to be considered. You can get a better idea by looking in the bioenzyme catalog from FMC Bioproducts and reading through their offers of agarose gels.

**Standard agaroses** can be used in concentrations of 0.5% to 2%, which cover fragment lengths of 0.2 to 20 kb (Table 3-4). To expand the separation range to include shorter fragments, **sieving agarose**, a special agarose that can be used at especially high concentrations (2% to 4%), makes it possible to resolve DNA fragments in the range of 10 to 1000 bp, which is otherwise possible only with polyacrylamide gels. The bands are not very sharply defined, and the whole procedure is very expensive. These agaroses are so brittle that they must be mixed with normal agarose so that they do not deteriorate in the event of hasty movement. At concentrations less than 2%, sieving agaroses are practically not functional. Nevertheless, they are still popular, because the agarose gels are simpler to manufacture and to use than polyacrylamide gels.

**Table 3-4.** Agarose Concentrations and Respective Fragment or Separation Lengths

| Fragment Length (kp) | Agarose Concentration (% w/v) | Bromophenol (bp) | Xylene Cyanol (kp) |
|---|---|---|---|
| 1 to 30 | 0.5 | 1000 | 10 |
| 0.8 to 12 | 0.7 | 700 | 6 |
| 0.5 to 7 | 1.0 | 300 | 3 |
| 0.4 to 6 | 1.2 | 200 | 1.5 |
| 0.2 to 3 | 1.5 | 120 | 1 |
| 0.1 to 2 | 2.0 | <100 | 0.8 |

The lengths of the fragments, which co-migrate (migrate to equal distances) with bromophenol and xylene cyanol dyes, are indicated for orientation.

Another specialty includes the **low-melting-point agarose** (LMP agarose). The agaroses melt at 65°C (149°F), not at 90°C (194°F) like normal agaroses, and at this lower temperature, DNA does not become denatured. The gelation temperature, at around 29°C (84°F) is also very low, lying below the incubation temperature of restriction enzymes, a feature that allows DNA fragments to be cut from a gel and to be digested without having to first purify them. LMP agaroses offer the possibility of very carefully purifying large fragments of the gel (see Section 3.2.2). LMP agaroses exist in two different forms. One type allows the separation of fragments up to 1 kb long and is substantially more expensive than the standard agarose, and the other allows a separation of fragments up to 20 kb long and is extremely expensive.

Agaroses with a high gel concentration are more stable than normal agaroses and therefore permit lower agarose concentrations, although they can be handled comparably. This is useful for the separation of very large DNA fragments by means of pulse-field gel electrophoresis.

## Agarose Concentration

Large DNA fragments migrate through agarose gels more slowly than do the smaller fragments. The reason lies in the sievelike structure of the agarose, because the pores in this material offer less resistance to the smaller fragments. With a sheet of logarithmic paper and the aid of a molecular-weight marker, the length of a fragment of DNA can be determined easily, because the length of migration is inversely proportional to the logarithm of the fragment length. This relationship, however, is valid only within a certain range. If you plot the distance of migration of known fragments against the logarithm of their respective length, you can see that the initially straight line makes a discrete kink upward in the event of larger fragments, until it eventually extends into uselessness (Figure 3-2). A reason for this kink is the tendency of DNA to align itself parallel to the electric field after some time, so that it can then migrate unhindered through the pores. A separation according to size is then no longer possible, and you are better off using pulse-field electrophoresis for fragments of more than 20 kb (see Section 3.2.5). The kink is found somewhere between 3 and 8 kb, and the exact localization is influenced by different factors such as the current or the electrophoresis buffer employed and most of all by the concentration of the agarose. The lower the concentration, the later the kink appears.

For short DNA fragments, there is the problem of DNA being too greatly diffused, resulting in indistinct bands that frequently cannot be distinguished from RNA contaminants (in plasmid DNA mini-preparations) or primer artifacts (in PCR products). To obtain more sharply defined bands, the agarose concentration can be increased, or polyacrylamide gels can be used instead.

To achieve a clear differentiation of the fragments, agarose concentrations of 0.8% to 2% are used, which can easily separate DNA fragments with a base length of 300 to 5000 bases. The optimal concentration depends on the fragment length (Table 3.5). In daily laboratory work, 1% gels are primarily employed.

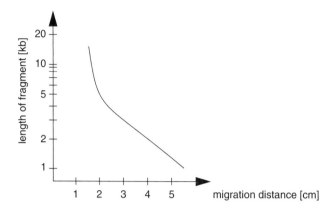

**Figure 3-2. Migration behavior of DNA fragments in agarose.** The fragment length is given in kilobases, and the migration distance is given in centimeters.

**Table 3-5.** Range of Fragment Size Separation in Polyacrylamide Gels

| Acrylamide Concentration (%) | Optimal Range of Fragment Size Separation (bp) | Apparent Migration Distance of Bromophenol Blue (bp) |
|---|---|---|
| 3.5 | 100 to 1000 | 100 |
| 5 | 100 to 500 | 65 |
| 8 | 60 to 400 | 45 |
| 12 | 50 to 200 | 20 |
| 20 | 5 to 100 | 12 |

# The Dye

The most common method for making DNA visible in agarose gels involves dyeing them with **ethidium bromide** (EtBr). The classic method is to dye the gel with $0.5\,\mu g/mL$ of EtBr for 10 to 30 minutes in a small amount of electrolysis buffer (or water). Although the entire situation usually ends in a large mess and the laboratory abounds in a multitude of invisible EtBr spots within a short period, you can obtain a beautiful and evenly colored gel whose contrast can be increased through additional destaining with water for up to another 30 minutes, which is especially advantageous for providing evidence of smaller or weaker bands.

A simpler alternative consists of adding some EtBr to the already melted agarose. The advantage is that you are spared the lengthy dyeing time and can take a brief look at the gel at any time to decide whether to interrupt the electrophoresis, which is an inestimable advantage for the always impatient experimenter. Standard books of methods recommend an EtBr quantity of $0.5\,\mu g/mL$, although $0.1\,\mu g/mL$ is sufficient. Concentrations of more than $0.5\,\mu g/mL$ are not recommended, because they increase the background under UV light. Although EtBr is photosensitive and is degraded in UV light, agarose solution with EtBr can be stored on the shelf for 2 weeks and then again brought to boiling. However, make sure that nothing boils over, because the mess with EtBr will spread within the microwave. Show consideration for those who are hoping to use this device after you.

Although EtBr is considered to be the ultimate example of a dangerous substance used in laboratory procedures, the danger appears to be less than is generally assumed. EtBr has been proved to be substantially mutagenic and is therefore considered to be carcinogenic, teratogenic, and poisonous, although studies are lacking to prove this assumption. You should take advantage of the usual security precautions (i.e., gloves, goggles, and a fume hood) to avoid any contact with this substance. If contact nevertheless occurs, there is no reason to panic. EtBr is not really poisonous nor carcinogenic, even if your neighbor insists that this is so! Remain quiet, carefully wash off the EtBr, and in the event that some of it has been swallowed, consult a physician. Further information about the potential danger of chemical substances can be found on the Internet (http://www.practicingsafescience.org).

There are more dyes than EtBr, such as **SYBR Green I**, **SYBR Green II**, **OliGreen**, or **PicoGreen** (Molecular Probes, Inc.,) whose far higher prices (e.g., 500-fold higher cost per gel) are attributed to their higher sensitivity, with a limit of detection of approximately 20 pg per band, compared with 5 ng per band for EtBr. The dangerous effects of these dyes are considered to be somewhat reduced, although the manufacturers of dyes such as these generally warn of their intercalating effects on DNA, and they must all be considered to be dangerous.

For the sake of completeness, the **methylene blue** dye must also be mentioned, which is visible without UV light but is substantially less sensitive, with the limit of detection of approximately 40 ng per band. The gel must be stained for 15 minutes in methylene blue solution (0.02% methylene blue in $H_2O$) and subsequently destained for 15 minutes in $H_2O$.

## Loading Buffer

Because DNA solutions demonstrate a density that is almost equal to that of the electrophoresis buffer, it is difficult to pipette such solutions directly into the sample wells of the gel (although not entirely impossible, as was once demonstrated to me unintentionally by a laboratory trainee). It is easier if you first add some loading buffer, which increases the density so that the solution sinks softly into the well.

A solution with sucrose (40% [w/v]), glycerin (30% [v/v]), or polysucrose 400 (Ficoll 400, 20% [w/v]), each with 50 mM EDTA, is suitable as a fivefold loading buffer solution. To better follow where the DNA migrates during the course of electrophoretic transfer, add one or more stains, usually xylene cyanol, bromophenol blue, or orange G (each at a concentration of 0.001% [w/v]). These stains migrate within the agarose gel like DNA fragments of approximately 2000, 500, or 200 bp, respectively (see Table 3-5).

Weighing such small quantities proves to be somewhat difficult in practice, because you rarely prepare more than 10 to 50 mL of a loading buffer, which serves well for a number of years. For that reason, these dyes are added as the experimenter thinks best. Concentration of the dye is usually too high at the end (which is also the case for commercially available blue dyes), ultimately allowing the experimenter to view these bromophenol bands from a distance of 10 m. Under UV light, these bands cover all of the DNA bands that migrate an equal distance and essentially serve to make them invisible. It is therefore recommended that the concentration of the dye be adjusted so that you can just barely see where the blue spot is in the gel.

## Molecular-Weight Marker

Formerly, every laboratory made its own molecular-weight markers, and each laboratory had its special favorites. The most common practice was to use λ DNA, which usually was digested with *Hind*III, *Eco*RI, or with both enzymes. If you can find it, take a look at the 1993/94 catalog from New England Biolabs, which provides a review in the appendix of the band patterns of λ DNA, adenovirus 2, and pBR322 DNA after their digestion with some common and less common restriction enzymes—a feast for the eyes that may inspire you to produce your own molecular-weight markers.

Today, molecular-weight markers are available from most of the companies that deal in molecular biologic accessories. The simple variants include offers of items that can be produced in the laboratory (e.g., λ × *Hind*III, λ × *Eco*RI/*Hind*III, λ ×*Bst*EII, ΦX74 × *Hae*III), although it is increasingly necessary to find so-called DNA ladders, which contain DNA fragments at regular distances, usually of 100 bp, 500 bp, or 1 kb. These DNA ladders are available for the lower range of sizes (up to 1500 bp) and for larger fragments (500 bp upward). The costs are higher than if you buy DNA and perform the digestion process personally. The patterns are also too regular, so that problems in orientation are frequently experienced. Molecular-weight markers with characteristic, irregular band patterns have an advantage because you can interpret them even if the migration in the gel has run poorly. Interesting offers for molecular-weight markers can be found from BioRad, AGS, or Roche.

## 3.2.2 Isolating DNA Fragments from Agarose Gels

After successfully separating the band of your dreams from all of the other indefinable DNA fragments, you must remove it from the agarose. There are many methods available, which do not deliver equally good results in all situations.

It is best to separate the DNA fragments in a TAE-buffered gel, because the borate in the TBE buffer interferes with most purification methods. Cut out the DNA band of interest from the gel under UV light and isolate the DNA from the piece of agarose using one of the following methods.

When cutting, work as quickly as possible, because DNA is very sensitive to UV light. Exposure results in depurination, in the formation of base dimers, and in cleavage of the strands and in other fine matters, which greatly complicates cloning and leads to mutations. In isolating DNA from agarose gels, keep the UV exposure as short as possible, limiting it to the time necessary to cut out the specific band.

MWG-Biotech markets a TAE buffer that is effective against this problem. This UV-safe TAE has an additive that protects the DNA from damage brought about by UV light (presumably, it does do this). A small disadvantage, however, is that the miracle cure absorbs the UV radiation, and the DNA is no longer visible. This method is therefore recommended only for very intensive bands, or you can dilute the stuff as far as necessary with normal TAE buffer until it finally works.

Another possibility, which is primarily of interest for longer fragments, is staining with **crystal violet** instead of with ethidium bromide. In this way, the DNA is made visible under normal light, and strand cleavage can be avoided. Invitrogen sells a corresponding kit under the name of S.N.A.P. UV-Free Gel Purification Kit. **Methylene blue** probably functions as well (see Section 3.2.1).

## Glass Milk

Silica material (e.g., glass) binds DNA in the presence of high concentrations of chaotropic salts (e.g., sodium iodide, guanidine isothiocyanate, sodium perchlorate) and can, after a washing step with a salt-ethanol buffer, be eluted with the use of a solution with a slight saline concentration (i.e., $H_2O$ or TE). The eluate can then be used directly (i.e., without any further precipitation). The method is also rapid and versatile: DNA can be purified from solutions and from agarose pieces, because the chaotropic salts dissolve the agarose. The yield is approximately 70%, although it is smaller for smaller DNA fragments (<500 bp).

In all of the available kits, the DNA solution or the agarose piece is diluted and mixed with a threefold to fivefold volume of high-salt buffer (i.e., 6 M sodium iodide, guanidine thiocyanate, or sodium perchlorate). The agarose piece must be shaken for about 5 minutes at 50°C (122°F), until the agarose has completely dissolved. Then, 5 to 10 μL of a well-resuspended silica suspension (i.e., glass milk) is added. It is incubated at room temperature for 5 minutes and centrifuged, and the pellet is then resuspended in 500 μL of wash buffer (0.1 M NaCl/10 mM Tris, pH 7.5/2.5 mM EDTA/50% EtOH) and centrifuged again. The pellet is then dried, and the DNA is eluted with 20 to 50 μL of $H_2O$ or TE. Instead of a silica suspension, silica membranes are frequently found in the kits that are packaged in small columns, which are much easier to handle.

There are different kits available on the market, whereby the most expensive does not always prove to be the best. Probably the most favorable variant is to produce your own silica suspension. To do so, 10 g of diatomaceous earth (SIGMA D-5384) is brought to a volume of 50 mL with $H_2O$; 50 µL of concentrated hydrochloric acid is added, and it is then mixed thoroughly. Before use, be sure to resuspend this solution well. Another advantage is that the quantity of the suspension to be used can be adjusted to suit the individual needs. One can therefore use them for plasmid DNA mini-preparations or midi-preparations.

**Disadvantages:** The different sizes of the particles complicate the washing steps and elution because the silica material and the supernatant poorly separate. Small filter sets for 1.5-mL tubes (e.g., Ultrafree-MC, Millipore) can help, because they allow the fluid but not the silica particles to pass through. Another disadvantage is that large DNA fragments (>5000bp) can be destroyed as a result of the developing shearing forces, and small nucleic acids (<100 bp) bind nearly irreversibly.

**Advantage:** The method is versatile and provides easily reproducible results.

**Literature**

Vogelstein B, Gillespie D. (1979) Preparative and analytical purification of DNA from agarose. Proc Natl Acad Sci U S A 76:615–619.

## Freeze and Squeeze Method

The piece of gel is frozen at −20°C (−4°F) and subsequently squeezed between two fingers of the hand (protected with a glove). The liquid pressed out is collected and centrifuged intensively for 5 minutes to obtain a pellet from the gel remnants. The supernatant can be used directly for additional procedures. To be certain, perform a purification using phenol-chloroform followed by ethanol precipitation. The disadvantage of this method is that it proves to be quite a mess.

**Literature**

Thuring RW, Sanders JPM, Borst P. (1975) A freeze-squeeze method for recovering long DNA from agarose gels. Anal Biochem 66:213–220.

## Centrifugation Method

The centrifugation method is a variant of the freeze and squeeze method, but it is faster and cleaner. A mini-sieve is prepared by making a hole in the bottom of a 0.5 mL plastic tube with a needle (Figure 3-3). The floor of the tube is covered with a piece of silanized glass wool that should fill less than one fourth of the tube. The piece of gel is put into the 0.5 mL tube, which is then placed into

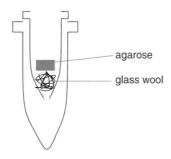

**Figure 3-3. Centrifuging DNA from a piece of agarose.** With a self-made filter, DNA can be centrifuged from a piece of agarose within 5 minutes. The glass wool forms a plug for a large pore and suffices to retain the agarose.

a 1.5-mL vessel and subsequently centrifuged in a table-top centrifuge at 8000 rpm for 5 minutes. The centrifugate in the 1.5-mL vessel contains approximately 70% of the DNA. The success of this procedure can easily be controlled through the use of UV light. The solution can be used directly, or the DNA can be precipitated with ethanol. Millipore offers a complete filter unit with a built-in shredder and filter (Ultrafree-DA Centrifugal Filter Units), which functions in the same manner but looks better. The method is very fast, it requires only a few actions, and the mini-sieve can be prepared in advance.

**Literature**
Heery DM, Gannon F, Powell R. (1990) A simple method for subcloning DNA fragments from gel slices. Trends Genet 6:173.

## Electroelution

The piece of gel can be placed in a dialysis tube and then subjected to further electrophoresis. The procedure is described later because it is used more for the purification of polyacrylamide gels than for agarose gels.

A nice variant is the **salt trap** (Figure 3-4), a method that makes use of the fact that DNA naturally runs into a high-salt solution quite easily but runs out again very slowly. The necessary device can be built in a workshop. The slab of gel is electrophoresed for 45 minutes in a V-shaped system of tubes. This system is filled at the deepest point with a 3 M sodium acetate solution (pH 5.2, stained clearly with a small amount of bromophenol blue). The blue solution is pipetted completely from the system, diluted with a twofold volume of $H_2O$, and the DNA precipitated with ethanol.

## Electrophoresis on an NA-45 Membrane

Electrophoresis on an NA-45 membrane is a variant of electroelution, although the DNA does not have to be cut out of the gel. Instead, a cut is made in the gel above and below the band of interest, and a piece of DEAE membrane (NA-45, Schleicher & Schüll) of the appropriate size is placed in each of these two slits. This is performed most easily by spreading the slits slightly with a pair of tweezers and inserting the DEAE membrane using a second pair of tweezers or by cutting out a

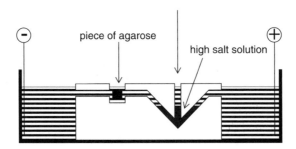

**Figure 3-4. Construction of a salt trap.** The DNA is eluted from the piece of agarose by means of electrophoresis. After the DNA has arrived in the high-salt solution, it continues to migrate further so slowly that it practically remains caught in the solution. It can easily be removed in its entirety if it is first diluted with bromophenol blue.

narrow piece of agarose. Continue the electrophoresis for about 15 minutes, until the DNA band has run completely onto the lower membrane and has become bound there. The upper membrane prevents other fragments from binding to the lower membrane when the electrophoresis has been performed for too long.

The selection of the correct acrylamide concentration is especially important for polyacrylamide gels, because larger DNA fragments are not differentiated when the concentration is above the optimal range.

The lower membrane is washed three times in TE buffer and then heated to 70°C (158°F) for 15 minutes in a 1.5-mL tube containing 400 μL of elution buffer (20 mM Tris, pH 8.5/2 mM EDTA/1.2 M NaCl) for fragments smaller than 500 bp and up to 1 hour for fragments larger than 1500 bp. The solution can be purified with phenol-chloroform and precipitated with ethanol, or you can use it as it is.

## Low-Melting-Point Agarose

Two other methods of purification are available for those who use low-melting-point (LMP) agarose for electrophoresis instead of normal agarose.

In one method, the slab of gel is melted at 65°C (149°F), diluted with TE buffer until the agarose concentration is below 0.4%, and then purified in a phenol solution. To this, add an equal volume of phenol at pH 7.4 (not phenol-chloroform solution!), mix vigorously, and then centrifuge. The aqueous phase is transferred into a new vessel, and the phenol phase is diluted with an equal volume of TE buffer, mixed, and centrifuged. The two aqueous phases are united and then precipitated with ethanol. For some applications, such as restriction digestion or ligations, you can dispense with the phenol purification, because LMP agarose gels only at a temperature below 30°C (86°F).

The second possible method makes use of the capabilities of β-**agarase** to disassemble long-chain polysaccharides and form short oligosaccharides. In this way, a DNA solution is produced from the piece of gel, which can be used directly, such as for restriction digestion or ligations, or precipitated with ethanol.

**Disadvantages:** LMP agarose is substantially more expensive than standard agarose. After the purification with phenol, the pellets still contain considerable quantities of agarose, which may have an inhibitory effect on some reactions. β-Agarase is not cheap and the digestion takes at least an hour.

**Advantages:** Unlike most other methods, the agarase digestion is very gentle. It also enables the purification of large DNA fragments that are many kilobases long.

## 3.2.3 Polyacrylamide Gels

The strength of polyacrylamide gels lies in their outstanding dissolution of small DNA fragments up to a size of 1000 bp. They also have a higher capacity (i.e., more DNA can be applied). For molecular biologists, however, they are less popular than agarose gels, because they are more difficult to cast, experimenters cannot attain an intermediate view of the development of the bands, and it is somewhat troublesome to isolate the DNA from the polyacrylamide gel.

The concentration of the acrylamide determines the range of separation of the specific gel. The range that is available by the use of 3.5% to 20% acrylamide is substantially larger than that available with agarose. At low concentrations, however, you should cast sufficiently thick gels, because the handling of such slimy objects is much more difficult the thinner and the larger they are.

To cast an acrylamide gel, you need two glass plates, two spacers, a comb, a large number of clamps, and an experienced colleague who can show you how to do this procedure. You can also read the corresponding chapters in a standard book with a protocol of this procedure, although it is certainly easier with the aid of a colleague.

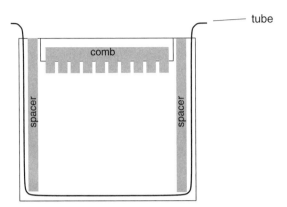

**Figure 3-5. Sealing an acrylamide gel with the aid of a rubber tube.**

One of the most frustrating experiences is to invest half an hour of time to set up the casting apparatus and mix the solutions, only to discover that the device has a leak and that the acrylamide solution is being distributed over your entire workplace. The fact that acrylamide in a nonpolymerized state is a neurotoxin, which can be absorbed by the skin, does not help to make cleaning up any more cheerful. Anyone who routinely casts polyacrylamide gels develops his or her own technology over the course of time, which to a great degree depends on the local conditions and the dimensions of the gel. A popular method is the application of agarose to seal the gel. A clever alternative involves the insertion of a rubber or silicon tube of a suitable diameter (i.e., a little bit thicker than the size of a spacer) between the glass plates, outside of the region occupied by the spacer (Figure 3-5). When the gel has been fully polymerized, the tube can be pulled out.

Many companies (e.g., BioRad, Hoefer, AGS) offer flawless casting apparatuses for a suitable amount of money, with which you can solve the casting problem more or less elegantly. Some experience is required nevertheless, and I have not discovered any foolproof device. For the totally lazy, hasty, untalented, or the rich or for diagnostic laboratories that are able to transfer their costs to the bills of the patients, several firms (e.g., Stratagene, Pharmacia, Novex) offer *precast gels*. The format of these gels corresponds to the equipment sold by these firms.

Usually, nondenatured gels in which the DNA remains double stranded are used. A special form is denatured polyacrylamide gels, which permit the electrophoresis of single-stranded DNA according to a specific length, such as is required for sequencing. This can be achieved through the addition of large quantities of urea (see Chapter 8, Section 8.1.1). Without such additives, single-stranded DNA (and RNA) demonstrates a diabolical effect of hybridizing with something during electrophoresis, most preferably with itself. The resulting formations are anything but linear, and they demonstrate migrational characteristics that are based somewhat on chance. This is generally an undesirable effect, except for in the case of single-strand conformation polymorphism (SSCP) electrophoresis (see Chapter 8, Section 8.2.2).

## Electrophoresis Buffer

As with agarose gels, the capacity is important for the selection of a buffer. Experimenters preferably use TBE, because this buffer is more stable and therefore permits higher currents. However, other buffers may be employed.

# Dyes

As with agarose gels, the experimenter chiefly stains polyacrylamide gels with ethidium bromide. To achieve good results, he or she must stain subsequently (after electrophoresis), which is a rather fast process due to a reduced gel thickness; 10 minutes usually is sufficient for a solution of electrophoresis buffer containing $0.5 \mu g/mL$ of ethidium bromide. Success with polyacrylamide gels requires some practice, because they are quite slimy and fairly thick, and they tend to tear easily.

An interesting alternative is provided by **silver dyes**, which demonstrate a substantially higher sensitivity than others. The method is more common in protein chemistry, although it functions equally well for DNA. This method involves the formation of a complex between silver ions ($Ag^+$) and DNA; the silver is subsequently reduced with alkaline formaldehyde and precipitates out. The bands that thereby develop appear brown to black. There are many protocols for this procedure (Bloom et al., 1987), but you can purchase the whole thing in the form of a complete kit, such as the one from Pharmacia. Silver stain can also be used after staining with ethidium bromide.

**Advantages:** Silver stain is substantially more sensitive than ethidium bromide. It is suitable for long-term documentation because the color does not disappear. You can shrink-wrap the gel in cellophane, or you can dry it. To do so, place the gel on filter paper, pour a little 10% (v/v) glycerin solution over it, cover it with a normal cellophane, and dry it in a commercially available gel dryer under a vacuum for 1 to 2 hours.

**Disadvantages:** The time required for the silver staining is substantially higher than for staining with ethidium bromide, and it is hard to avoid decorating the entire workplace with blackish-gray spots over time.

**Literature**

Bloom H, Beier H, Gross HS. (1987) Improved silver staining of plant proteins, RNA and DNA in polyacrylamide gels. Electrophoresis 8:93–99.

## 3.2.4 Isolating DNA Fragments from Polyacrylamide Gels

### Elution by Means of Diffusion

The simplest method to regain DNA from polyacrylamide gels is to make use of the laws of diffusion for your own purposes. This is very time-consuming, and the yield is limited. The larger the quantity of elution buffer used, the more DNA is diffused from the gel, although the risk of losing material through the subsequent precipitation is also higher.

The bands should be cut out, sliced into small pieces, and then put in a 1.5-mL tube. The gel pieces are covered with elution buffer (0.5 M ammonium acetate/1 mM EDTA) and incubated at 37°C (99°F). For fragments smaller than 250 bp, 2 to 3 hours of elution are sufficient, but it is better to incubate large fragments (>750 bp) overnight. The supernatant is transferred to a new tube, the pieces of gel are washed again with buffer, and the combined supernatants are precipitated and centrifuged in a twofold volume of ethanol. The pellet is then placed in water, and the precipitation is performed again. This mixture is then washed and dried and dissolved in TE buffer.

### Electroelution

Electroelution is somewhat more elegant. The DNA band is cut from the gel, the piece of gel is placed in a small dialysis tube, and the smallest possible volume of a 0.1-fold electrophoresis buffer is pipetted into it. The tube is then placed in a migration chamber, and it is electrophoresed at 2 V/cm interelectrode distance for 2 hours. The buffer is pipetted from the dialysis tube, the tube is washed with some buffer, and the DNA is precipitated from the combined buffers with sodium acetate and ethanol.

Many manufacturers (e.g., Schleicher & Schüll, BioRad, AGS, Hoefer) offer comparable devices, which usually permit a smaller elution volume than that used with dialysis tubes.

This method functions for agarose gels as well. An advantage is that even larger DNA fragments can be isolated with a good yield.

## 3.2.5 Pulse-Field Gel Electrophoresis

Although agarose gels are sufficient for normal use, their resolution for fragments of larger than 15 kb is unsatisfactory. The reason for this lies in the size. An agarose gel is essentially a sieve with various pore sizes. Whereas smaller fragments can easily pass through the pores, larger fragments must deform themselves more or less extensively to pass through. More time is required for larger fragments, although the larger fragments can also pass through the matrix to some extent after they have oriented themselves properly. For 20-kb and larger fragments, the difference in the time required for the fragments to find the correct orientation is quite short, so that they all run the same distance. A physical trick permits the separation of fragments that are larger than this. If the direction of the electric field is changed routinely during electrophoresis, the fragments must repeatedly reorient themselves. Even if the differences in time are only slight, the total number of directional changes allows larger fragments to be separated.

This approach presupposes the use of a special apparatus. There are different methods (and devices), but the simplest and presumably the most common is to make use of reversing the field of current (**field inversion electrophoresis**). With this, fragments that are 10 to 2000 kb long can be separated. Other methods allow the separation of substantially larger fragments. The resolution depends on different parameters such as current, pulse time, temperature of the gel, agarose, and buffers. The run time for such a unit is a multiple of that observed for normal gel electrophoresis.

Handling DNAs of this size is difficult. A chromosome is composed of only two molecules, which in humans are as long as 1.7 to 8.5 cm per chromosome and so narrow that they are best seen with an electron microscope. Consequently, such a molecule can break or tear very easily, and special protocols are mandatory for isolating DNA fragments of this length and working with them. The trick is to embed the material in small agarose blocks and to then work with them. In this way, you can avoid unnecessary shearing forces that occur in the course of pipetting. Because all buffers and enzymes in the agarose must be diffused, their preparation and digestion occurs over a period of several days. The enzyme quantities required are considerably larger than those used in other procedures.

**Literature**

Birren B, Lai E, Clark SM, et al. (1988) Optimized conditions for pulsed field electrophoretic separations of DNA. Nucleic Acids Res 16:7563–7581.

Birren B, Lai E. (1994) Rapidly pulsed field separation of DNA molecules up to 250 kb. Nucleic Acids Res 22:5366–5370.

Schwartz DC, Cantor CR. (1984) Separation of yeast chromosome-sized DNAs by pulsed-field gradient electrophoresis. Cell 37:67–75.

## 3.2.6 Capillary Electrophoresis

A new technique, which is little known but gaining popularity, is capillary electrophoresis. The gel matrix is poured into a capillary tube with a internal diameter of 20 to 100 μm. Because the surface in this system is very large compared with the volume of the gel, the warmth that develops in the course of the electrophoresis can very rapidly be directed away. Currents of up to 800 V/cm can be applied without overheating the gel, whereas a maximum current of 10 to 40 V/cm is possible for standard agarose gels. The run times for such a gel with this technique are reduced to a matter of minutes. With the corresponding apparatus, the experimenter can detect fluorescently labeled DNA fragments and still recognize extremely small quantities (as small as 10 fg). The modern sequencing units function for the most part with the aid of capillary electrophoresis, because the very difficult gel casting and sequencing procedures can be accomplished on a large scale and almost fully automatically in this manner.

**Literature**
Landers JP. (1993) Capillary electrophoresis: Pioneering new approaches for biomolecular analysis. Trends Biochem 18:409–414.

## 3.3 Blotting

Blotting involves an attempt to fix "something" long-term that has been separated onto a membrane gel electrophoretically (Figure 3-6). If the item is DNA, this process is known as a Southern blot, in honor of Ed M. Southern, who made this procedure public in 1975. An equivalent procedure is called a Northern blot if it involves RNA, and because scientists are cheerful individuals, a Western blot is designated to represent the blotting of proteins.

For a **membrane**, you can choose between the classic nitrocellulose and nylon membranes; the latter exist in two variants: neutral and positively charged. The choice of the membrane depends on the application and on personal preference. There is not much difference in cost between the various types of membranes. The different modes of application, however, are quite varied; because nitrocellulose tears easily in a dry form and crumbles easily, it can be labeled only with difficulty,

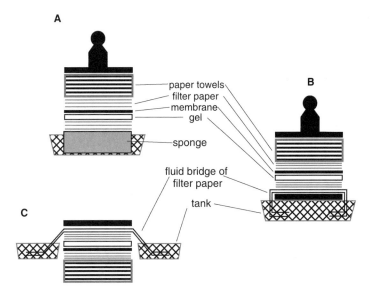

**Figure 3-6. Different possibilities for the construction of blots. A,** Blotting with a sponge. **B,** Blotting with a filter-paper bridge. **C,** Upside-down blotting.

whereas nylon membranes are able to withstand almost everything. This stability is the main reason that nylon membranes can be hybridized very frequently, whereas nitrocellulose usually develops a form similar to a puzzle after using it a third time. DNA can be bonded covalently to the nylon, whereas it adheres only very firmly to nitrocellulose, and a portion is lost after each hybridization. Fragments smaller than 500 bp bind to nitrocellulose very poorly from the start.

These characteristics have led molecular biologists to prefer working with nylon membranes. Nylon, however, frequently results in a substantially higher background, especially when using nonradioactive methods of detection. Some manufacturers also offer strengthened ("supported") nitrocellulose, which demonstrates a stability similar to that of nylon, but it has not become popular.

## 3.3.1 Southern Blots

A goal of the Southern blot is to fix DNA fragments on a membrane, which have previously been separated according to their length by means of gel electrophoresis. Hybridization with labeled probes can provide evidence of specific, individual DNA fragments. The amount of work required for such a transfer of DNA, which has been separated in agarose gel, is not extensive and can be accomplished without special instruments. Several methods are available and are all based on the fact that the grid of the agarose is very large. For polyacrylamide gels, an electroblotter is required.

The gel must be pretreated before blotting. To efficiently transfer fragments more than 5 kb long to the membrane, the gel should first be deposited in 0.25 M HCl for 30 minutes. These and the subsequent incubations are best performed on a see-saw or a shaker to ensure that the movements are not too rapid. The gel should always *float* in the solution to permit diffusion on both sides. The volume of this fluid must be at least threefold that of the gel volume. A look at the side of the gel makes it possible to view the progress of this diffusion process, because xylene cyanol turns green in HCl, and bromophenol blue turns yellow. If the color markers have fully changed their color—the time necessary for this procedure depends on the thickness of the gel—the incubation should still be carried out for another 10 minutes. Through the step involving HCl, the DNA is depurinated in part and then breaks into smaller fragments. For shorter DNA fragments, no depurination is necessary.

After that, the DNA (i.e., each strand individually) is denatured over 30 minutes of incubation in a denaturing solution (0.5 M NaOH/1.5 M NaCl). This material is neutralized by 30 minutes of incubation in a neutralization solution (0.5 M Tris HCl, pH 7.0/1.5 M NaCl). Now, the blotting can begin.

### Blotting with High-Salt Buffer

The most common method is blotting with high-salt buffer. Under the exploitation of capillarity, 10× SSC buffer (1.5 M NaCl/0.15 M Na-citrate, pH 7.0) passes through the gel and the membrane into a sandwich of paper towels due to the elevation. The DNA migrates along with this buffer and adheres to the membrane.

The construction of the transfer sandwich is interesting. It is most practical to obtain a fine-pored sponge with sufficient size and stability. Sponge rubber is well suited for this purpose. The sponge should not be too high nor too soft, because it may break apart as a result of the weight of the construction. It is fully soaked in a small tank with 10× SSC. You then add three layers of 10× SSC-saturated filter paper (absolutely free of bubbles), which have been cut to suit the size of the gel; then add the gel itself, a membrane that has been soaked in water, and three more layers of 10× SSC-soaked filter paper, followed by a neat package of paper towels (which may not be soaked). On top of this is placed a glass plate and a small weight (200 to 400 g [approximately 7 to 14 ounces] are sufficient). Check the fluid level in the tank, which should be about 2 cm (3/4 inch). It is important that the transfer puffer does not run beside the gel. For that reason, all components, including the paper towels, are cut to suit the size of the gel, or the area around the gel is covered with cellophane. After 3 to 14 hours (depending on the particular transfer efficiency desired), you disassemble the construction and fix the DNA to the membrane (discussed later).

If you do not have a sponge, you can construct a "fluid bridge." Place a glass plate over the tank, which is about as wide as the gel, and put four layers of soaked filter paper strips on top of it, whose two ends hang in the fluid of the tank. The remaining construction is carried out as described earlier.

Another alternative, which is not cheap, involves the use of a well-made capillary blotting apparatus from Scotlab. This proves to be worthwhile for those who must perform a great number of blots, because it obviates messing around with temporary arrangements such as those previously described.

Because the gel can collapse if the weight of the construction becomes too large, and the transfer efficiency thereby decreases, there is also an upside-down variant that avoids this problem. Begin with a pile of paper towels, and pack the filter paper, membrane, gel, filter paper, and four layers of filter-paper strips on top, which are left to hang on the left and on the right into a tank containing the transfer buffer. For a balanced pressure, place a glass plate on top of the sandwich. The transfer times with this construction should be somewhat shorter than with other approaches.

When setting up the blot, it is important in every case that the transfer progress only as long as the fluid is still in the tank and dry paper is still to be found in the sandwich. If either requirement is no longer fulfilled, no further waiting will help.

## Blotting with Alkaline Buffer

Blotting with alkaline buffer represents a rapid protocol that can avoid preparation of the gel as described previously, because the DNA becomes denatured by the transfer buffer, which should be used jointly with positively loaded nylon membranes. Nevertheless, the effectiveness is reduced with neutral nylon membranes and nitrocellulose, because the highly alkaline pH causes the membrane to disintegrate into brownish crumbs during the transfer.

The transfer is accomplished as described previously; the only difference is that 0.4 M sodium-hydroxide solution is used instead of $10\times$ SSC. Be careful with the use of this agent, because NaOH is a caustic alkali. Following this protocol, the transfer time is somewhat shorter, and satisfactory results can be achieved after 2 hours.

## Quick and Dirty Method

To carry out blotting quickly and minimize the expense when high effectivity is not important, select the extremely fast, semi-dry variant. Denature the gel for 15 minutes; place it on a sheet of cellophane; cover this with a membrane, filter paper, and paper towels; and then blot for 2 hours. This method is highly suitable for all gels on which you observe broader bands, because these would otherwise produce signals likely to be too intensive during hybridization.

## Vacuum and Pressure Blotters

Because the blotting methods described so far will not earn a huge profit, resourceful minds have designed apparatuses that provide equivalent results but that function even better, although only if an electrical outlet is available. A vacuum blotter from any of several manufacturers (e.g., Hoefer, Pharmacia, BioRad) draws the transfer buffer up through the gel and membrane, whereas a pressure blotter (Stratagene) presses this buffer through these materials. These instruments should function more rapidly (according to the manufacturers, 60 minutes is a sufficient blotting time) and offer better transfer efficiencies. However, the gel mask that is required usually disappears, or it proves to be inadequate for keeping the apparatus from leaking. Cleaning these instruments is even more troublesome. The acquisition of such an apparatus is primarily beneficial for individuals who perform a great deal of blotting.

## Electroblotters

Because capillary transfer does not function with polyacrylamide gels, it is difficult to avoid purchasing an electroblotter. Although the traditional tank system that is commonly used for Western blotting also functions for DNA, *semi-dry blotting* devices help to make this procedure easier. AGS, BioRad, and Pharmacia market corresponding instruments at a cost of approximately $1000 to almost $2750. If you have access to a workshop, you can have such a device built. It is essentially composed of two graphite plates, an anode and a cathode (the commercial versions use platinum alloys and stainless steel instead) with dimensions of about $20 \times 20$ cm (almost $8 \times 8$ inches), which is connected to a part of the base and to a cover made of plastic, and it is furnished with a plug to connect it to an electrical outlet.

With this method, you can construct a small transfer sandwich on one of the electrodes, similar to that used for blotting with agarose gels. Five layers of filter paper, soaked in $0.5 \times$ TBS buffer, are placed on the cathode (free from air bubbles); onto this is placed the gel, the membrane, and another five layers of filter paper soaked in $0.5 \times$ TBS buffer. All components must be cut to suit the size of the gel. The cover of the device, to which the anode is attached, is then placed on top. Electrophoresis is most effectively performed at 30 V for 4 hours at a temperature of 4°C (39°F). The optimal duration depends on the thickness of the gel, the tension, and the size of the DNA fragments.

## Fixation of the DNA on the Membrane

Once blotted onto the membrane, the DNA adheres relatively tightly, although not irreversibly. To achieve this, treat the membrane by placing nitrocellulose between two layers of filter paper and then bake for 2 hours at 80°C (176°F), preferably in a vacuum oven. Nylon membranes are best when irradiated with UV light so that the DNA is covalently and irreversibly bonded to the membrane, a process known as *crosslinking*. This is performed most easily with the use of a crosslinker (e.g., from Stratagene), which permits irradiation of the membrane using a defined amount of energy. Irradiation performed at 120 mJ, for instance, is sufficient for crosslinking. The membrane used should be dry, at least for the most part. A normal source of UV light can be used, as long as you take the trouble to perform a test hybridization beforehand to determine what duration of exposure provides the best results. Any experimenter who only wishes to hybridize his or her membrane within a single cycle can forget about using crosslinking.

**Literature**
Southern EM. (1975) Detection of specific sequences among DNA fragments separated by gel electrophoresis. J Mol Biol 98:503–517.

## 3.3.2 Northern Blots

The principle of Northern blots is equivalent to that for Southern blots, with the small difference being that this procedure is concerned with RNA. In practice, it does not matter, but it is certainly advantageous to have separate instrumentation for RNA—an electrophoresis cabinet and a gel tank to minimize contamination with RNases. Aside from that, all solutions are made with the use of DEPC-treated water (see Chapter 5). The composition of the gel differs somewhat, because RNA under denatured conditions must be separated because the individual strands tend to form secondary structures that may influence the transfer characteristics. Two different kinds of denatured gels have gained general acceptance.

In **method 1**, 1 volume of RNA (3 to 10 μL usually with 0.2 to 10 μg of RNA all together or 0.1 to 3 μg of Poly A+ RNA) is mixed with a twofold volume of RNA sample buffer (10 mL of deionized formamide/3.5 mL of 37% formaldehyde/2 mL of 5× RNA electrophoresis buffer; the latter composed of 200 mM MOPS/50 mM Na-acetate/5 mM EDTA in DEPC-treated water, pH 7.0, autoclaved) and heated for 5 minutes at 65°C (149°F). This is put on ice, and 2 μL of RNA gel buffer (50% glycerin/1 mM EDTA/0.4% bromophenol blue in DEPC-treated water) is added. The RNA is applied to an agarose gel (1% agarose/6.5% formaldehyde in 1× RNA electrophoresis buffer), which had previously been run for 10 minutes and separated using a current of 5 to 10 V/cm interelectrode distance for 2 to 5 hours (do not forget the RNA molecular-weight marker, which can be obtained from Roche or Promega). Electrophoresis should be performed under a fume hood, because formaldehyde is poisonous and carcinogenic, and it stings dreadfully in the nose. The gel can be stained with ethidium bromide, although its sensitivity is only mediocre.

**Method 2** makes use of **glyoxal** for denaturation. The bands become sharper, and you can avoid the use of the unpleasant and unhealthy formaldehyde. The disadvantage is that the buffer has only a small capacity, and it must be recirculated with the aid of a pump during the course of this procedure. Consequently, this protocol is only rarely used.

Burnett (1997) presented a variant that is easier to employ and that immediately combines the denaturation with the staining. Use BTPE as an electrophoresis buffer (10 mM PIPES/30 mM Bis-Tris/10 mM EDTA, pH 6.5/10× solution: 6 g Bis-Tris, 3 g PIPES, 2 mL of 0.5 M EDTA in 90 mL of H₂O [DEPC treated; see Chapter 5]), 2 μL of RNA and 10 μL of glyoxal mix (6 mL of DMSO, 2 mL of 6 M glyoxal [deionized]), 1.2 mL of 10× BTBE, 0.6 mL of 80% glycerin, 0.2 mL of ethidium bromide (10 mg/mL). After mixing, 1-mL aliquots can be stored at −70°C (−94°F). The solution should not be frozen again after use. It is incubated for 1 hour at 55°C (131°F), applied to an agarose gel (1% to 1.5% agarose in 1× BTPE buffer), and separated electrophoretically at 5 V/cm for 3 hours. Some bromophenol or xylene cyanol solution is pipetted into an extra well to assess its migration. The RNA can be viewed under UV light without any further staining.

The blotting proceeds with both methods, as with DNA gels, using 10× SSC or 7.5 mM NaOH. With a methylene blue dye (staining for 2 minutes in 0.02% methylene blue solution, destained with H₂O; see Herrin and Schmidt, 1988). The RNA can be seen on the membrane substantially better than in the gel.

An interesting alternative involves commercially available Northern blots, as offered by Clontech (Multiple Tissue Northern blots [MTN]) or Invitrogen (Northern Territory blots). For mice and rat tissues, it simplifies the work, whereas Northern blots of human tissues are primarily of interest because upstanding citizens rarely have a chance to obtain the appropriate tissues in sufficient qualities.

For those who have no desire to dirty their fingers, a completely different option is available. A flood of articles published over the past few years is concerned with such techniques with which the entire work is carried out on a computer workstation. An example is the **electronic Northern**, which makes use of a large number of **expressed sequence tags** (ESTs), with which databases are currently bustling. ESTs are fragmentary sequences of clones that had been blindly fished out from cDNA banks and then sequenced, a method that was introduced in the early 1990s by Craig Venter (Adam et al., 1991, 1993) with the goal of discovering a maximum number of genes in a minimum amount of time through the use of such a sequencing overkill. The sequences are mostly evaluated only rudimentarily before they disappear into the databases. Many cDNA banks from different organisms, tissues, and tumors have been sequenced in this manner, and it is now possible to carry out a quantitative evaluation. Through the use of statistical evaluation of the frequency of sequences in different cDNA databanks, especially strongly expressed genes can be used to obtain an expression profile in various tissues. Consequently, these can be related to different developmental stages and can be employed to investigate the upregulation or downregulation of genes in different tumors (Schmitt et al., 1999) or to compare different genes with one another (Rafalski et al., 1998). The information is usually examined experimentally at the end, for instance by using *DNA arrays*. How much longer will this continue to be the case?

**Literature**

Adams MD, Kelley JM, Gocayne JD, et al. (1991) Complementary DNA sequencing: Expressed sequence tags and human genome project. Science 252:1651–1656.

Adams MD, Soares MB, Kerlavage AR, et al. (1993) Rapid cDNA sequencing (expressed sequence tags) from a directionally cloned human infant brain cDNA library. Nature Genet 4:373–380.

Burnett WV. (1997) Northern blotting of RNA denatured in glyoxal without buffer recirculation. Biotechniques 22:668–671.

Herrin DL, Schmidt GW. (1988) Rapid, reversible staining of Northern blots prior to hybridization. Biotechniques 6:196–200.

McMaster GK, Carmichael GG. (1977) Analysis of single- and double-stranded nucleic acids on polyacrylamide and agarose gels by using glyoxal and acridine orange. Proc Natl Acad Sci U S A 74:4835–4838.

Puissant C, Houdebine LM. (1990) An improvement of the single-step method of RNA isolation by acid guanidine thiocyanate-phenol-chloroform extraction. Biotechniques 8:148–149.

Rafalski JA, Hanafey M, Miao GH, et al. (1998) New experimental and computational approaches to the analysis of gene expression. Acta Biochim Pol 45:929–934.

Schmitt AO, Specht T, Beckmann G, et al. (1999) Exhaustive mining of EST libraries for genes differentially expressed in normal and tumour tissues. Nucleic Acids Res 27:4251–4260.

## 3.3.3 Dot Blots and Slot Blots

If you can dispense with separation according to size, you can obtain results much more quickly with the aid of dot blots (Kafatos et al., 1979). The DNA is applied directly to the membrane without any previous electrophoresis. You can also perform dot blots by hand, but this is very troublesome and can be recommended only in an emergency, because you must apply the DNA in 2-μL portions to limit the diffusion of the solution throughout the entire membrane. It is substantially easier to use an apparatus as offered by Hoefer, BioRad, or Scotlab. This device is made up of an upper part with holes and a hollow, lower part to which a vacuum pump is connected. The membrane (best to use nylon) and a piece of filter paper of equal size are soaked in 6× SSC and tightened within the apparatus, with the membrane on top and the filter paper below. The DNA is dissolved in 200 to 400 μL of 6× SSC, the vacuum pump is started, and then 500 μL of 6× SSC and a DNA solution are successively drawn off through each hole (the holes not required should be closed with Scotch tape). The membrane is incubated for 5 minutes each in denaturing and in neutralizing solution (see Chapter 3, Section 3.3.1), and the DNA is fixed as in Southern blotting.

If you use positively charged nylon membranes, you can instead employ an SSC solution with 0.4 M NaOH and 10 mM EDTA and then bring the DNA solutions to the same concentrations. In this way, subsequent denaturation is unnecessary.

Slot blots differ only from dot blots in that their form is a slot rather than a dot. Slot blots may possibly be more advantageous, because the characteristic form of their hybridization signals cannot be confused with hybridization artifacts. In both cases, you should not forget to apply control substances aside from the DNA of interest, because the background may prove to be quite high during hybridization.

In their program, Clontech offers an interesting RNA dot blot (Human Multiple Tissue Expression Array) that contains 76 poly A+ RNA dots from various human tissues and cell lines. The membrane is very practical in case you need to identify in which human tissue the cDNA, which has just been found, is expressed.

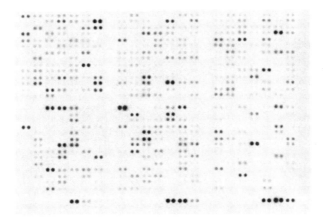

**Figure 3-7. Macroarray.** In the *cDNA expression array*, each cDNA has been applied doubly to facilitate the distinction between artifacts and genuine signals. Comparison of the signal intensities of the two hybridization experiments demonstrates upregulation and downregulation of the expression of individual genes. (Courtesy of Clontech.)

A relatively new method, which could be called **reverse Northern**, functions in an opposite manner. Instead of blotting the RNA, known cDNAs are dotted onto a membrane and hybridized with a probe produced from the poly A+ RNA that stems from a specific tissue or cell line (Gress et al., 1992). Based on the intensity of the hybridization signal, you can obtain an idea of the intensity of expression of the specific genes in the tissue being examined. You can also carry out such cDNA dot blots yourself if you have the corresponding cDNAs (which frequently presents a problem in practice), or you can purchase them. Such membranes are known as **DNA arrays** or as **macroarrays** to distinguish them from the related, but technically different **microarrays** (see Chapter 9, Section 9.1.6). The hybridization usually is carried out with radioactively labeled cDNAs (Figure 3-7). The membranes can be stripped (i.e., the probes can be removed) up to three times and used repeatedly. Probes that are not labeled radioactively continue to cause problems, although systems for such probes are in preparation or have been made available on the market recently.

The principal manufacturers offering different membranes with cDNA sequences from humans, mice, rats, or yeast are Genome Systems, Research Genetics, and Clontech. The price for these membranes is approximately $1400 to $2800 each. The offer is constantly being expanded and includes complete cDNA sets (e.g., Clontech offers all of the 5000 known human genes on three membranes) and filters specific for individual purposes, such as with the apoptosis-relevant genes that have been identified. Because these companies have extensive cDNA clone collections, the experimenter can order custom membranes as long as a sufficient quantity is purchased.

**Literature**

Gress TM, Hoheisel JD, Lennon JG, et al. (1992) Hybridization fingerprinting of high density cDNA library arrays with cDNA pools derived from whole tissues. Mam Genome 3:609–619.

Kafatos FC, Jones CW, Efstratiadis A. (1979) Determination of nucleic acid sequence homologies and relative concentrations by dot hybridization procedure. Nucleic Acids Res 24:1541–1552.

# 4 The Polymerase Chain Reaction

*Heil dem Wasser! Heil dem Feuer!*
*Heil dem seltnen Abenteuer!*

*Safely the water! Safely the fire!*
*Safely the rare adventure!*

The polymerase chain reaction (PCR) should be found in the chapter on tools. However, it is the method of choice in molecular biology, which is reason enough to dedicate a chapter to this subject alone.

The principle is wonderfully simple, and it is reasonable to ask why it was discovered long ago (Figure 4-1). The answer is perhaps no one knew what it could be used for. Since its beginnings in 1985 (Saiki et al.), resourceful heads have continuously discovered new applications. For a different reason, the PCR has incited much attention in general. It has rather quickly become an exemplary case for economizing. Never before has a patent in the field of molecular biology triggered such extensive discussions concerning the action of patents in the free field of research. Sentiments have been allayed because patents have increased the price of the enzyme only to a small degree and everything else has remained as it was.

## 4.1 Standard Polymerase Chain Reaction

The requirements for a PCR are a thermostabile DNA polymerase, a small amount of initial DNA (**template**), and two suitable oligonucleotide primers. Add buffers and nucleotides to this, place the tube in the PCR machine, and the reaction begins. It sounds simple, as does nearly everything that the experimenter begins, but it turns out to be extremely difficult, like everything in life. I begin with how it functions theoretically and then explain why this does not work.

The typical PCR program consists of a **denaturing step**, an **annealing step**, and an **extension step**. Denaturing, whereby the two strands of the DNA template are separated from one another, is carried out at 94°C (201°F). The temperature is subsequently lowered to 55°C (131°F) so that a hybridization of the oligonucleotide primer, which is available in a massive excess, is carried out on the single-stranded DNA template. After that, the temperature is increased to 72°C (162°F), which is the optimal temperature for Taq polymerase. The primer is thereby elongated until a double-stranded DNA is formed, which is exactly equivalent to the original DNA template. Because complementation occurs along both strands of the DNA template, the number of template DNAs is doubled in the course of one cycle. If this cycle is repeated, the experimenter obtains a fourfold quantity and so on.

The mother of all PCR programs demands the following conditions: an initial denaturation of 5 minutes at 94°C (201°F); 30 cycles, consisting of 30 seconds at 94°C (201°F), 30 seconds at 55°C (131°F), and 90 seconds at 72°C (162°F); and an extension for 5 minutes at 72°C (162°F) to be sure that the polymerase has completely finished its work. The reaction is subsequently cooled to 4°C (39°F) until somebody can be found who removes this tube from the machine. This protocol

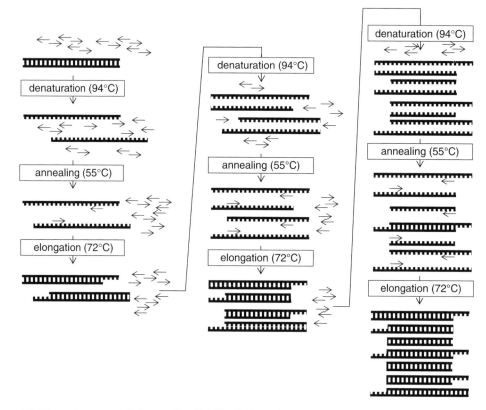

**Figure 4-1. The polymerase chain reaction (PCR) principle.** Step 1 (denaturing): The template DNA is heated to 94°C (201°F), and the strands are separated from one another. Step 2 (annealing): The temperature is reduced to give the primers a chance to hybridize on the DNA. Step 3 (extension): The temperature is increased to a level that is optimal for the function of the polymerase to permit synthesis of the second strand. At the end of one cycle, the quantity of DNA is almost doubled. In the three cycles portrayed, several problems associated with PCR can be recognized. In the first cycle, the products are usually longer than expected. The DNA fragments of the expected length proliferate after four to five cycles. Amplification occurs only so long as there is a sufficient amount of nucleotides and primer. As the concentration of amplification products increases, they hybridize more frequently with one another instead of with the primer. The replication rate per cycle therefore decreases drastically toward the end of the PCR, and additional amplification is uncommon, even in the presence of primers and nucleotides.

is presumably used today by hundreds of thousands and principally functions quite well. With some consideration and practice, you can improve these results (Figure 4-1).

**Denaturation** is a very rapid process that begins at a temperature of 70°C (158°F). All of the components are susceptible to heat. Polymerase is denatured, the nucleotides are degraded, and the DNA template and the primer are depurinated. These are sufficient arguments for keeping the denaturation as short as possible. A time of 5 seconds is adequate for separating the DNA strands from one another. Fortunate are those individuals who have a PCR machine that measures the temperature in the tube or can at least calculate it (e.g., one from Perkin-Elmer). All others must be satisfied with estimated values. As a matter of orientation, a 50-μL preparation covered with oil requires approximately 20 seconds until it has reached a temperature equivalent to that set on the heating unit.

The **annealing temperature** primarily conforms with the primers used. There are many programs on the market that calculate the (theoretical) melting temperature ($T_m$) of the primer and that of potential secondary structures. Various equations are available for the calculation of the melting temperature. They demonstrate a different efficacy and are calculated differently. With the simplest variant, you can calculate an approximate value of $T_m$ from the GC content of the primer:

$$T_m = 4 \times (\text{number of G or C}) + 2 \times (\text{number of A or T}) \tag{4-1}$$

It is useful only for short primers with a length of about 20 bases. An equation that is more laborious to calculate (Baldino et al., 1989) was developed to determine the melting temperature of oligonucleotides under the conditions of in situ hybridization:

$$T_m = 81.5 + 16.6 \times (\log_{10} [J^+]) + 0.41 \times (\%G + C) - (675/n)$$

$$- 1.0 \times (\% \text{ mismatch}) - 0.63 \times (\%FA) \tag{4-2}$$

In Equation 4-2, J is concentration of monovalent cations (in mol/L), % G + C is the proportion of the guanine and cytosine bases given as a percentage, n is the amount of the bases in the oligonucleotide, % mismatch is the proportion of bases with a faulty pair given as a percentage, % and FA is the proportion of formamides in the buffer given as a percentage (which is normally lacking in PCR).

Equation 4-2 is valid for cation concentrations of up to 0.5 M, GC concentrations of between 30% and 70% and oligonucleotide lengths of up to 100 bases. A disadvantage is that it is not particularly useful for oligonucleotides smaller than 30 bases long. The most acknowledged method is the *nearest-neighbor* method, as described by Breslauer and colleagues (1986). It is based on the observation that the melting temperature depends on the relationship of the bases and on the sequences of the oligonucleotides, because adjacent bases mutually influence one another. This process is the most exact, although it can no longer be managed manually. Fortunately, practically every program that analyzes sequences or primers also calculates the melting temperature of a sequence today.

The melting temperature alone is not enough, because it only provides information concerning the temperature at which 50% of the primer no longer binds to the template, but not concerning when the primer is reliably hybridized. Consider the annealing temperature of 5°C to 10°C (41°F to 51°F) to be lower. A quaint, more scientific dispute concerning this topic is to be found in the article by Rychlik and coworkers (1990), who came to the conclusion that the optimal annealing temperature depends on the primer and the amplified fragment. From this, they developed the following equation:

$$T_a^{opt} = 0.3 \times T_m^{Primer} + 0.7 \times T_m^{Product} - 14.9 \tag{4-3}$$

In Equation 4-3, $T_m^{Product}$ represents the product from Equation 4-2. Because many have difficulties with this equation initially, a small arithmetic example is provided. A fragment that is 632 bp long and has a GC content of 45% is amplified in a buffer with 50 mM KCl. The two primers have respective melting temperature of 55.3°C (131.5°F) and 48.6°C (119.5°F).

$$T_m^{Product} = 81.5 + 16.6 \, (\log 0.05) + 0.41 \times 45 - (675/632) = 77.3° \, C \, (171° \, F)$$

$$T_a^{opt} = 0.3 \times 48.6 + 0.7 \times 77.3 - 14.9 = \mathbf{53.8°C \, (129°F)}$$

This equation is the best available, although it does not provide completely reliable values. Like it or not, the optimal annealing temperature must be determined empirically. Extremely helpful in this process is a PCR machine that makes temperature gradients possible, and it is likely to be available from every manufacturer of PCR units, such as from Applied Biosystems, Stratagene, Eppendorf, or Biometra.

The **extension time** should be selected to suit the length of the awaited product. If it is too short, the polymerase cannot finish its job, and if it is too long, it has too much time to produce nonsense. Calculate 0.5 to 1 minute per 1 kb when using Taq and 2 minutes per 1 kb with the use of Pfu.

**Oil immersion** is another useful technique. With classic PCR machines, the preparation must be immersed in a mineral or paraffin oil to eliminate the possibility that the contents of the tube evaporate (otherwise, fluid will condense under the lid). Most newer machines have a heating cover that heats the lids of the tubes to more than 100°C (212°F) and consequently prevents the development of condensation, making the application of oil superfluous. Evaporation occurs nonetheless. The droplets accumulate at the edge of the tube between the central heating unit and the heating cover. Although this has no influence on larger volumes, smaller volumes ($<20\,\mu L$) can extensively alter the relationship of concentrations, preventing a successful amplification. In such cases, despite using the heating cover, the preparation should be immersed in oil.

The better the quality of the **template** DNA, the better the results. Sometimes, the PCR works when it is not expected to. A selection of functional templates includes plasmid DNA, cosmid DNA, phage DNA, DNA from a previous PCR, genomic DNA, cDNA, bacteria (from a colony transferred directly into the tube), phages (eluted from individual plaques), and DNA in polyacrylamide gel. Fundamentally, there is nothing too crude to serve as a template; give it a try if you doubt it.

The **multiplication factor** should be considered. Many sources explain that the DNA molecules double themselves from cycle to cycle. This would be wonderful, because a billion molecules could be obtained from a template molecule within 30 cycles or a trillion after 40 cycles; for a length of 1 kb, that rate would make an impressive quantity of at least $1\,\mu g$ of DNA. However, this is not the case. The average multiplication factor per cycle actually is about 1.6 to 1.7, with variations somewhat above or below. The difference appears to be small, but at the end, it is decisive for whether the PCR is a success or a failure. The situation is much more complicated, because the multiplication rate is smaller at the beginning of the PCR, presumably because the chance is relatively small that the template, primer, and enzyme meet one another when necessary. This probability then increases with an increase in the concentration of template, decreasing again at the end of the reaction because of the inhibition of multiplication through the increase in pyrophosphate, crumbling nucleotides, and the rehybridizing products.

The **template concentration** is an important factor. Theoretically, a single template molecule is sufficient to make an experimenter happy, but in practice, at least 10,000 DNA template molecules are needed to perform an amplification without any problems. Determination of the amount of molecules from a particular DNA concentration is difficult for many, and the following standard values may provide some help: 100,000 molecules of DNA template correspond to approximately 0.5 pg of plasmid DNA, 300 ng of human genomic DNA, or the cDNA in $1\,\mu g$ of total RNA. In most cases, these quantities can be reduced substantially, and the experimenter should not use template quantities that are too large; otherwise, the result may be exactly the opposite of that desired. The probability of faulty annealing increases, and the reaction can attain saturation too early, resulting in an unpleasant effect on the quality of the product.

Table 4-1 shows that a little cause produces a large effect. Even slight changes in the average rate of multiplication per cycle have a substantial effect on the yield. The DNA concentration at the end of the amplification can be calculated using the following equation:

$$X - x \cdot V^n$$

**Table 4-1.** Multiplication of DNA in the Polymerase Chain Reaction Tube

| Cycles before Multiplication | Average Multiplication Factor (V) per Cycle | | | | |
|---|---|---|---|---|---|
| | 2 | 1.8 | 1.7 | 1.6 | 1.5 |
| 10 cycles | $10^3$ | 357 | 201 | 110 | 58 |
| 20 cycles | $10^6$ | $1.3 \times 10^5$ | $4.1 \times 10^4$ | $1.2 \times 10^4$ | $3.3 \times 10^3$ |
| 30 cycles | $10^9$ | $4.6 \times 10^7$ | $8.2 \times 10^6$ | $1.3 \times 10^6$ | $1.9 \times 10^5$ |
| 40 cycles | $10^{12}$ | $1.6 \times 10^{10}$ | $1.6 \times 10^9$ | $1.5 \times 10^8$ | $1.1 \times 10^7$ |

In the equation, x is the initial concentration, V is the multiplication factor per cycle, and n is the number of cycles.

An interesting idea is the reuse of template DNA, as described by Sheikh and Lazarus (1997). The authors describe how they use pieces of unloaded nylon membranes (Duralon-UV membranes, Stratagene), on which DNA had previously been blotted, as a PCR template. Such a template can be used repeatedly, although it proves to be more unsatisfactory for the amplification of longer fragments (>850 bp).

A **buffer** must be chosen. The activity of Taq polymerase is maximal at a pH above 8. In the face of that, much experimentation has taken place with many different buffers, although evidently without any resounding success, because tris(hydroxymethyl)aminomethane hydrochloride (Tris HCl) remains the standard buffer used by molecular biologists, even though it is quite unfavorable for this purpose. It changes its pH extensively depending on the temperature ($\Delta pK_a$ of $-0.021/°C$), and the frequently used buffer, with a pH of 8.3, demonstrates a pH of 7.8 to 6.8 at 20°C (68°F) during the PCR—a situation that is less than favorable for the activity of the Taq polymerase. A Tris buffer with a pH of 8.55 or even of 9.0 is therefore frequently used.

For the **salts**, the selection is somewhat larger. One uses KCl (up to 50 mM) or $(NH_4)_2SO_4$ (up to 20 mM). In certain situations, it may even prove to be profitable to test which salt concentration is optimal. The yield, for example, gradually increases with the use of KCl at a concentration of between 10 and 50 mM, and it then falls dramatically at concentrations of more than 50 mM. NaCl inhibits the amplification.

The **$MgCl_2$** concentration is important. $Mg^{2+}$ influences the annealing of the primer, the separation of the strands during denaturation, the product specificity, the formation of primer dimers, and the rate of errors. The enzyme requires free $Mg^{2+}$ for its activity. Because primer, nucleotides, and EDTA bind $Mg^{2+}$, which may eventually be present, you must determine the optimal $Mg^{2+}$ concentration for each PCR; it normally lies between 0.5 and 2.5 mM. In most cases, you will be successful with 2 mM $Mg^{2+}$. In the event that this is not so, it is beneficial to optimize the concentration, and in some cases, the window of the optimal $Mg^{2+}$ concentration is very small. Pwo and Tfl prefer $MgSO_4$.

If the PCR does not function as it should and a change in the $Mg^{2+}$ does not bring about the desired effect, you can make an attempt with a number of different **additives**. Dimethyl sulfoxide (DMSO) (up to 10% [v/v]) and formamide (up to 5% [v/v]) can increase the specificity and facilitate the amplification of GC-rich sequences. Glycerin (10% to 15% [v/v]), PEG 6000 (5% to 15% [w/v]), and Tween 20 (0.1% to 2.5% [v/v]) can accelerate the reaction. There is no rule concerning which method is best; you must simply give it a try.

Heparin (>5u), SDS (>0.01%), Nonidet P40 (>5%), and Triton X-100 (>1%) inhibit the reaction. Although these substances would not be pipetted to the reaction, you should be sure that the DNA template does not contain disturbing amounts of any of them.

The quality of **nucleotides** has a great influence on the success of these procedures. If necessary, try the reaction with different nucleotides obtained from various manufacturers.

The **nucleotide concentration** is an important factor. In the standard protocol, a concentration of 200 μM of each nucleotide is cited. The quantity is adequate to synthesize 13 μg of DNA from a 50-μL preparation, a quantity the experimenter cannot expect to attain. A large proportion of the nucleotides ends instead as a nucleotide diphosphate or monophosphate, so that only one fourth of this quantity is normally sufficient. There is no reason to be fearful, because the yield remains the same even if the concentration of the nucleotides is small. The frequency of errors with Taq polymerase tends to be reduced when associated with a lower concentration of nucleotides, at least up to a limit of 10 μM per nucleotide, although study authors do not seem to completely agree with this point.

A concentration of dNTP that is too high can inhibit amplification, possibly resulting in complete failure. This effect may result from the fact that the nucleotides bind free $Mg^{2+}$ ions, although it also may reflect the quantity of inhibitory pyrophosphates that rapidly increase during the course of PCR in the event of high dNTP concentrations. The experimenter should beware, because some manufacturers label their nucleotide mixtures with the confusing designation of "dNTP 100 mM", a sign that the concentration of each nucleotide is present at a concentration of only 25 mM.

A trick to increase the yield consists of displacing the balance of the reaction in favor of the product by removing one of the reaction products, the **pyrophosphate**. This has been feasible, because it has been possible to successfully isolate a thermostabile pyrophosphatase from thermophilic bacteria following the polymerase. Perkin-Elmer also uses this as a part of their *Cycle-Sequencing* and *Thermosequenase* kits, although thermostabile pyrophosphatase has been difficult to obtain. Currently, only CHIMERx (www.chimerx.com) is available.

**Primers** must be made for the desired amplification, and they are most decisive for the success of the amplification (Figure 4-2). As unpredictable as their conduct may be, there are some rules that can serve to substantially increase their probability of functioning. PCR primers normally have a length of 18 to 30 bases and demonstrate a proportion of 40% to 60% guanidine and cytosine (G + C). Primers for the amplification of extra-long products should be 25 to 35 bases long. The primer should contain not more than four subsequent similar bases (e.g., AAAA) to avoid faulty hybridization and frameshifts. The melting temperature should be 55°C to 80°C (131°F to 176°F) to permit sufficiently high annealing temperatures.

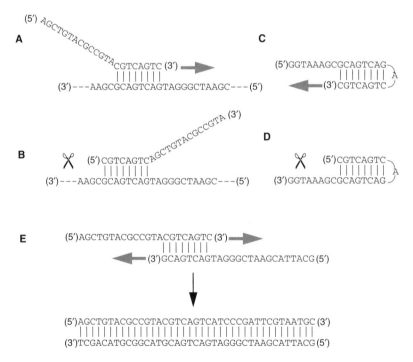

**Figure 4-2. Problems in the primer design. A,** If the 3′ end of the primer hybridizes to another DNA, the polymerase can lead to the ligation of additional bases so that nonspecific amplification products are formed. **B,** If the 5′ end hybridizes on another DNA instead, the polymerase can remove bases from the 5′ end because of its 5′-3′ exonuclease activity. **C,** If hairpin structures are found at the 3′ end, the polymerase can bind additional bases so that the primers no longer bind specifically. **D,** In the event of hairpins at the 5′ end, the polymerase can eliminate bases from the 5′ end. The shortened primer then binds to the template DNA with a reduced probability. **E,** Two primers whose 3′ ends are complementary can use themselves as templates, resulting in primer dimers (i.e., hybrids of the two primers), which are no longer of any use for amplification.

Two G or C bases should be located at the 3' end to obtain a better bonding and extension; a maximum of three G or C bases should be located at the 3' end, because faulty hybridized primer is otherwise stabilized and increases the danger of nonspecific amplification products.

The sequence should be as specific as possible for the desired amplification product so that the primer only hybridizes at the proper site. The more sequences employed in the template DNA, the higher is the probability of obtaining nonspecific amplification products. The more complex the template DNA (i.e., the more varied the different DNA sequences that are contained, such as in cDNA, to say nothing of genomic DNA), the higher is the probability of amplifying the artifacts. You can best approach the problem with the highest possible annealing temperatures (approximately 55°C to 65°C [131°F to 149°F]), as long as the primer sequence permits this. With increasing temperature, the probability of annealing and the yield decrease. Inserting mutations in the primer for the purpose of mutagenesis is best performed at the 5' end.

The primers should form no internal secondary structures, such as *hairpins*, because these reduce the probability of a hybridization with the DNA template. The possibility of forming *hairpins* is best controlled with an appropriate computer program.

The primers may not hybridize with one another and should demonstrate the smallest possible complementarity at their 3' ends; otherwise, one primer may serve as a template for the other. In this way, instead of the desired amplification product, the fearful primer dimers are likely to appear. The quantity of useful primer is thereby drastically reduced, and the amplification efficiency plummets as far as possible.

The purer the primer, the more favorable is the amplification. You should always use HPLC or PAGE—purified buffers—in the event of critical applications; the normal, desalinated primer is adequate for the standard PCR.

In contrast to the official statements, it is possible to store buffers and primers at 4°C (39°F) if the work is done cleanly and free of nucleases, as expected from a good experimenter. The nucleotides are not able to withstand such storage (stabilized nucleotide solutions are available, such as those from Peqlab, and it is better to store the polymerases at −20°C (−4°F), although an enzyme that can be boiled for 1 hour should be considered robust.

Oligonucleotides can be obtained on almost every second street corner. Among the larger companies, Pharmacia, MWG, and Roth manufacture synthetic primers, along with other chemicals. Many biotechnology companies also have specialized in this field. Consequently, experimenters only have to look through the laboratory mail attentively.

The **primer concentration** can have a considerable influence on the yield. Usually, about 10 pmol of primer are required for a 50-μL preparation; with this method, 1.0 pmol of primer is sufficient for the synthesis of 6.6 μg of a 1-kb fragment. If the yield is small, an attempt can also be made with substantially higher quantities of primer. Be careful, because the yield also drops again when more than the optimum is used (depending on the experimental preparation, that can be approximately 50 to 200 pmol per 50 μL).

The **number of cycles** is a factor in the outcome. For a specific quantity of the product (approximately 0.3 to 1 pmol), a plateau effect is to be observed, for which the rate of multiplication is greatly reduced. There are several reasons for this. The accumulation of end products (e.g., DNA, pyrophosphate) slows down the synthesis, the probability that two completed DNA strands reanneal reduces the neosynthesis, the substrate concentration is reduced (especially the concentration of intact nucleotides), the quantity of intact polymerase is reduced, and a competition with nonspecific products (e.g., primer dimers) and faulty hybridization develops. The number of faulty products increases at this point and, along with the desired product, the background disappears. The product tends to become somewhat "smeared" after too many cycles, because a continuous increase in the faulty products develops along with the desired product, which differs from the correct product primarily through its length, and it is seen as a smear on the agarose gel above and below the correct band. For the best results, you must select the number of cycles so that the PCR is completed when the plateau has been reached. An increase at this point does not increase the yield!

**Temperature-dependent activity** of the Taq polymerase must be considered. Although the thermostabile polymerases have an optimal effect at about 70°C (158°F) (for Taq, it is 74°C [165°F]), they are not fully inactive at other temperatures. The Taq polymerase demonstrates an insertion of approximately 2800 nucleotides/min at 70°C (158°F), 1400 nucleotides/min at 55°C (131°F), 90 nucleotides at 37°C (99°F), and about 15 nucleotides/min at 22°C (72°F). This makes it possible, for instance, to make use of only a single annealing or extension step at 60°C (140°F). Although the residual activity at low temperatures is not a problem with the application of Taq polymerase and the preparations can remain overnight in the PCR machines at room temperature after the amplification without showing any changes, it can be a problem for polymerases demonstrating error correction activity. Pfu, Pwo, and Tli degrade superfluous primer at room temperature, and they nibble at the PCR product. PCR preparations demonstrating polymerases with error correction activity must be cooled in the machine after the completion of the amplification until they are used again.

The classic **polymerase** for PCR is the Taq DNA polymerase isolated from *Thermus aquaticus*, a thermostabile bacterial strain that grows in hot springs at a temperature of 70°C (158°F). This miraculous substance demonstrates an activity maximum at 74°C (165°F), at a pH above 8, and aside from 5'-3' DNA polymerase activity, it also has a 5'-3' exonuclease activity, but no 3'-5' exonuclease activity. Taq has a template-dependent (normal) polymerase activity and a template-independent activity. The later is responsible for the fact that an additional base, which is not to be found on the template strand and is usually adenosine, is frequently attached to the end of the newly synthesized strand. The rate of DNA synthesis is approximately 2800 nucleotides per minute. In the meantime, the situation involving patent law and the inquiring mind have presented us with thermostabile polymerases, which demonstrate different characteristics. The fewest of these demonstrate synthesis rates that are equivalent to that of the Taq polymerase, so that its popularity continues to carry on today.

For the most part, the polymerases can be differentiated into two groups. The first group is **Pfu** (from *Pyrococcus furiosus*), **Pwo** (from *Pyrococcus woesei*), **Tma** or **UlTma** (from *Thermotoga maritima*), and **Tli** or **Vent** (from *Thermococcus litoralis*). These polymerases demonstrate a substantially higher temperature stability than that of Taq (i.e., the half-life of Pfu at 95°C [203°F] is approximately 12 hours) and a 3'-5' exonuclease activity, which permits error correction activity (i.e., *proofreading*). Their amplification products have no base overlap like the Taq-generated products, but they instead have smooth ends. What is the price for such a product? Handling this substance is somewhat more difficult (because the enzymes also "correct" single-stranded primer), and it has a substantially lower rate of synthesis; for example, with 550 nucleotides per minute, Pfu activity is far below that demonstrated by Taq polymerase. The reason is the different processivity of the enzyme (Table 4-2).

*Processivity* is the number of nucleotides that a polymerase attaches to the 3' end of a DNA strand in a single cycle. This amount represents the rate of polymerization and the disassociation constant ($K_d$) of

**Table 4-2.** Characteristics of Thermostabile Polymerases

| Feature | Taq | Tth | Pfu, Pwo |
|---|---|---|---|
| 5'-3' DNA polymerase | + | + | + |
| Processivity | High | High | Low |
| 5'-3' exonuclease | + | + | + |
| 3'-5' exonuclease (proofreading) | − | − | + |
| Unspecific overlap | + | + | − |
| Reverse-transcriptase activity | − | + | − |
| Estimated error frequency of inserted bases | $10^{-5}$ | $10^{-5}$ | $10^{-6}$ |

the polymerase. The processivity of Pfu is about 9 to 12 bases (per cycle), whereas that of Taq is 35 to 100. This explains why the two polymerases demonstrate such different rates of synthesis.

To compensate for the degradation of primers by exonucleases, you should use higher concentrations of primer. The manufacturers usually recommend that a substantially higher amount of enzymes be used.

The second group is **Tth** (from *Thermus thermophilus*) and **Tfl** (from *Thermus flavus*). The profile of characteristics is very similar to that of Taq, although these polymerases demonstrate a unique property. Their reverse transcriptase activity, which can be observed to a small degree in all thermostabile polymerases, is very high. The enzyme can be used for cDNA synthesis and for PCR (usually after an alternative buffering, because the reverse transcriptase activity is $Mn^{2+}$ dependent, whereas the normal DNA polymerase activity is $Mg^{2+}$ dependent). Although the RT activity is not as high as that of the normal reverse transcriptase and the rate of synthesis of the normal polymerase activity is not as high as that of a Taq, they are very interesting enzymes, especially if a large number of reverse transcriptase–polymerase chain reactions (RT-PCRs) must be carried out.

For polymerases with error correction activity, the **rate of failure** is approximately one error every $10^6$ bases (the values vary somewhat depending on the particular test used), and it is higher by a factor of about 10 for those without any correction. If you would like to investigate this in more detail, a more extensive examination has been performed by Cline and colleagues (1996). Years of investigation have shown that Pfu demonstrates the lowest rate of error among all the thermogelatin-stable polymerases.

What does this mean? If a PCR is performed with Taq polymerase and with 10,000 template molecules, producing a product that is 1000 bp long, 1% of the newly generated fragments will have a mutation in the first cycle. In the second cycle, another 1% of mutated fragments will be observed, and so on. After 30 cycles, 25% of the fragments demonstrate one or more mutations. The same PCR with a correctable proofreading polymerase would result in mutated fragments at a rate of about 3%. These mutations can cause a massive problem; for example, for a 2-kb template, these methods translate to rates of 45% and 7%, respectively. For this reason, when cloning new DNA fragments with the aid of PCR, at least three clones should be sequenced independently to ensure that there is ultimately at least one error-free clone.

The rate of error can increase, depending on the particular conditions used. As a result, Taq polymerase reveals different rates of error under various buffer conditions (Table 4-3) (see also Chapter 9, Section 9.2).

The **enzyme concentration** must be taken into account. Typically, a preparation with 50 μL of a *unit* of Taq polymerase is used. Especially with polymerases demonstrating a proofreading activity, substantially larger quantities are required. It is best to follow the instructions of the manufacturers.

### Literature

Baldino F, Chesselet M-F, Lewis ME. (1989) High-resolution in situ hybridization histochemistry. Method Enzymol 168:761–777.

Breslauer KJ, Frank R, Blocker H, Marky LA. (1986) Predicting DNA duplex stability from the base sequence. Proc Natl Acad Sci U S A 83:3746–3750.

**Table 4-3.** Buffers

| Composition of the Buffer | Rate of Error |
|---|---|
| 20 mM Tris HCl, pH 9.2/60 mM KCl/2 mM $Mg^{2+}$ | $42 \times 10^{-6}$ |
| 20 mM Tris HCl, pH 8.8/10 mM KCl/10 mM $(NH_4)_2SO_4$/2 mM $MgSO_4$/100 μg/mL BSA | $21 \times 10^{-6}$ |
| mM Tris HCl, pH 8.8/50 mM KCl/1.5 $MgCl_2$/0.001% gelatin | $9.6 \times 10^{-6}$ |

BSA, bovine serum albumin; Tris, tris(hydroxymethyl)aminomethane.

Cline J, Braman JC, Hogrefe HH. (1996) PCR fidelity of Pfu DNA polymerase and other thermostabile DNA polymerases. Nucleic Acids Res 24:3546–3551.

Innis MA, Myambo KB, Gelfand DH, Brow MA. (1988) DNA sequencing with *Thermus aquaticus* DNA polymerase and direct sequencing of polymerase chain reaction-amplified DNA. Proc Natl Acad Sci U S A 85:9436–9440.

Rychlik W, Spencer WJ, Rhoads RE. (1990) Optimization of the annealing temperature for DNA amplification in vitro. Nucleic Acids Res 18:6409–6412 (erratum: Nucleic Acids Res 19: 698).

Saiki RK, Scharf S, Faloona F, et al. (1985) Enzymatic amplification of beta-globin genomic sequences and restriction site analysis for diagnosis of sickle cell anemia. Science 230: 1350–1354.

Sheikh SN, Lazarus P. (1997) Re-usable DNA template for the polymerase chain reaction (PCR). Nucleic Acids Res 25:3537–3542.

## 4.2 Suggestions for Improving the Polymerase Chain Reaction

PCR is a miracle weapon—if it functions. With this method, it becomes very clear that every experiment demonstrates its own, independent existence. This is not surprising, because the number of templates is enormous and the number of possible primers is almost infinite.

For the most part, there are two essential possibilities for the use of PCR. You always amplify the same fragments, sometimes from one DNA and sometimes from another. In this way, it is possible to perform tests to determine the proper conditions and to avoid these problems. As a full-fledged researcher, you may instead amplify one thing today and another thing tomorrow, until a standard protocol is developed, and you then hope that it functions from experiment to experiment. If it does not work, you must search for the error.

**No product may be attained.** Typically, you have forgotten to add a component while pipetting. Sometimes, the nucleotides are not functional, and one of the primers may be lacking; the latter error may occur despite storage at $-20°C$ ($-4°F$), although this is rare. Occasionally, the concentration of the template is too small, such as when you are working with rare mRNAs, and you must increase the number of cycles. More than 40 cycles, however, should be employed only in exceptional cases.

**In addition to the band seen from the product, many unexplainable bands are seen.** In a classic case of PCR artifacts, the primer hybridizes at the wrong site. All of the parameters that can be changed have been mentioned. The $Mg^{2+}$ concentration must be adjusted more accurately; DMSO, glycine, formamide, and other components must be added (Sarkar et al., 1990); and the annealing temperature is increased. Another possibility is the *hot start*. The underlying idea is simple. At the beginning of the PCR, because of the initially still low temperatures, there are undesirable effects such as a faulty hybridization of the primer or primer dimer formation, which in the further course of this procedure can lead to a poor yield or to the development of unspecific products. That should be counteracted by a hot start so that one or more important components are first added after the correct working temperature has been reached. Although the first realization of this idea was rather elaborate (i.e., pipetting of polymerases), a whole host of alternatives are now available for the experimenter. Many manufacturers offer wax beads that are used instead of oil and that liquefy only at a temperature of $60°C$ ($140°F$). In a first step, the first half of the preparation (i.e., without polymerase and nucleotides) is pipetted together and added to the wax until the wax solidifies to form a solid layer. In the second half of this preparation, nucleotides and polymerase are pipetted onto this hardened layer, and the PCR is started. As soon as the wax melts, the upper portion of the preparation sinks through the wax, and the amplification begins. Be careful; if the volume of the upper layer is too small ($<10\,\mu L$), it will not sink, and nothing at all will occur. Newer alternatives inactivate the Taq polymerase with a specific antibody, which is denatured at higher temperatures and sets the Taq free (TaqStart and TthStart, Clontech) or makes use of oligonucleotides that block the DNA binding sites of the Taq at low temperatures (Dang and Jayasena, 1996). Invitrogen offers HotWax $Mg^{2+}$ beads that first release the cofactor magnesium at a temperature of 60°C (140°F), and Promega imbeds the Taq polymerase in wax beads (Kaijalainen et al., 1993).

A cheap alternative, which is frequently successful, is the *cold start*, a pompous term that indicates that all of the components up to the start of the PCR are put on ice. If none of this helps, use another primer. Not infrequently, a hybridized primer, located 10 or 20 bases further along, reacts completely differently and produces sparkling white, clean bands.

**A single band is obtained, but the yield is very poor.** This problem results from the primers (i.e., their melting temperature is below that of the annealing temperature) or from a sequence section in the amplified fragment that forms secondary structures that can only be unraveled with difficulty (e.g., persistent, hair-needle structures). The addition of DMSO or formamide can help, or you can use one of the many polymerase mixtures for extra-long PCRs (discussed later) instead of Taq polymerase; they are frequently more effective with such secondary structures. If the PCR reaction functions in principle, the yield can occasionally be raised through the use of higher primer concentrations (up to 50 pmol per primer for 50-μL preparations).

**A smear is obtained instead of a band.** A smear can appear if the planned product is just barely too long for a successful amplification. If too much template DNA has been inserted or the number of cycles has been too high when the PCR reaction comes to an end (i.e., the Taq is not used to full capacity), it begins to produce nonsense. You can attempt to amplify a PCR product again with the same primers, a venture that should theoretically represent no real problem, although it frequently does not function in practice because there is an accumulation of nonspecific product or too much template is used. Even if the product contains repetitive sequences, such as in the amplification of genomic fragments, a smear may develop.

**Bands are obtained although there is no template in the preparation.** The problem is contaminants. Contamination is the constant bogeyman of experimenters. They cannot be avoided, and experimenters may be lulled into a false sense of security after many problem-free amplifications, until suddenly, nothing works. Because most molecular biologists work with the same genes, it cannot be avoided that the traces of older PCRs and plasmid preparations distribute themselves throughout the entire laboratory and occasionally land in the primers, the nucleotides, or the polymerase. The most important rule is to perform a negative control in the form of a PCR preparation with no template DNA before every important amplification. I know that negative controls do not make any academics happy, but the failure to perform them can be guaranteed to come back and haunt the experimenter sooner or later.

There are various ways to limit the probability of contamination. The most important is having your own set of PCR pipettes. If you are unable to afford them, you can use displacement tips, which help to avoid contamination from being passed from the pipette to the PCR preparation. However, over the long run, they cost as much as the pipettes. To be especially safe, use your own pipettes and displacement tips.

If the space is available in the laboratory, set up a PCR workplace where only PCR preparations are pipetted. Otherwise, be sure to have a clean surface. You must be able to differentiate between activities before and after PCR. Consider exactly with whom you wish to share solutions. Too many cooks spoil the broth and the PCR.

**Literature**

Dang C, Jayasena SD. (1996) Oligonucleotide inhibitors of Taq DNA polymerase facilitate detection of low copy number targets by PCR. J Mol Biol 264:268–278.

Kaijalainen S, Karhunen PJ, Lalu K, Lindstrom K. (1993) An alternative hot start technique for PCR in small volumes using beads of wax-embedded reaction components dried in trehalose. Nucleic Acids Res 21:2959–2960.

Sarkar G, Kapelner S, Sommer SS. (1990) Formamide can dramatically improve the specificity of PCR. Nucleic Acids Res 18:7465.

Sharkey DJ, Scalice ER, Christy KG Jr, et al. (1994) Antibodies as thermolabile switches: High temperature triggering for the polymerase chain reaction. Biotechnology 12:506.

## 4.2.1 Nested Polymerase Chain Reaction

A single template molecule should suffice theoretically to cover the world ankle-deep in a PCR product, but in practice, the experimenter always seems to be unsuccessful exactly then when he or she relies on it most. After about 40 cycles, the quantity of faulty hybridization continues to become stronger so that, after a greater number of cycles, you end up with a smear from the products and with little satisfaction.

A trick to demonstrate even the smallest of template concentrations involves the use of *nested* PCR. You perform a PCR and use the product as a template for a second amplification with other primers. The second primer pair lies nested between the outer limits of the first pair (Figure 4-3). In this way, the wrong amplification products are selected during the first round. By means of nested PCR, you can demonstrate extremely small quantities of template. This method allows the amplification of almost every cDNA from any desirable tissue, because each cell of the body expresses most genes, even though only to an extremely slight degree—a process known as *illegitimate transcription* (Chelly et al., 1989).

**Literature**

Chelly J, Concordet JP, Kaplan JC, Kahn A. (1989) Illegitimate transcription: Transcription of any gene in any cell type. Proc Natl Acad Sci U S A 86:2617–2621.

## 4.2.2 Multiplex Polymerase Chain Reaction

That molecular biology has something to do with true life can be seen in the fact that *multiplex* is found in the laboratory as well as the movies. The idea was born from laziness: to make several

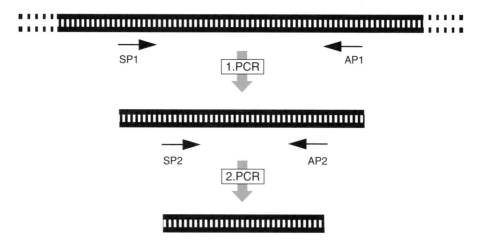

**Figure 4-3. Nested polymerase chain reaction.** AP1, antisense primer 1; SP1, sense primer 1.

amplifications on the same DNA template, why not add a respective amount of primer in the process? Chamberlain and colleagues (1988) showed that this approach works by using as many as six primer pairs. This is by no means the possible limit; 36 fragments have been amplified successfully in a tube, and it appears that amplification may be feasible with 100 primer pairs.

A prerequisite is that the primer amplifies no homologous sequences, because it would otherwise easily come to a faulty annealing of the primer or even of the products, and the result would be a smear in the agarose gel. Evidence of the individual fragments becomes increasingly difficult with a larger number of primer pairs. With only a few fragments, it is still possible to see that the amplified fragments have different lengths. If the number of the fragments increases, you can use fluorescence-labeled primer and separate the PCR products on an automatic sequencer. For demonstrating many fragments, it is best to detect them by means of hybridization, such as using *hybridization strips* (Cheng et al., 1999) or *microarrays* (see Chapter 9, Section 9.1.6).

The *multiplex* PCR frequently causes problems initially, and you must invest some time in optimization of the primer and PCR conditions. The method is therefore suitable for screenings, in which the same primer is always employed. Zangenberg and coworkers (1999) have provided good instructions for beginners.

**Literature**

Chamberlain JS, Gibbs RA, Ranier JE, et al. (1988) Deletion screening of the Duchenne muscular dystrophy locus via multiplex DNA amplification. Nucleic Acids Res 16:11141–11156.

Cheng S, Fockler C, Barnes WM, Higuchi R. (1994) Effective amplification of long targets from cloned inserts and human genomic DNA. Proc Natl Acad Sci U S A 91:5695–5699.

Cheng S, Grow MA, Pallaud C, et al. (1999) A multilocus genotyping assay for candidate markers of cardiovascular disease risk. Genome Res 9:936–949.

Zangenberg G, Saiki RK, Reynolds R. (1999) Multiplex PCR: Optimization guidelines. In: Innis MA, Gelfand DH, Sninsky JJ (eds): PCR Applications. New York, Academic Press.

## 4.2.3 Amplification of Longer DNA Fragments

Taq polymerase is very efficient, although it permits only short amplifications (up to approximately 5 kb). The reason for this lies in the mode of operation of Taq polymerase, which begins to stutter after installing an incorrect base, and it continues very slowly with further extension or not at all. In the next PCR cycle, faulty hybridization occurs, which leads to a long-term accumulation of incorrect amplificates. At the end, the agarose gel presents a large, impenetrable smear, and the experimenter knows that it has not worked.

In 1994, a large breakthrough occurred. Because the Taq polymerase was fast but only briefly effective and therefore permitted the amplification of only a few kilobases, whereas the Pfu was correct and effective for longer but resulted in a slight yield so that it was not particularly suitable for the amplification of long fragments, a small group of clever scholars came up with the idea of uniting both of their positive characteristics by mixing the two substances (Barnes, 1994; Cheng et al., 1994). A mixture of 11 parts of Taq and 1 part of Pfu accomplished true miracles. Since then, it has been possible to amplify fragments of up to 50 kb (at least occasionally). Most suppliers of thermostabile polymerases have several such mixtures to offer (with original names such as TaqPlus, ProofSprinter, or PowerScript). Buy them, and try them. However, you should limit your expectations. The yield tends to be smaller for longer fragments, and the amplification of long fragments of genomic DNA—which is the most interesting application—is the most difficult process, because the risk of finding repetitive sequences is high, a situation that can make an amplification difficult or impossible. Nevertheless, the existence of mixtures of polymerases are justified because they amplify longer products, frequently degrade unpleasant secondary structures better, and can considerably increase the yield of shorter fragments that can be amplified only with difficulty.

Aside from a polymerase mixture, a longer primer is needed for the amplification of long products, higher nucleotide concentrations, and other (higher) salt concentrations than those required for standard PCRs. Take the advice of the manufacturer concerning this topic.

**Literature**

Barnes WM. (1994) PCR amplification of up to 35-kb DNA with high fidelity and high yield from lambda bacteriophage templates. Proc Natl Acad Sci U S A 91:2216–2220.

Cheng S, Fockler C, Barnes WM, Higuchi R. (1994) Effective amplification of long targets from cloned inserts and human genomic DNA. Proc Natl Acad Sci U S A 91, 5695–5699.

# 4.3 PCR Applications

The uses of PCR appear to be endless, at least there does not seem to be any end to this ingenuity at the present. To obtain at least an impression of the possibilities, a list follows which hopefully includes at least the most frequent applications.

## 4.3.1 Reverse Transcription-Polymerase Chain Reaction

Reverse transcription-polymerase chain reaction (RT-PCR) is a long term for a simple technology. cDNA is synthesized from an arbitrary RNA, and it is used as a template for a PCR.

The palette of applications is large. For instance, you can determine the transcription of a gene in certain tissues or cells, find splice variants, or manufacture cDNA probes for hybridization. Various other applications, such as quantifying the expression of a gene, are clever modifications of the RT-PCR protocol and are discussed later in more detail.

The reverse transcription is accomplished mostly with AMV or M-MLV-RT (see Chapter 5, Section 5.4), although thermostabile DNA polymerases with reverse transcriptase activity can also be used, such as Tth or Tfl, which can also be used for the PCR amplification directly afterward (see Section 4.1). The advantage lies in the lower expense, although the sensitivity is higher if the cDNA synthesis and PCR are performed separately. Oligo(dT) primer, hexamers, or specific primers may be used, although the results with hexamers are usually the best.

cDNA frequently causes greater problems in the PCR than other templates such as plasmids or genomic DNA. Sometimes, problems are caused by the occasionally quite varied cDNA yield with reverse transcription, or the cDNA contains considerable quantities of substances such as primers, nucleotides, and salts, which are also found in PCR preparations. In unfavorable cases, the concentration of such a substance brought in with the cDNA is changed to leave the area of optimal amplification so that the yield is reduced. A purification, such as by means of gel chromatography or *spin columns* (see Chapter 7, Section 7.1.1), may help. Reverse transcriptase also has an inhibitory effect on the Taq polymerase (Sellner et al., 1992).

Sometimes, the cDNA does not appear to be very gelatin stable, even when stored properly. Everything still works well during the first amplification, whereas the bands are rather weak 1 week later, and nothing functions some time later. The cause is unknown, because cDNA is not any more unstable than other nucleic acids; it may help to use an enzyme or kit from another manufacturer.

**Literature**

Sellner LN, Coelen RJ, MacKenzie JS. (1992) Reverse transcriptase inhibits Taq polymerase activity. Nucleic Acids Res 20:1487–1490.

## 4.3.2 Rapid Amplification of cDNA Ends

After producing a cDNA clone, the experimenter may discover that it is not complete. Usually, the cDNA bank is screened a second time, but what if the second run also delivers no complete clone? The solution is to use rapid amplification of cDNA ends (RACE), which is also known as *anchored* PCR or *one-sided* PCR.

RACE is an RT-PCR by another means. Depending on whether the 5′ or the 3′ end is to be elongated, the method differs somewhat (Figure 4-4). With the **3′ RACE**, cDNA synthesis is performed using a modified oligo(dT) primer (frequently known as an anchor primer), the 5′ half of which consists of a self-selected sequence that is long enough to permit the annealing of two primers. The cDNA is amplified with a *nested* PCR (see Section 4.2.1) through the use of a transcript and an anchor-specific primer.

For the **5′-RACE**, a transcript-specific primer is used for the reverse transcription, and with the aid of a terminal transferase, a poly A (or poly C) tail is added to the synthesized cDNA in a second step. Nested PCR is used as a substitute for three primers in the first amplification: slight quantities of anchor primers for the formation of amplifiable templates and normal quantities of anchor and transcript-specific primers for the actual amplification. The second amplification runs as usual with

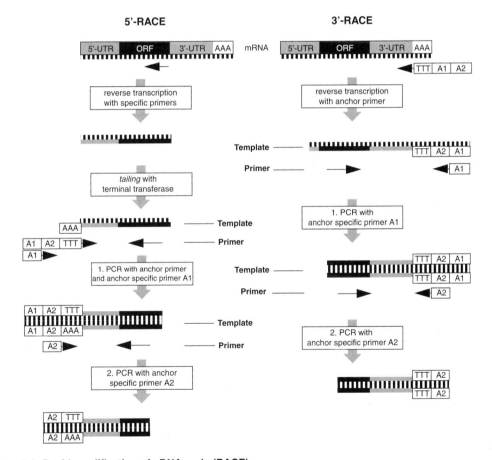

**Figure 4-4. Rapid amplification of cDNA ends (RACE).**

the use of two primers. Clontech offers a kit (Marathon cDNA amplification kit), which permits implementation of 5′ and 3′ RACE. If that is too much work, you can purchase RACE cDNA from them as a finished product for the trifling amount of about $800.

RACE is frequently used if you are interested in the exact start of transcription of the mRNA, such as in promoter studies. Exactly then, however, some weak points of the 5′ RACE will be observed. The poly A tail added is not very suitable for amplifications because of the low melting point, and the addition of a poly C tail demonstrates a melting point that is too high. Another problem is that mRNAs frequently have GC-rich sequences at their 5′ end, which causes difficulties for the reverse transcriptase so that the DNA synthesis is interrupted prematurely. The consequence is that there are some bases missing at the 5′ end. A remedy is seen in a modified protocol, which demands much work, but at least it promises to deliver complete 5′ ends.

In a first step, the mRNAs are dephosphorylated with alkaline phosphatase. The intact mRNAs remain unchanged because their 5′ ends are protected by a 7-methylguanosine cap, where the 5′ ends of the partially degraded mRNAs are dephosphorylated. After that, the cap is removed with tobacco acid pyrophosphatase (TAP), and a phosphorylated 5′ end remains. In a third step, an RNA oligonucleotide on the 5′ end of the (phosphorylated) mRNA is ligated using a T4 ligase. The required RNA primer (or linker) is manufactured through in vitro transcription of a linearized plasmid. Ultimately, reverse transcription is performed with specific primers or *random* hexamers, and a *nested* PCR is added, each with a gene and a linker-specific primer.

A detailed description of classic and modified RACE with protocols can be found in articles by Frohman (1994) and Schaefer (1995).

**Literature**

Frohman MA. (1994) On beyond classic RACE (rapid amplification of cDNA ends). PCR Methods Appl 4:S40-S58.

Schaefer BC. (1995) Revolutions in rapid amplification of cDNA ends: New strategies for polymerase chain reaction cloning of full-length cDNA ends. Anal Biochem 227:255–273.

## 4.3.3 Amplification of Coincidental Products

*Ein schöner, süßer Zeitvertreib!*

*A sweet and seemly pastime that!*

This section proves that PCRists are lunatics, for whom every problem is a nail, simply because they have only a hammer. The title of this section is really a contradiction because the amplification of DNA demands two specific primers, with which a defined product becomes amplified. Instead of this, protocols were designed by the dozen in the beginning of the nineties, which caused this idea to be thrown out. The first were the *arbitrarily primed PCR* (AP-PCR) (Welsh and McClelland, 1990) and *random amplified polymorphic DNA* (RAPD) (Williams et al., 1990), which were originally designed for DNA fingerprinting for the purpose of genetic analyses. Soon, the process called *RNA arbitrarily primed PCR* (RAP-PCR) (Welsh et al., 1992) and *differential display* (DD) (Liang and Pardee, 1992) were developed for the fingerprint analysis of RNA. McClelland and Welsh (1994a, 1994b) provide a deeper insight into the endless breadth of this method and offer a few sample protocols.

One questions addresses the basic idea: How do I analyze a large quantity of DNA, whose sequence is not known to me? Instead of using two specific primers, usually of 17 to 23 bases long, the answer is to use random primers of about 10 bases long. The annealing temperature must be reduced to less than 40°C (104°F) for this procedure, so that the primer is also able to be hybridized on only partially complementary sequences. If you have chosen the proper conditions, you will receive a jumble of bands of different sizes, whose sequence continues to be unknown, although this band pattern is reproducible for the template used. If you amplify different DNAs under the same conditions, you can identify differences between individuals, phyla, or species through a comparison of the band patterns.

The skill is to figure out the conditions so that the number of developing bands is precisely large enough that you can evaluate them. If there are too many, you can no longer distinguish the individual bands, and if there are too few, the chance of finding a polymorphism is reduced. In this process, you can change the length of the primer, use degenerated primer or mixtures of several primers, or play around with the amplification protocol. In each case, the result is a matter of luck. You should also be aware that the band pattern may differ among laboratories and, even worse, from day to day. The article by Meunier and Grimont (1993) is very interesting in this regard. In the daily routine of researchers, many differences in patterns are caused by artifacts. For these reasons, this method cannot be used for critical applications, such as for paternity tests.

The RNA variant of random amplification (differential display PCR [DD-PCR]) is very popular because, at least in principle, it can provide evidence concerning differences in gene expression, such as between two different tissues or between induced cell lines and those that are not induced. It is performed with a reverse transcription using an anchor primer that, aside from the usual oligo(dT) portion and a specific adaptor sequence at the 5′ end, can demonstrate two defined bases (e.g., 5′-NNNNNN(T)$_{16}$AG-3′) at the 3′ end. The aim is to rewrite only a part of the mRNAs in cDNA in this way and to thereby reduce the complexity of the cDNA. Subsequently, you amplify with an anchor-specific and a short, random primer under reduced stringency and look to see whether there are differences in the band patterns between the RNAs being examined. If such differences are found, the bands are then isolated, cloned, and sequenced, and you then have it where you want it.

Although the theory of such a preparation is quite logical, it is in practice quite laborious, because it deals less with a systematic search for differences and represents a refined form of poking around in a haystack. If a difference is identified, the difficulties begin with characterization of the band to ascertain whether it is only an artifact. The method is only successful if the difference in the expression of a gene under two different conditions amounts to some orders of magnitude. More subtle differences are frequently lost in this kind of molecular biologic lottery. It is nevertheless very exciting, because the experimenter is likely to find something that provides hope.

There are laboratories where satisfactory and interesting results have been achieved using this method. *Differential display* also demonstrates strength. You do not have to be familiar with what you are searching for beforehand, unlike the situation with *microarrays*, with which you can investigate only known sequences (see Chapter 9, Section 9.1.6).

More positive information, which may incite your interest in *differential display*, can be found on the Internet (http://www.genhunter.com). However, it must be pointed out that a doctoral candidate will most likely be overburdened with this information.

*Das Erdetreiben, wies auch sei,*
*Ist immer doch nur Plackerei.*

*This earthly life, howe'er 'tis passed,*
*Proves but a sorry jest at last.*

**Literature**
Liang P, Pardee AB. (1992) Differential display of eukaryotic messenger RNA by means of the polymerase chain reaction. Science 257:967–971.
McClelland M, Welsh H. (1994a) DNA fingerprinting by arbitrarily primed PCR. PCR Methods Appl 4:S59–S65.
McClelland M, Welsh H. (1994b) RNA fingerprinting by arbitrarily primed PCR. PCR Methods Appl 4:S66–S81.
Meunier JR, Grimont PA. (1993) Factors affecting reproducibility of random amplified polymorphic DNA fingerprinting. Res Microbiol 144:373–379.

Welsh J, Chada K, Dalal SS, et al. (1992) Arbitrarily primed PCR fingerprinting of RNA. Nucleic Acids Res 20:4965–4970.

Welsh J, McClelland M. (1990) Fingerprinting genomes using PCR with arbitrary primers. Nucleic Acids Res 18:7213–7218.

Williams JG, Kubelik AR, Livak KJ, et al. (1990) DNA polymorphisms amplified by arbitrary primers are useful as genetic markers. Nucleic Acids Res 18:6531–6535.

## 4.3.4 Classic Quantitative Polymerase Chain Reaction

Many questions revolve around the quantification of nucleic acids. For instance, you may want to know how much of a certain mRNA is expressed in a specific tissue. Employing the PCR for quantification is senseless when you consider that the method was originally conceived for the ruthless replication of the smallest traces of DNA (i.e., purely for providing qualitative evidence). However, because it deals with an exponential increase, which follows mathematical principles, quantification also is possible.

This approach can be carried out only with difficulty. The process involves an exponential increase in DNA, although on closer inspection, you will notice that the rate at which the DNA is increased constantly changes in an almost uncontrollable manner. In the first phase of the amplification, the template concentration is still very limited so that the probability of template, primer, and polymerase coming together is suboptimal. In the third phase, the quantity of products (i.e., DNA, pyrophosphate, and monophosphate nucleotides) increases to such an extent that it inhibits product production. The product fragments increasingly hybridize with the primer and with other product fragments. The substrates (i.e., primer and nucleotides) ultimately disappear, and the polymerases and nucleotides, despite the stability they may demonstrate, slowly give up the ghost.

A more or less exponential increase can be found in the intermediate phase. It is assumed that an exponential increase proceeds up to a product concentration of $10^{-8}$ M (as far as up to $10^{-7}$ M, the increase remains linear, and after that, almost no increase is observed). Only in this phase is a comprehensible relationship seen between the concentration of the product and that of the template.

The central problem is that a standard is missing to enable determination of how much template was available at the beginning. Various attempts have been made to get a grip on this problem.

**Application of an external standard** is important. In addition to an actual test amplification, another gene, which is present in the template DNA, is amplified using multiplex PCR to be used as a standard. In the case of quantifying viral DNA in a tissue, this may, for example, be an area from the genomic DNA of the tissue. The quantity of genomic DNA in the preparation can be determined and consequently used to calculate the quantity of template molecules to be used as a standard. A defined quantity of DNA, such as a plasmid, also can be added and amplified, and the product can be used as reference material. Subsequently, the quantity of the standard is compared with that of the test product, and the quantity of the test template is calculated from that.

The main problem of this method is that two different fragments are only rarely amplified to exactly the same extent. Not all primers hybridize with equal effectiveness, and longer fragments increase more poorly than shorter ones, producing drastic differences in the rate of multiplication (see Section 4.1) and making a comparison of the product concentrations impossible. A requirement for quantification therefore is that the primer pairs and the amplified products must resemble one another as far as possible in regard to annealing temperature, size, and GC concentration to make a homogeneous amplification possible. This method, however, is always open to criticism, because truly satisfactory conditions are rare.

Critical in this preparation is the number of cycles, because you do not want to leave the exponential area of the PCR. This depends primarily on the number of template molecules that are present at the beginning, and this number must ultimately be determined experimentally. For guidance, you can use the following values:

3,000 to 50,000 template molecules: 20 cycles
200 to 3000 template molecules: 25 cycles
10 to 400 template molecules: 30 cycles

**Application of an internal standard (competitive PCR)** is the solution to the problem just described. Instead of co-amplifying a second DNA fragment from the same test template, add a so-called internal standard to the preparation through a small deletion or cleavage site that is easy to recognize. The internal standard is a defined quantity of a standard template that differs as little as possible from the template that you are interested in. Because the remainder of both templates is identical and the same primer pair is used, you can proceed on the assumption that the amplification conditions are identical for both templates. You then perform several amplifications, which make use of the same test template concentrations but employ different quantities of the standard template. After the PCR, the quantities of the standard and test product are compared with one another, such as by means of electrophoresis or by using Southern blots. At sites where the test and standard bands are equally intense, the quantity of the test and standard DNAs are identical.

**Advantage:** Because the amplification conditions are the same for both fragments during the entire PCR and because the relationship of the quantities of the two substances remains constant until the end, there is no necessity to remain in the exponential area.

**Disadvantages:** For every fragment that is to be quantified, you must first construct a fitting standard. For every quantification, you must perform several attempts with different quantities of the standard DNA, because a realistic appraisal is possible only at a ratio of the test to the standard DNA between 1 to 10 and 10 to 1; most precise is the use of a ratio of 1 to 1. An exact quantification of the PCR product therefore demands quite a bit of work.

**Quantitative RT-PCR** can be used in other situations. Difficulties are encountered in quantifying mRNA instead of DNA from cells or tissues. You can accomplish a reverse transcription and quantify the desired cDNA by means of an external or internal standard, but this provides information only about the quantity of cDNA in the preparation, telling you nothing about the quantity of mRNA in the initial material.

The problem is that the cDNA synthesis can proceed in varied forms, and the effectiveness usually lies somewhere between 5% and 90%. A quantification performed under these conditions provides a reference to the minimal amount of mRNA in the preparation, but no absolute value, because nothing is known about the effectiveness of the cDNA synthesis or the integrity of the cDNA. You can quantify the cDNA synthesis in principle, but only with a fair amount of work and not particularly accurately.

An attempt to solve this problem is performed with an external standard. A *housekeeping gene* often is selected as a standard, because it can be expected to be constitutively expressed at a constant quantity during the course of the investigation of the preparation. The quantity of the specific product is then related to this housekeeping gene, whose quantity is considered to be constant. β-Actin is very popular for this purpose, although it has been shown that the quantities of β-actin in dividing cells can fluctuate considerably, and glyceraldehyde 3-phosphate dehydrogenase (GAPDH) currently is preferred as a standard (Apostolakos et al., 1993, Zhao et al., 1995). Housekeeping genes are also upregulated and downregulated to a certain extent and must therefore be considered to be problematic as a standard. Another problem is that housekeeping genes are frequently expressed more intensely than the gene being examined and that it is not very easy to make a quantitative comparison of an intensely and a weakly expressed gene. More information about the problem of quantifying mRNA is available elsewhere (McCulloch et al., 1995; Souaze et al., 1995).

**Quantifying the product** can be achieved with competitive PCR (Figure 4-5). It is possible to estimate the quantity of the test and the control bands in agarose gel with the naked eye; anyone with a video documentation system and a program for the quantification of DNA bands can record

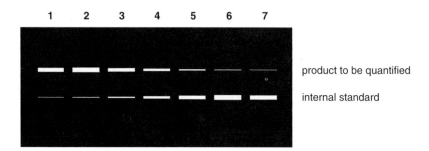

**Figure 4-5. Principles of competitive polymerase chain reaction (PCR).** At sites where the bands of the product and the standard are equally intense, equal quantities of the test and the standard DNA were found at the beginning of the PCR.

this value in figures. Otherwise, radioactively labeled nucleotides are added to the PCR, so that the bands can be visualized with the aid of autoradiography (with the known problem that these are easily affected by saturation) (see Chapter 7, Section 7.3.1), or they are cut out and measured in the scintillation counter or phosphorus imager (see Chapter 7, Section 7.3.1), a device that is sensitive and easy to operate but quite expensive. Nucleotides that are not labeled radioactively can also be inserted, although it is more problematic to provide evidence of these substances and the methods for quantification have not been fully developed.

**Literature**

Apostolakos MJ, Schuermann WH, Frampton MW, et al. (1993) Measurement of gene expression by multiplex competitive polymerase chain reaction. Anal Biochem 213:277–284.

Gaudette MF, Crain WR. (1991) A simple method for quantifying specific mRNAs in small numbers of early mouse embryos. Nucleic Acids Res 19, 1879

Hart C, Chang SY, Kwok S, et al. (1990) A replication-deficient HIV-1 DNA used for quantitation of the polymerase chain reaction (PCR). Nucleic Acids Res 18:4029–4030.

Kellogg DE, Sninsky JJ, Kwok S. (1990) Quantitation of HIV-1 proviral DNA relative to cellular DNA by the polymerase chain reaction. Anal Biochem 189:202.

McCulloch RK, Choong CS, Hurley DM. (1995) An evaluation of competitor type and size for use in the determination of mRNA by competitive PCR. PCR Methods Appl 4:219–226.

Murphy LD, Herzog CE, Rudick JB, et al. (1990) Use of the polymerase chain reaction in the quantitation of mdr-1 gene expression. Biochemistry 29:10351–10356.

Souaze F, Ntodou-Thome A, Tran CY, et al. (1996) Quantitative RT-PCR: Limits and accuracy. Biotechniques 21:280–285.

Zhao J, Araki N, Nishimoto SK. (1995) Quantitation of matrix Gla protein mRNA by competitive polymerase chain reaction using glyceraldehyde-3-phosphate dehydrogenase as an internal control. Gene 155: 159–165.

Zimmermann K, Mannhalter W. (1996) Technical aspects of quantitative competitive PCR. Biotechniques 21:268–279.

# 4.3.5 Real-Time Quantitative Polymerase Chain Reaction

The most modern method of quantifying nucleic acids is *real-time quantitative PCR*. Unfortunately, some people refer to this new method succinctly as real-time PCR and then, out of laziness, abbreviate it as RT-PCR, an acronym that has stood for *reverse transcription-PCR* for many years. An official acronym has not been decided on, but in the meantime, anyone who wants to abbreviate this term should try it RTQ-PCR or RTD-PCR (for *real-time quantitative* or *real-time detection PCR*).

Attempts to experience anything "live" that occurs in a PCR tube were made quite early. The most unadventurous method consisted of stopping all the apparatuses every five cycles and removing an aliquot. Applied to a gel, these samples provide a nice overview of the developments that take place in the tube, and if this is followed by a hybridization, they can be used to quantify the history of this procedure quite well. However, this approach provides only a limited picture of what is actually taking place, apart from the work required for the procedure.

One development consisted of providing the PCR apparatus with a UV lamp and a CCD camera and adding some ethidium bromide to the PCR reaction. The ethidium bromide intercalates in double-stranded DNA, and the fluorescence is increased by a measurable amount (Higuchi et al., 1992, 1993). This method is still used, although other dyes, such as Hoechst 33258, YO-PRO-1 and especially SYBR$^{RM}$ Green I (Molecular Probes), have been used that deliver a better signal-to-background ratio. The advantage is their universal applicability, because every PCR reaction can be performed. The signal strength is high, because every DNA molecule binds several molecules of the dye. A disadvantage results from this advantage, because it is not possible to distinguish between the correct product and artifacts—a considerable problem because artifacts are omnipresent in PCR.

The problem of lacking specificity has been solved by Holland and collegues (1991). A largely unnoticed characteristic of the Taq polymerase, its 5'-3' exonuclease activity, is made use of by adding a radioactively labeled oligonucleotide, in which the 3' end has been blocked (by means of a dideoxynucleotide or a phosphate group) to the PCR reaction, so that it cannot function as a primer. The third oligonucleotide is chosen in such a manner that it hybridizes between the two primers. During the synthesis of a new strand, if the polymerase stumbles on the oligonucleotide, it disassembles the strand and sets the radioactively labeled structures free, which can later be distinguished from the intact oligonucleotides by means of thin-layer chromatography. Decisive flaws are observed, which require quantification, removal of aliquots, and considerable revisions to be performed.

The current solution involves the use of **fluorescence resonance energy transfer** (FRET), as defined by Förster and introduced by Cardullo and colleagues (1988). A fluorescent dye (i.e., flu-orochrome) can be excited at a certain wavelength (A1) and subsequently emits the energy taken up in the form of light of a different wavelength (E1). The excitation and emission spectra of the fluorescent dyes are characteristic. However, if a fluorochrome molecule (F1) is brought sufficiently close to a second molecule (F2), whose excitation spectrum (A2) corresponds with the emission spectrum of the first fluorochrome (E1), a "spark" leaps between the two. Instead of being emitted as light of the wavelength of E1, the energy is transmitted directly to fluorochrome 2, which makes light with a wavelength of E2 (Figure 4-6). Depending on the experimental design, the light intensity of a wavelength of E1 or E2 can be followed during the PCR, and the experimenter can see whether the two fluorochromes are spatially located far apart (measurement of E1) or closely together (mea-surement of E2). If E1 is measured, fluorochrome 1 is a **reporter**, fluorochrome 2 is a **quencher**, and the measurement of E2 results in a **donor** and an **acceptor**.

Through the combination of PCR instruments with fluorescence detection, specific oligonucleotides, and FRET, three more or less similar methods of detection, known as *TaqMan, molecular beacons*, and *hybridization probes*, were designed. The differences in these methods are not very easy to understand, but they all essentially lead to the same results. The TaqMan principle is the oldest (Livak et al., 1995) and perhaps also the best known. Here, the reporter and quencher are found on the same oligonucleotide, preferably at the 5' and the 3' end. As long as the oligonucleotide is intact, the intensity of the light at E1 is slight, although the light production increases at this wavelength when the reporter is set free as a result of an approaching polymerase. The more DNA that is synthesized, the larger the quantity of reporter molecules that is released, so that the signal intensity increases. Another development of the TaqMan primer is the molecular beacons (Tyagi and Kramer, 1996); in addition to the specific region of the oligonucleotide located centrally, there are complementary sequences to be found at the 5' and 3' ends, which cause the oligonucleotide to assume a hairpin structure. This guarantees that the *reporter* and *quencher* come together sufficiently to

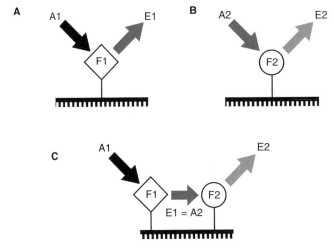

**Figure 4-6. Principles of fluorescence resonance energy transfer (FRET).** Fluorochrome 1 (F1) has a characteristic excitation and a typical emission spectrum (A1 and E1). The same is true for fluorochrome 2. If E1 and A2 overlap, the emitted light from F1 can excite the second fluorochrome, and light with a wavelength of E2 occurs. This presupposes, however, that both fluorochromes are localized close to one another. If the fluorochrome is excited with light of a wavelength of A1, the experimenter can determine whether the two fluorochromes are separate from or adjacent to one another through the measurement of E1 or E2.

prevent an emission of light, whereas TaqMan primers, according to the sequence used, occasionally form unfavorable structures, and the distance between the two fluorophores becomes too large for a complete suppression of the light emission.

**Hybridization probes** are made of two oligonucleotides. One is associated with an acceptor at the 3′ end, and the other is associated with a donor at the 5′ end. The oligonucleotides are selected in such a manner that they both bind to the same DNA strand, in which the distance between the acceptor and the donor must total 1 to 5 nucleotides. In this case, the light intensity is measured at E2; light of this wavelength is demonstrable only if both oligos are bound to the DNA.

How does the quantification function? The principle is fundamentally different from that of a normal quantification, because the absolute quantities of the PCR product are not measured. Instead, the experimenter takes advantage of the kinetics of the PCR reaction. During the early cycles of PCR, there is an exponential increase in the reproduction of DNA fragments, even if this is not so apparent in the initial stages. If the product accumulates, the disturbing influences also increase. Whether the primer or nucleotide tend to disappear, polymerase or nucleotides give up the ghost because of the persistently high temperatures, or the reaction is inhibited through the accumulation of products (pyrophosphate!), the rate of the process decreases to a linear growth curve and then comes to a standstill, as occurs with bacteria (Figure 4-8). As a guideline, you should use the number of cycles at which the fluorescence signals can just be distinguished from the background ($C_T$ value), because the growth is still exponential at this point. Theoretically, you could draw a conclusion concerning the original amount of template, if you knew the rate of growth (see Table 4-1). Unfortunately, amplification of a specific fragment is influenced by so many elements that it proves to be to amplify and compare known template quantities with their respective $C_T$ values. A standard curve can be plotted to obtain the template quantity as based on the respective $C_T$ value (see Figure 4-8).

The use of specific oligonucleotide probes allows amplification of several fragments in the same tube (i.e., multiplex PCR, see Section 4.2.2) and provides evidence for them individually. However, you should not expect too much. Because the emission spectra of the measured fluorophores should

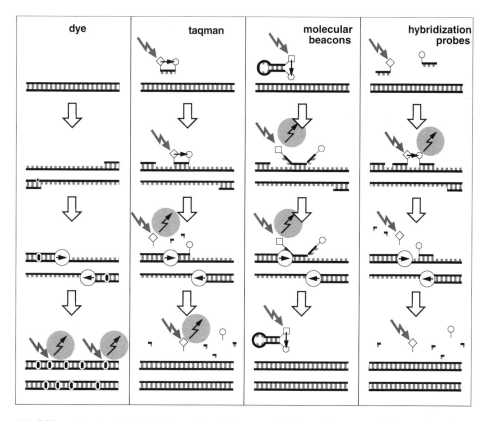

**Figure 4-7. Different methods of detection of real-time quantitative polymerase chain reaction.** Dyes such as SYBR green I are installed nonspecifically in double-stranded DNA. TaqMan probes and molecular beacons fluoresce if the oligonucleotide has been disassembled from the polymerase, whereas hybridization probes fluoresce as long as both oligonucleotides are bound to the template.

overlap as little as possible for clear evidence, the number of fragments that can be followed in practice is limited to two. This still facilitates the work considerably, because you can easily compare the expression of two genes directly or can immediately demonstrate two mutations.

Four manufacturers (Applied Biosystems, Roche, BioRad, and Cepheid) offer RTD-PCR–suitable instruments, although they will surely be followed by others in the near future. Before purchasing such an instrument, be clear about what method of detection you want to use. Because of the different types of construction, not every type of fluorophore can be used in all cases, nor can several oligonucleotide probes be employed simultaneously. The apparatus from Roche is unusual, because it is not based on the conception of a normal PCR unit; it instead functions with glass capillary tubes (Wittwer et al., 1997). This makes the work with such an instrument somewhat more difficult, but it also achieves somewhat more reproducible results and has fantastically short amplification times, so that a PCR requires just 45 minutes.

**Advantage:** You can see the final results of a PCR and what occurs beforehand. This is a relatively simple method with the most precise results.

**Disadvantage:** The instruments cost about $37,000 to $75,000, in addition to the costs for the labeled oligonucleotides, which are expensive because of the small number of suppliers. The evaluation is not simple either, in contrast to what the high-gloss brochures indicate.

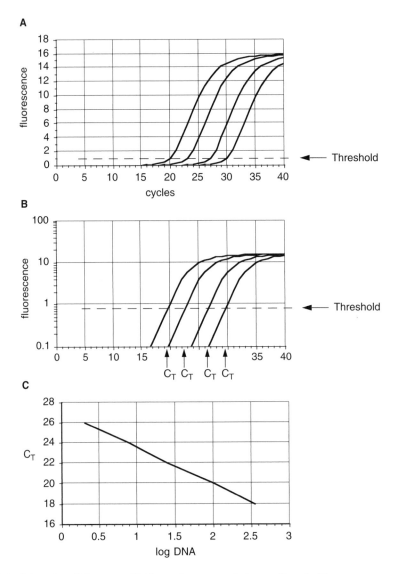

**Figure 4-8. Principles of real-time quantitative polymerase chain reaction. A,** Fluorescence curves for four amplifications with different template DNA quantities. **B,** The same fluorescence curves plotted using a logarithmic scale. The $C_T$ values have been determined in an area of the curve in which the amplification proceeds exponentially (i.e., straight area of the curve). **C,** If the $C_T$ values are plotted against the logarithm of the DNA quantities initially used, a standard curve results from which the $C_T$ values can be used to determine the quantity of DNA initially present.

**Suggestion:** Make efforts to see that the data can be evaluated using another type of instrument. If instruments of this type are to be used frequently (and that is usually the case), you will experience incalculable bottlenecks as a result of your other colleagues, which will make the evaluation appear to take forever.

**Literature**

Cardullo RA, Agrawal S, Flores C, et al. (1988) Detection of nucleic acid hybridization by nonradiative fluorescence resonance energy transfer. Proc Natl Acad Sci U S A 85:8790–8794.

Higuchi R, Dollinger G, Walsh PS, Griffith R. (1992) Simultaneous amplification and detection of specific DNA sequences. Biotechnology 10:413–417.

Higuchi R, Fockler C, Dollinger G, Watson R. (1993) Kinetic PCR: Real time monitoring of DNA amplification reactions. Biotechnology 11:1026–1030.

Holland PM, Abramson RD, Watson R, Gelfand DH. (1991) Detection of specific polymerase chain reaction product by utilizing the 5′-3′ exonuclease activity of Thermus aquaticus DNA Polymerase. Proc Natl Acad Sci U S A 88:7276–7280.

Livak KJ, Flood SJA, Marmaro J, et al. (1995) Oligonucleotides with fluorescent dyes at opposite ends provide a quenched probe system useful for detecting PCR product and nucleic acid hybridization. PCR Methods Appl 4:357–362.

Tyagi S, Kramer FR. (1996) Molecular beacons: probes that fluoresce upon hybridization. Nat Biotechnol 14:303–308.

Wittwer CT, Herrmann MG, Moss AA, Rasmussen RP. (1997) Continuous fluorescence monitoring of rapid cycle DNA amplification. Biotechniques 22:130–138.

Wittwer CT, Ririe KM, Andrew RV, et al. (1997) The Light Cycler: A microvolume multisample fluorimeter with rapid temperature control. Biotechniques 22:176–181.

## 4.3.6 Inverse Polymerase Chain Reaction

Always amplifying sequences between two known DNA sequences is quite boring. How about reversing this situation and amplifying unknown sequences that lie to the left and to the right of a known sequence? This is known as inverse PCR.

The principle is illustrated in Figure 4-9. The DNA template is digested with a restriction enzyme, and the fragments are religated; nevertheless, the DNA concentration must be so small that the fragments ligate preferably with each other and result in ring-shaped DNAs. These can be used as a template for PCR, in which the primers are oriented away from one another. The product's internal regions are composed of unknown sequences that lie on the outer side in the original template—a somewhat complicated situation. The method is very clever, although it primarily functions with templates of only moderate complexity, such as cosmid clones.

**Literature**

Ochman H, Gerber AS, Hartl DL. (1988) Genetic applications of an inverse polymerase chain reaction. Genetics 120:621–623.

Triglia T, Peterson MG, Kemp DL. (1988) A procedure for in vitro amplification of DNA segments that lie outside the boundaries of known sequences. Nucleic Acids Res 16:8186.

## 4.3.7 Biotin-RAGE Method and Supported Polymerase Chain Reaction

Biotin–rapid amplification of genomic DNA ends (RAGE) and supported PCR are additional methods for amplifying unknown sequences in the vicinity of known sequences. The DNA template is digested with a restriction enzyme. A reduced PCR reaction is performed with only a single primer and a single cycle, with the addition of biotin-labeled dUTP, and the PCR product is then fished out of the reaction preparation with the aid of streptavidin-agarose or using magnetic particles. In this way, it is possible to concentrate the product by several orders of magnitude. A poly A tail is then removed

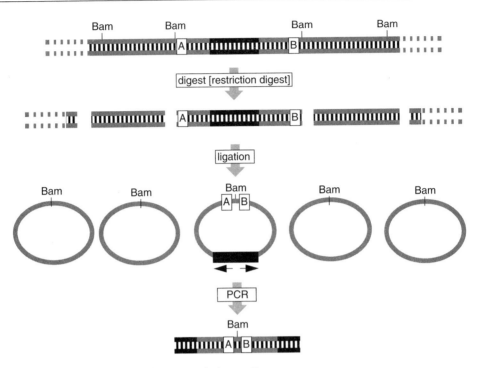

**Figure 4-9. Principles of inverse polymerase chain reaction.**

using a terminal transferase (Bloomquist et al., 1992), or a linker fragment is ligated (Rudenko et al., 1993), and the product modified in this manner serves as a template for a proper PCR, with a gene-specific and an oligo(dT)- or linker-specific primer.

**Literature**

Bloomquist BT, Johnson RC, Mains RE. (1992) Rapid isolation of flanking genomic DNA using Biotin-RAGE, a variation of single-sided polymerase chain reaction. DNA Cell Biol 11:791–797.

Rudenko GN, Rommens CM, Nijkamp HJ, et al. (1993) Supported PCR: An efficient procedure to amplify sequences flanking a known DNA segment. Plant Mol Biol 21:723–728.

## 4.3.8 Mutagenesis with Modified Primers

Because you can order a tailor-made primer for any purpose, the PCR is particularly well-suited for undergoing mutagenesis. An ordinary amplification is carried out, and one of the two primers contains the desired mutation. The mutations should be localized as near as possible to the 5′ end of the primer, because the probability that the polymerase will not tolerate a faulty pairing is higher as the mutation approaches the 3′ end (a characteristic that can be made use of in ARMS; see Section 4.3.9), and as a consequence, the amplification does not function. Polymerases with a corrective activity primarily demonstrate an additional risk of degrading the primer up to the site of faulty pairing so that the mutation is then eliminated. If you wish to insert a restriction cleavage site, it should not lie completely at the 5′ end of the primer, because many restriction enzymes cut poorly at such locations (see Chapter 6, Section 6.1.6). More can be found concerning the topic of mutagenesis in Chapter 9, Section 9.2.

**Literature**
Kaufman DL, Evans GA. (1990) Restriction endonuclease cleavage at the termini of PCR products. Biotechniques 9:304–305.

## 4.3.9 Amplification Refractory Mutation System

The *amplification refractory mutation system* (ARMS) can provide evidence of point mutations. Two primers are designed whose sequences correspond to the two variants that you would like to demonstrate. The primer must end at the 3′ end of the base, where both sequences differ (Figure 4-10). As a reverse primer, you can use any arbitrary primer, as long as its melting temperature is approximately on the same order of magnitude.

To identify the two variants in an arbitrary DNA template, two PCR amplifications must be performed, one with the reverse primer and one on one of the two specific primers. If one chooses the conditions correctly (i.e., the fact that the annealing temperature is sufficiently high is particularly decisive), the amplification is then successful only if the sequence of the primer and the template correspond with one another exactly. If these two substances differ by only one significant base, no PCR product can be expected.

The entire process does work, although not without problems. For it to function, the primer should be sufficiently long and have a melting temperature of approximately 55°C (131°F), so that extremely high annealing temperatures can be attained. The annealing temperature during the amplification should be higher than the melting temperature of the primer. The temperature range at which the differential amplification functions is relatively narrow, and only a few degrees can be decisive for success or failure. The optimal temperature must therefore be approached gradually in the test PCRs if no PCR apparatus is available with a temperature gradient.

Newton and colleagues (1989) report that evidence can be attained better for some sequences if another mutation could additionally be inserted close to the 3′ end of the primer; the faulty pairing destabilizes the linking to a certain extent and may possibly provide a clearer yes or no answer.

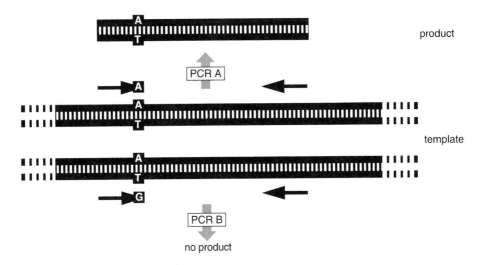

**Figure 4-10. Amplification refractory mutation system (ARMS).**

**Literature**

Newton CR, Graham A, Heptinstall LE, et al. (1989) Analysis of any point mutation in DNA. The amplification refractory mutation system (ARMS). Nucleic Acids Res 17:2503–2516.

## 4.3.10 In Situ Polymerase Chain Reaction

An extremely fascinating application is the in situ PCR of thin layers of tissues or of cells. The process combines the sensitivity of PCR with the resolution offered by histology. The fixed material is amplified directly on the slide. The technical difficulties brought about by this system, such as exact control of the temperature and the danger that the preparation may dry up, have been eliminated to a large degree through the introduction of special PCR machines or attachments by the manufacturers (e.g., Hybaid, Perkin-Elmer,), along with mandatory accessories. The conditions deviate somewhat from the standard PCR conditions, so that higher $Mg^{2+}$ concentrations and bovine serum albumin (BSA) are required in the reaction preparation (Nuovo, 1995).

The method is frequently used for providing evidence of viral DNA, although it can be used to recognize RNA if a reverse transcription is performed before the amplification. The procedure has been described in more detail by Nuovo and coworkers (1992).

Evidence of the amplification products is provided directly by inserting radioactively labeled or nonradioactively labeled nucleotides during the PCR or indirectly through the use of in situ hybridization (see Chapter 9, Section 9.1.3). Long and colleagues (1993) have studied and described this topic in more detail. Direct evidence is more comfortable, because radioactive products can be portrayed through a simple autoradiography, fluorescence-labeled products are visible under the microscope, and biotin- and digoxigenin-labeled probes can be made visible using antibodies without much difficulty. The weakness of the direct evidence lies in a problem that is typical for PCR; the correct amplification products cannot be distinguished from the artifacts. Attaining indirect evidence by means of a classic in situ hybridization is more difficult, although it is substantially more specific. For routine examinations, it is beneficial to carry out both procedures until the PCR conditions have been optimized to such a degree that you can dispense with any indirect evidence.

**Literature**

Haase AT, Retzel EF, Staskus KA. (1990) Amplification and detection of lentiviral DNA inside cells. Proc Natl Acad Sci U S A 87:4971–4975.

Long AA, Komminoth P, Lee E, Wolfe HJ. (1993) Comparison of indirect and direct in situ polymerase chain reaction in cell preparations and tissue sections. Detection of viral DNA, gene rearrangements and chromosomal translocations. Histochemistry 99:151–162.

Nuovo GJ. (1995) In situ PCR: Protocols and applications. PCR Methods Appl 4:S151–167.

Nuovo GJ, Gorgone GA, MacConnell P, et al. (1992) In situ localization of PCR-amplified human and viral cDNAs. PCR Methods Appl 2:117–123.

## 4.3.11 Cycle Sequencing

Taq polymerase has been employed for some time in sequencing. In contrast to the normal PCR, only a primer is used, so only linear growth (not exponential) is observed.

Taq functions at higher temperatures than a classic DNA polymerase and, in part, even permits better sequencing results, because the GC-rich structures can be broken down better. Perhaps it is also the somewhat reduced pipetting or merely the tendency of the PCR lover to solve all problems with their PCR apparatus, which has helped to make *cycle sequencing* so popular. More on the topic of sequencing can be found in Chapter 8, Section 8.1.

## 4.3.12 cDNA Synthesis

RNA occasionally forms secondary structures that can be unraveled only poorly by an ordinary reverse transcriptase. This problem may be solved by performing the cDNA synthesis at a higher temperature. The **avian myeloblastosis virus reverse transcriptase** (AMV-RT), for instance, is active at temperatures of up to 58°C (136°F), whereas **Moloney murine leukemia virus reverse transcriptase** (MMLV-RT) is inactivated at 42°C (108°F). This is nothing compared with our beloved thermostabile polymerases. In addition to their DNA-dependent DNA-polymerase activity, most **thermostabile polymerases** demonstrate a more or less distinctive, RNA-dependent, DNA-polymerase

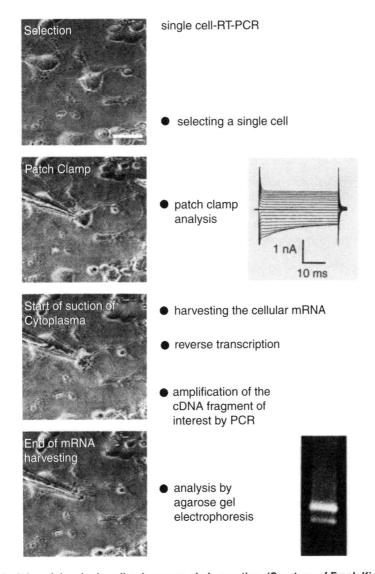

**Figure 4-11. Principles of the single-cell polymerase chain reaction. (Courtesy of Frank Kirchoff)**

activity, meaning that they can synthesize cDNA from RNA if the experimenter can provoke them sufficiently so that their hidden talents can be employed. Good candidates are the Tth and the Tfl polymerases, which perform RNA-dependent polymerization in the presence of manganese ($Mn^{2+}$) and DNA-dependent polymerization in the presence of magnesium ($Mg^{2+}$). The temperature-determining component is then no longer the polymerase, but rather the primer, because *random* hexamer primers must be given a chance to hybridize at room temperature before gradually increasing the temperature to the desired value, or nothing will occur. Frequently, it is better in this case to use a specific primer for the cDNA synthesis with a higher annealing temperature. The thermostabile polymerases are highly inferior to the classic reverse transcriptases with regard to sensitivity, yield, and transcript length.

## 4.3.13 Single-Cell Polymerase Chain Reaction

*Nun sind wir schon wieder an der Grenze unsres Witzes,*
*da wo euch Menschen der Sinn überschnappt.*

*Here we are again at the end of our wits,*
*The point at which the minds of you mortals are strained to breaking.*

With consultation with an electrophysiologist, you can make an attempt using an extreme PCR variant, the single-cell PCR (Figure 4-11). This represents an excursion into the marginal zones of what is feasible.

The contents of an individual cell are sucked out, and this little bit of nothing is then used for cDNA synthesis. Subsequently, a normal PCR is performed; you should try to do this with a *nested* PCR (see Section 4.2.1) because of the small amount of template. Under these extreme conditions, evidence of transcripts is seen in only 50% of cases, and the bands do not necessarily look as good as in the normal PCRs. However, the fact that it functions at all is fascinating.

# 5 RNA

RNases are found everywhere. All organisms produce RNases and eliminate them. These enzymes are even found in our sweat.

Working with RNA should not be any more complicated than with DNA because the chemical difference between the two substance classes is slight—but that would be the case only if there was no such thing as an RNase. Because all RNases, in contrast to the DNases, require no cofactors such as $Mg^{2+}$ to be activated and because they are mostly stable (e.g., RNase A solution is made free of DNase by cooking it for 20 minutes), they can make the experimenter's work more complicated.

The beginner should prepare well in advance for the use of RNA. It is sensible to have a separate set of pipettes, pipette boxes, tubes, solutions, and other materials; to label them with *RNA*; and to stow everything in a separate cabinet. The researcher must then convince his or her colleagues that the letters RNA signify "Montezuma's revenge is nothing compared with the trouble you will encounter if you do not keep your hands off these things," and success will be ensured, at least halfway.

All work should be performed while wearing gloves. Experience will teach which points are crucial and which are insignificant. When handled properly, the use of RNA is not much more problematic than work with DNA. Until you have acquired this experience, however, you must work in an extremely fastidious manner.

## 5.1 Inactivating RNases

Considering the broad distribution of the RNases and their great stability, their inactivation is one of the largest problems related to RNA. The most common method used to free solutions from RNase is to treat them with diethyl carbonate. To do so, a 1/200 volume of **diethyl pyrocarbonate** (DEPC) is added and mixed until the DEPC beads have dissolved. The solution is incubated overnight at room temperature and then autoclaved. The DEPC is degraded to $CO_2$ and ethanol, and the solution smells mildly like schnapps.

DEPC binds to primary and secondary amines (e.g., histidine) through covalent bonds, and the activity is destroyed by RNases, along with the activity of all other enzymes. Because all substances that contain primary amines are modified, this method cannot be employed with many different buffers, such as Tris. There is only one possibility: Purchase RNase-free substances, maintain them in safety away from other colleagues who do not work with RNA, handle these substances using utensils that are free of RNase, and mix these solutions with DEPC-treated water. In practice, this means that a separate set of chemicals must be used for working with RNA.

DEPC is not completely free of problems. It is carcinogenic and poisonous, and it is very unstable if it becomes damp. Only smaller quantities of this substance should be ordered, and each order should be used completely before replenishing supplies.

Glassware and other utensils that can withstand high temperatures are baked at 180°C to 200°C (356°F to 392°F) for 4 hours. Autoclaving is not sufficient in this case, although it is better than nothing. Utensils that cannot withstand heat are treated with hydrogen peroxide solution (1% [v/v]) and then rinsed with DEPC-treated water. Rinsing with chloroform can effectively inactivate RNases, although chloroform also dissolves all types of plastic. It is a mess and therefore not employed very often.

Molecular Bio-Products (Fisher Scientific) offers a solution (RNase AWAY) that purports to free plastic and glassware from RNases by means of wiping their surfaces with this solution. The manufacturer states that this product is less harmful than DEPC but does not describe how the solution functions. Another alternative is the use of as many disposable, sterile utensils as possible, an approach that is popular but quite costly.

## 5.2 Methods of RNA Isolation

In isolating RNA, the cells are lysed, which inactivates the RNases, and the RNA is then isolated from this soup. The result is some amount of total RNA, a lively mixture of ribosomal RNA (rRNA), transfer RNA (tRNA), messenger RNA (mRNA), and other forms.

The mRNA makes up only about 2% of the total. For most applications, such as reverse transcriptase–polymerase chain reaction (RT-PCR) and Northern blot hybridization, this amount is sufficient. For more demanding applications, such as production of cDNA banks or for higher sensitivity when using a rare mRNA, polyA+ RNA should be selected from the total RNA.

### 5.2.1 Single-Step Method

The *single-step method* requires more than one step, but through the clever combination of two methods, the essential work can be performed in a single step. The material is first lysed in a guanidine isothiocyanate (GIT) solution. GIT is a chaotropic salt that effectively denatures and inactivates proteins, even RNases. Phenol is added to remove the proteins. Because the pH is decreased, even smaller DNA fragments dissolve in the acidic phenol, and the larger fragments collect at the interphase during centrifugation. RNA is found in the liquid supernatant, from which it must be precipitated. Because the inactivation of the RNases occurs immediately after the lysis of the cells, the method is very stable and delivers a qualitatively high-grade RNA. This has become the most important method for RNA isolation.

In the protocol, 100 mg of tissue is homogenized in 1 mL of denaturing solution (4 M guanidine isothiocyanate/25 mM sodium citrate, pH 7.0/0.5% [w/v] of $N$-lauryl sarcosine/0.1 M β-mercaptoethanol, which is imperishable for 3 months at room temperature without β-mercaptoethanol and only 4 weeks with it) with the aid of an Ultraturrax or a Potter homogenizer. Very RNA-rich tissues should be worked with in a fresh state; otherwise, deep-frozen material can be used. Cultivated cells are required only in the denaturing solution (1 mL per $10^7$ cells) to resuspend them, and they then lyse on their own.

The experimenter adds 100 μL of 2 M sodium acetate (pH 4.0), 1 mL of phenol, and 200 μL of chloroform (shaking intermittently) per 1 mL of lysate and incubates the solution for 15 minutes on ice. After centrifugation for 20 minutes at 10,000 g, the aqueous phase is transferred to a new tube, and an equal volume of isopropanol is added. After 30 minutes at −20°C (−4°F), the solution is centrifuged for 10 minutes at 10,000 g, the pellet is dissolved in 0.3 mL of denaturing solution and precipitated a second time using 0.3 mL of isopropanol. The pellet is washed with 70% ethanol, dried, and dissolved in 100 μL of $H_2O$, which is best done at 60°C (140°F) for 15 minutes. If the RNA is too dry, it will dissolve poorly.

A large number of commercial kits for isolating RNA are available from various manufacturers (e.g., Qiagen, Macherey-Nagel, Promega, Stratagene, Roche). They are based on the *single-step* method described (Chomczynski and Sacchi, 1987) or on a variety of methods described by Chomczynski and Mackey (1995). Several of these kits allow production of RNA and DNA from the same probe (TRIzol from Gibco, TriStar from AGS.

**Advantage**: The method can be used without any problem for larger and even for the smallest quantities of material. It provides good yields and is very reliable.

## 5.2.2 Lysis with Nonidet P40

Another alternative is available. First, $2 \times 10^7$ cells are washed in PBS, centrifuged (5 minutes, 1000 rpm), and the pellet is then resuspended in 375 µL of cold lysis buffer (50 mM Tris HCl, pH 8.0/100 mM NaCl/5 mM $MgCl_2$/0.5% [v/v] Nonidet P40). After 5 minutes on ice, the cells are lysed. The cell nuclei are centrifuged off (2 minutes, 3000 g), because this can eliminate much of the genomic DNA. The supernatant is combined with 4 µL of 20% (w/v) SDS and 2.5 µL of proteinase K (20 mg/mL) in a new vessel, mixed, and incubated for 5 minutes at 37°C (99°F). Then, 400 µL of phenol-chloroform solution is added, which is vortexed well and centrifuged in a table-top centrifuge for 5 minutes at 13,000 rpm. The upper layer is transferred into a new tube without taking along any contaminants from the interphase, and the phenol-chloroform purification is carried out again. The upper layer is subsequently mixed with 400 µL of chloroform and centrifuged, and it is then transferred to a new tube and precipitated with ethanol. The pellets are dried and then dissolved in approximately 100 µL of DEPC-treated water.

**Disadvantage**: Inactivation of the RNases is less efficient, because the method is more appropriate for RNase-poor tissues or cells.

## 5.2.3 General Information

The yield in RNA isolation differs extensively from tissue to tissue. As a consequence, almost three times as much RNA can be obtained from the liver as from the kidney.

RNA that is truly free of DNA can be obtained through an additional DNase digestion. To this, 50 µL of RNA solution, 10 µL of a DNase mixture (0.2 µL of RNase-free DNase (2.5 mg/mL), and 0.1 µL of ribonuclease inhibitor (25 to 50 u/µL) in 100 mM $MgCl_2$/10 mM DTT), and 40 µL of TE buffer are mixed and incubated for 15 minutes at 37°C (99°F). Then, 25 µL of stop mix (50 mM EDTA/1.5 M sodium acetate/1% [w/v] SDS) is added, and a phenol-chloroform purification is carried out.

Clean RNA is just as stable as DNA and can therefore be stored at 4°C (39°F), although even minimal contamination can destroy the entire work. You should play it safe and store the RNA at −70°C (−94°F). For a short-term freezing, −20°C (−4°F) is adequate.

Instead of using $H_2O$, you can dissolve the RNA in formamide and store it at −20°C (−4°F) or −70°C (−94°F), and it will be protected from any RNase digestion in this manner (Chomczynski, 1992). For Northern blots, the RNA dissolved in formamide can be used directly, although it must previously be precipitated using four volumes of ethanol for a cDNA synthesis.

To avoid any undesirable digestion of the RNA during incubation, you must add RNAse inhibitors to the reaction preparation (e.g., RNAsin from Promega, RNase Block from Stratagene, ribonuclease inhibitor from Clontech), which are composed of mixtures of proteins that inhibit a more or less broad spectrum of RNases. Like the enzymes, they are stored at −20°C (−4°F) and are sensitive to heat. The latter feature should be considered when you heat RNA, such as in the synthesis of cDNA, so that you add the inhibitors. Unfortunately, the inhibitors do not offer perfect protection, because all of the available inhibitors inhibit only some part of the existing RNases. As a result, these substances cannot be handled without working cleanly.

## 5.2.4 Determination of the RNA Concentration

Usually, the concentration of RNA solutions can be determined by means of the $OD_{260}$ measurement. For determination of the concentration of DNA (see Chapter 2, Section 2.4 and Fig. 2-5), only OD values between 0.1 and 1 are significant, because values between 0.02 and 0.1 may be associated with substantial errors. Values less than 0.02 primarily show whether the experimenter has washed his or her fingers before the measurement, how old the photometer is, and whether the cuvette holders of the apparatus have too much play.

The experimenter frequently works with small quantities of tissues or cells, which leads to small yields. Quartz cuvettes with volumes from 1 mL to 10 μL are available, but the probability of encountering an error increases with a reduction in the measurement volume. If you use kits, you should pay attention to the statements made by the manufacturers concerning typical yields.

**Literature**

Chirgwin JM, Przybyla AE, MacDonald RJ, Rutter WJ. (1979). Isolation of biologically active ribonucleic acid from sources enriched in ribonuclease. Biochemistry 18:5294.

Chomczynski P, Sacchi N. (1987) Single-step method of RNA isolation by acid guanidine thiocyanate-phenol-chloroform extraction. Anal Biochem 162:156–159.

Chomczynski P. (1992) Solubilization in formamide protects RNA from degradation. Nucleic Acids Res 20:3791–3792.

Chomczynski P, Mackey K. (1995) Substitution of chloroform by bromochloropropane in the single-step method of RNA isolation. Anal Biochem 225:163–164.

# 5.3 Methods of mRNA Isolation

Things usually work well with the use of the total RNA, although mRNA only makes up about 2% of it. However, isolation of mRNA may be necessary if you want to establish a cDNA bank that should not predominantly consist of ribosomal sequences or if you work with weakly expressed transcripts that cannot be observed in normal Northern blots.

The designation of mRNA isolation is incorrect, because RNA with a poly A tail is selected for the purification. Unfortunately, other RNAs can also contain adenosine-rich sequences that are also isolated through this procedure. In contrast, there are mRNAs without an existing poly A tail, which are then lost. Nevertheless, it remains the best method for isolating mRNA from the total RNA.

The purification is based on a linkage of the poly A+ RNA to oligo(dT) nucleotides of about 20 bases, which are bound in a matrix. The matrix can consist of cellulose, magnetic, or latex beads. Purification can be done in a *batch* (i.e., suspending the matrix in the RNA solution), a situation that is especially profitable for large preparations, or by constructing a column by stuffing silanized glass wool into a silanized Pasteur pipette and then pipetting the matrix onto this. Magnetic pellets can be collected with the aid of a magnet.

In a protocol for cellulose, oligo(dT)-cellulose is washed and allowed to swell in 0.1 M NaOH, washed with $H_2O$, and ultimately equilibrated with binding buffer (0.5 M LiCl/10 mM Tris HCl, pH 7.5/1 mM EDTA/0.1% [w/v] SDS). In this case, 1 mL of bloated oligo(dT)-cellulose is sufficient for 5 to 10 mg of total RNA, but you should read the instructions of the manufacturer. The total RNA is denatured for 10 minutes at 70°C (158°F), an equal volume of 1 M lithium chloride is added, and the mixture is then poured onto the cellulose. This is washed twice with binding buffer and twice with wash buffer (0.15 M LiCl/10 mM Tris HCl, pH 7.5/1 mM EDTA/0.1% [w/v] SDS) before eluting the RNA with 2 mM EDTA/0.1% SDS. Because a relatively large volume of elution solution is required for performing the elution, the RNA is concentrated through ethanol precipitation.

After purification, the proportion of mRNA has been raised from 2% to about 50%. If you want to attain more, you must purify the RNA a second time. The quality of the purification can be controlled by applying an aliquot directly on a denatured agarose gel, as performed in a Northern blot, and then carrying out electrophoresis. You can dye the gel with ethidium bromide or, better still, blot it on a membrane and dye it with methylene blue (the quality of the evidence is better). Poly A+ RNA appears as a smear of almost 0- to 20-kb fragments, with a maximum in the region of 5 to 10 kb. After a twofold purification, the 18S and 28S rRNA bands that are typical for total RNA should be more evident.

For the poly A+ RNA isolation, it is easiest to use one of the many commercial kits that are available. All manufacturers of molecular biology products offer kits for RNA isolation.

## 5.3.1 Purchasing RNA

The simplest but most expensive method is to purchase a finished RNA. Clontech offers total RNA and poly A+ RNA from different organisms and tissues. A fat billfold is mandatory for this product, because 5 μg of mouse brain poly A+ RNA costs about $350. Nevertheless, you can acquire material in this way that cannot otherwise be attained easily, such as human poly A+ RNAs from the prostate gland or the caudate nucleus, which are more than twice as expensive as poly A+ RNA from some other tissues.

# 5.4 Reverse Transcription: cDNA Synthesis

*Complementary DNA* (cDNA) mandates a change in thinking about the direction of the sequence of DNA to RNA to protein. Viruses can make DNA from RNA; the responsible enzyme is reverse transcriptase. Molecular biologists primarily use cDNA as an initial material for the PCR and for the construction of cDNA banks.

For cDNA synthesis, you pipette RNA and primer together, heat the preparation to 70°C (158°F) to melt the secondary structure of the RNA, and cool it slowly to room temperature so that the primer can hybridize. After that, you add buffer, nucleotides, RNase inhibitors, and reverse transcriptase, and you incubate it for 1 hour at 37°C to 42°C (99°F to 108°F). In the protocol, there is not much that has to be revised, although the components may be altered.

Whether you use total **RNA** or poly A+ RNA depends on the purpose. If the cDNA is to serve as a template for a PCR, total RNA is usually sufficient, because the subsequent amplification is tremendous. If necessary, a small amount of template can be adjusted through the use of several more cycles; the efforts required are substantially less extensive than those required for poly A+ purification. If you are working with mRNA, which is only slightly expressed, you should add poly A+ RNA, and this method also works if the cDNA is to be used as a hybridization probe.

There are many **primers** to choose from. The classic primers are oligo(dT) primers, made up of 16 to 20 thymidines that bind to the poly A+ tail of mRNAs.

**Advantage**: You can synthesize complete cDNAs, beginning at the poly A+ tail and ending at the 5′ end of the mRNA.

**Disadvantage**: The reverse transcriptases used to synthesize cDNAs have an average length of 1 to 2 kb, whereas mRNAs can easily be 10 kb long. The protein coding portion of interest is usually in the vicinity of the 5′ end. As a consequence, it is frequently impossible to demonstrate long mRNAs, or they are poorly demonstrated. In this case, you can take advantage of a random hexamer primer (*random hexamers*), which hybridizes somewhere along the mRNA so that all mRNA segments are represented in the cDNA and the non-mRNA structures. If you are interested in a specific mRNA, you can also make use of a specific primer, although you must frequently optimize the conditions first, or the yields will be smaller than that observed from either of the two other types of primer.

Two **reverse transcriptases** (RTs) challenge one another's positions. The reverse transcriptase from the *avian myoblastosis virus* (AMV-RT) is a DNA polymerase that synthesizes DNA either RNA or DNA dependently. It also demonstrates DNA exonuclease and RNase activities. Its optimal working temperature is about 42°C (108°F), although the enzyme can withstand temperatures of up to 60°C (140°F). The reverse transcriptase from the *Moloney murine leukemia virus* (MMLV-RT) is an RNA-dependent DNA polymerase, likewise with RNase H activity, which precipitates more weakly than the AMV-RT; the optimal working temperature is about 37°C (99°F), with a maximum of 42°C (108°F). Which of the two you use is a matter of taste. The AMV-RT is frequently preferred, because the RNA is overcome somewhat better by the secondary structures at a higher working temperature, although the MMLV-RT produces longer cDNA transcripts because the mRNAs survive longer due to their lower RNase activity. The development of modified MMLV-RTs represents another step in this direction (e.g., Truescript from AGS, MMLV-RT RNase H Minus from Promega, SuperScript from Gibco BRL), using products that withstand higher temperatures and have no intrinsic RNase H activity, so that it is possible to obtain substantially longer transcripts. In the event of more difficult secondary structures, with which normal reverse transcriptases fail, you can make an attempt with thermostabile polymerases that have reverse transcriptase activity (see Chapter 4, Sections 4.1 and 4.3.12). However, they have substantially smaller yields and are therefore rarely used for standard applications.

You can **quantify** the amount of synthesized cDNA by adding radioactively labeled nucleotide (5 to 10 μCi) to the reaction, removing two 1 μL aliquots (aliquots 1 and 2) before adding reverse transcriptase, and then carrying out the cDNA synthesis. From the completed product, you remove another aliquot (aliquot 3). Aliquot 1 is thinned 1 to 100, and the total activity in the preparation is determined from 1 μL of this preparation. Aliquots 2 and 3 are precipitated with trichloroacetic acid, and the activity of the pellets is measured; the difference between the values is the quantity of additional activity. The equation is as follows:

$$\text{cDNA amount [ng]} = (\text{incorporated activity/total activity}) \times \text{ volume of preparation [μl]}$$

$$\times 4 \times \text{dNTP concentration [nmol/μl]} \times 330 \text{ [ng/nmol]}$$

# 5.5 In Vitro Transcription: RNA Synthesis

Occasionally, an experimenter is faced with the embarrassing situation of trying to synthesize RNA from DNA. Although the molecular biologist is not very fond of working with RNA, particularly because of the increased demands regarding working in a clean manner, in vitro transcription offers them some interesting possibilities:

- The manufacture of RNA probes. RNA probes bind more strongly than DNA probes and are therefore much more sensitive (i.e., produce stronger signals). They can also be employed for every type of hybridization (e.g., Northern, Southern, in situ hybridization, RNase protection assays).
- Microinjection of RNA into *Xenopus* oocytes. If the RNA codes for a protein, it will subsequently be expressed in the oocyte. If it is an antisense RNA, it will bind to the complementary RNA, whose translation can be suppressed in this way.
- In vitro transcription is the first step in the in vitro translation (see Chapter 9, Section 9.3).

Each DNA with an RNA polymerase binding site can be transcribed, although only a few them have one. The most common method is to clone the desired DNA in a vector that contains a binding site. Best suited is a vector that has an SP6 or a T7 or T3 binding site to the left and to the right of the

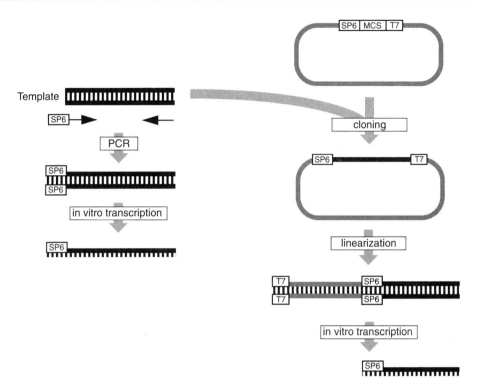

**Figure 5-1. In vitro Transcription.** Every DNA that has a promoter for an RNA polymerase is suitable for the synthesis of RNA in a test tube. The experimenter can clone the fragment from which to synthesize RNA in a vector using a fitting promoter or can amplify it by means of the polymerase chain reaction, in which one of the two primers on the 5′ end is required to demonstrate the promoter sequence. Typically, viral RNA polymerases are used for in vitro transcription; they generally stem from SP6, T7, or T3, and each has different promoters.

cloning site (see Chapter 9, Table 9-3). In this way, you can later transcribe both chains by selecting the corresponding RNA polymerase (Figure 5-1). Before the transcription, the clone is sliced using restriction enzymes at precisely the site where the future 3′ end of the RNA should be, eliminating the presence of vector sequences from the RNA and obtaining RNAs of a defined length. You should not use restriction enzymes that produce **overlapping 3′ ends** (e.g., *Aat*II, *Apa*I, *Ban*II, *Bgl*I, *Bsp*1286I, *Bst*XI, *Cfo*I, *Hae*II, *Hgi*AI, *Hha*I, *Kpn*I, *Pst*I, *Pvu*I, *Sac*I, *Sac*II, *Sfi*I, *Sph*I), because they can serve as starting points for RNA polymerase (Schenborn and Mierendorf, 1985). If one of these enzymes is necessary, however, you should smooth the end of the DNA with Klenow or T4 DNA polymerase (see Chapter 6) before performing the in vitro transcription. Alternatively, you can manufacture your own transcription template using the PCR. One of the primers should have about 20 more nucleotides on its 5′ end so that the polymerase will later accept such a template as readily as possible.

In each case, the DNA must be clean to guarantee a high yield; this usually means using cesium chloride gradient, ion exchange column, or glass milk purification. It also must be free of RNase because the transcription would function but the synthesized RNA would immediately be degraded. The same is true for all other solutions and products used.

The actual transcription is not problematic: 2.5 μg of linearized template DNA, 5 μL of 10× buffer, 5 μL of 50 mM DTT, 5 μL of nucleotides (10 mM each), 1.5 μL of RNase inhibitor, and 3.5 μL of RNA polymerase are pipetted together and incubated at 37°C to 42°C (99°F to 108°F) for 60 minutes. If you want to quantify the RNA synthesis or produce a radioactive probe, add a radioactive nucleotide and control its incorporation as in cDNA synthesis (for a nonradioactive probe, nucleotides that are not labeled radioactively can be inserted in the same way). It is certainly simpler and more sensible to apply an aliquot of the preparation to an agar gel, and in this way, you can estimate the approximate yield and see whether you have a single, lovely band. It must not necessarily be a denatured gel because a normal "DNA gel" is sufficient for this purpose.

If you want to dispose of the nucleotides that have not been inserted, the transcription is followed by a gel filtration (*spin column*). If the DNA template causes a disturbance in future processes, carry out another DNase digestion (use RNase-free DNase), purify the preparation with an acidic phenol-chloroform mixture (1 part phenol, pH 4.5, plus 1 part chloroform; the RNA remains in the supernatant, while the DNA remains in the phenol—the same principle used in RNA preparation), shake it with chloroform, and precipitate the RNA with salt and ethanol.

### Literature

Melton DA, Krieg PA, Rebaglialti MR, et al. (1984) Efficient in vitro synthesis of biologically active RNA and RNA hybridization probes from plasmids containing a bacteriophage SP6 promoter. Nucleic Acids Res 12:7035–7056.

Schenborn ET, Mierendorf RC. (1985) A novel transcription property of SP6 and T7 RNA polymerases: Dependence on template structure. Nucleic Acids Res 13:6223–6236.

## 5.6 RNA Interference

Because RNA is not very popular with molecular biologists, the number of applications is not large. Nevertheless, this field occasionally produces pioneering innovations. One such innovation is **RNA interference** (RNAi).

It began in 1990, as Napoli and colleagues reported the introduction of a pigment-producing gene, under whose control a strong promoter in petunias did not lead to the expected deep violet color of the blossoms. Its expression frequently resulted in speckled or white blossoms because the introduced gene and the homologous, endogenous gene suppressed one another, a phenomenon now known as **post-transcriptional gene silencing** (PTGS).

Later, it was shown that this effect could be induced in fungi and in animals up to mammals, awakening the interest of many researchers because the fast and simple suppression of individual genes (*loss of function*) is a tool that had been lacking. The production of transgenic organisms has become well established (see Chapter 9, Section 9.5), although extensive efforts are required.

Because of this great interest, there has been an extensive increase in the knowledge about PTGS, primarily for the RNAi, a special form of PTGS whose essential mechanisms have become well known.

Antisense oligonucleotides have been employed for a longer period to suppress the expression of individual genes. The idea is that the antisense oligo binds specifically to the mRNA, and the double-stranded RNA is cut by double-strand–specific nucleases. The quantities of oligos required for this are very large, and the results are not guaranteed.

In 1998, Fire and coworkers were able to show that the injection of very small quantities of **double-stranded RNA** (dsRNA) could trigger the development of RNAi. This silencing effect continues for several days and is apparently transferred to the daughter cells when the cell divides. In plants, *Caenorhabditis elegans*, and in *Drosophila*, the dsRNA is broken down into short fragments of only 21 to 23 nucleotides, the so-called **small interfering RNAs** (siRNAs), and these siRNAs are then incorporated in RNA protein complexes known as **RNA-induced silencing complexes** (RISC). A RISC is activated through the elimination of one of the strands of the siRNA; the activated complex can then bind to homologous mRNA and cut this strand approximately 12 nucleotides from the 3' end with the aid of the siRNA. This process is evidently very effective, because even a relatively few number of dsRNA molecules are sufficient for specific gene suppression in some organisms.

In mammalian cells, the process is unfortunately not as simple as described. The first attempts carried out in mammals were not particularly successful, until it was realized that dsRNA that is more than 30 bases long causes a nonspecific suppression of gene expression. Smaller dsRNAs lead to the development of RNAi, whereby fragments of 21 bases with a two-base, 3' overhang (as is generated in *Drosophila*, for instance, through the action of the RNase in their own cells) demonstrate the most effective action.

What is the origin of the different *gene silencing* mechanisms? Does it originally deal with a defensive mechanism of the cell against viruses or transposons? This may be the case; after all, the antiviral, nonspecific suppression of gene expression in embryonic stem cells from mice is not available because it is precisely in the developmental stage during which transposons can increase in the genome. There are also indications that it may involve the addition of new gene-regulatory mechanisms, through which individual, specifically targeted genes can be switched off.

siRNAs for laboratory purposes can be synthesized chemically or manufactured through in vitro transcription; the latter process is time consuming, but it is frequently less expensive. The transfer of siRNAs normally occurs by means of transfection or injection. Expression vectors have been developed for the prolonged expression of siRNAs in the cell (Brummelkamp et al., 2002; Miyagishi and Taira, 2002).

**Suggestions**: Ambion has an interesting Internet site on the topic of RNAi (http://www.ambion.com/RNAi), and the following are some of their **suggestions concerning the design of siRNAs**:

- The sequence should be absolutely specific, because individual base pairs can reduce the efficiency dramatically.
- Double-stranded RNAs of 21 nucleotides with a 2-nucleotide, 3' overhang are the most effective. The sequence should begin with two adenosines (AA) at the 5' end. A GC concentration of 30% to 50% functions better than higher GC concentrations.
- Making sequences from the nontranslated areas (5' and 3' end) or from the initial regions of the open reading frames (first 75 bases) is not advised, because these regions frequently contain binding sites for regulatory proteins, and these can disturb the linking to RISC complexes.
- Sequences with a high degree of homology to other genes should be avoided to enable later obtaining a specific answer.
- Do not rely on an individual sequence, but choose instead three to four and test them all, because not all sequences function equally well. In the best case, you can reduce the expression of the targeted gene by more than 90%.
- siRNAs, which have an equivalent nucleotide relationship to that of the selected siRNA but demonstrate no significant homology to any site in the genome of the selected organism, are considered to be suitable negative controls. You whirl the sequence and perform a search for homologs, such as with BLAST (http://www.ncbi.nlm.nih.gov/BLAST/).

**Literature**

Brummelkamp TR, Bernards R, Agami R. (2002) A system for stable expression of short interfering RNAs in mammalian cells. Science 296:550–553.

Elbashir SM, Harborth J, Lendeckle W, et al. (2001) Duplexes of 21-nucleotide RNAs mediate RNA interference in cultured mammalian cells. Nature 411:494–498.

Elbashir SM, Martinez J, Patkaniowska A, et al. (2001) Functional anatomy of siRNA for mediating efficient RNAi in *Drosophila melanogaster* embryo lysate. EMBO J 20:6877–6888.

Fire A, Xu S, Montgomery MK, Kostas SA, et al. (1998) Potent and specific genetic interference by double-stranded RNA in *Caenorhabditis elegans*. Nature 391:806–811.

Guru T. (2000) A silence that speaks volumes. Nature 404:804–808.

Miyagishi M, Taira K. (2002) U6-promoter-driven siRNAs with four uridine 3′ overhangs efficiently suppress targeted gene expression in mammalian cells. Nat Biotechnol 20:497–500.

Napoli C, Lemieux C, Jorgensen R. (1990) Introduction of a chalcone synthase gene into petunia results in reversible co-suppression of homologous genes in trans. Plant Cell 2:279–289.

# 6 Cloning DNA Fragments

*Des Löwen Mut,*
*des Hirsches Schnelligkeit,*
*des Italieners feurig Blut,*
*des Nordens Dau'rbarkeit.*

*The lion's dauntless mood,*
*The deer's swift pace,*
*The Italian's fiery blood,*
*The toughness of the northern race.*

The key word *cloning* should not to be confused with producing *clones*, the manufacture of an identical copy. Cloning is the introduction of a DNA fragment into a vector, which makes it possible to increase this DNA to an abundant quantity. Cloning has for the first time made it possible to obtain such large quantities of this fragment that they can be seen with the naked eye. In the past, the cloning procedure was practically obligatory to obtain sufficient quantities of DNA for an experiment. Since the introduction of PCR, it is possible to do without it. Nevertheless, PCR will not replace cloning, because DNA as a clone is so steady and unproblematic to use that experimenters cannot do without it, and these features are not available with PCR fragments.

The boundless replication is not the only important aspect DNA clones. Through the vector, it is possible to give a fragment additional characteristics that may expand its use considerably. RNA polymerase promoters in the vector, for instance, allow the problem-free in vitro transcription of RNA (see Chapter 5, Section 5.5), and viral promoters in a construction permit the expression of cDNAs in mammalian cells (see Chapter 9, Section 9.4.4).

Banks of genomic DNA or cDNA can be manufactured in this way from a great variety of organisms and tissues. In the past, this was the only possibility for obtaining new genes, because a wild mixture of clones had to be screened (i.e., sifted through) until the desired clone was isolated (see Chapter 7, Section 7.4). Because the clones can be increased again and again, researchers can get along with a small amount of initial material, completely unlike the isolation of new proteins, which previously required kilograms or even tons of tissues. In the meantime, because cDNA expression banks can be produced with the help of such cloning techniques, which allow screening of the clone for functional proteins, the previous method is almost a thing of the past. Entire genomes are being sequenced, and the extent to which the manufacture of DNA banks will play a role remains to be seen.

## 6.1 The Basics of Cloning

Cloning is a simple procedure. You digest the vector and DNA fragments, purify them, ligate them with one another, and transform the wild mixture that emerges within the bacteria. You then select the desired clone among the bacteria, and the cloning procedure is finished. The difficulties lie in the details.

## 6.1.1 The Vector

Usually, at the beginning of the work, the experimenter decides in favor of a specific cloning vector and remains with it (Figure 6-1). The researcher also uses the same cutting sites again and again. Although only a small amount (25 to 50 ng) of the vector is required for the individual ligation, larger quantities (1 to 5 μg) should be prepared (i.e., well digested) and, if necessary, dephosphorylated (discussed later), purified, adjusted to a concentration of 25 ng/μL, and then frozen. Initially, this means more work, but you can attain an attractive set of ready-to-use vectors within a short time, which many of your colleagues will envy. You must work carefully, or you will quickly end up with a large number of poorly defined preparations, which you are best advised to throw away.

If the vector is digested with two restriction enzymes, a small piece of a polylinker will be released that disturbs the following ligation, because such small DNA fragments are far better ligated into the vector than the fragment that you would like to insert. If the pieces of polylinker are shorter than 10 to 15 bases, it can be eliminated during the ethanol precipitation. Vector fragments demonstrating longer pieces should be separated by means of gel electrophoresis, and the vector can then be purified from the gel.

If the vector is cut with a single restriction enzyme, two compatible ends emerge that can ligate with one another again. In this way, the enemy is in the vector DNA, and the proportion of the desired clones usually is minimal. The problem can be reduced (although it cannot be eliminated) by **dephosphorylating** the vector DNA. Colloquially, experimenters speak of CIP reactions, because they use **calf intestine alkaline phosphatase** (alkaline phosphatase from calf intestines [CIP or CIAP]). After restriction digestion, phosphate remnants remain at the 5′ ends of the DNA fragment that are required for ligation. If they are removed with a phosphatase, the vector will no longer undergo any self-ligation. The fragment you want to insert still has both phosphate remnants and can therefore ligate with the vector DNA, at least with one of the two strands (Figure 6-2). Surprisingly, this is sufficient for successful cloning.

The dephosphorylation is as simple as a restriction digestion. CIP buffer and 1 μL of CIP are added and then incubated for 1 hour at 37°C (99°F). The enzyme functions very efficiently with fragments demonstrating a 5′ overhang, but it is recommended to incubate for 30 minutes at 37°C (99°F) and 30 minutes at 56°C (133°F) for smooth ends and 3′ overhangs.

The CIP is a remarkable enzyme. It is chemically stabile, can be stored for years at 4°C (39°F), and almost always functions as it should. This stability, however, is also a problem at the end of the reaction. The enzyme must be inactivated, because it would otherwise dephosphorylate the DNA

**Figure 6-1. Capacity for accepting heterologous DNA of different cloning vectors.**

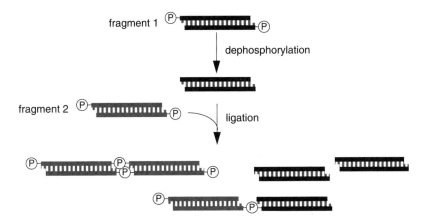

**Figure 6-2. Ligating with dephosphorylated DNA fragments.** The number of possible combinations in a ligation reaction can be reduced by dephosphorylating one of the DNA fragments that is used. This technique most frequently is used in cloning to prevent vector fragments from ligating with themselves, because they have two ends that are compatible with one another.

fragments that you would like to insert during the ligation, ending its brilliance as a cloning agent. This problem is primarily solved with the aid of phenol-chloroform or glass milk purification.

To facilitate the life of the experimenter, the use of **shrimp alkaline phosphatase** (SAP) was launched a few years ago. Its action can be switched off through normal heat inactivation (20 minutes at 65°C [149°F]). Whereas some are very pleased with this enzyme, others consider the SAP to be poorer than CIP. The primary advantage of CIP is its great stability. It suffices to test the batch employed only once, and you can be sure that it will also function in the future. Because the quantities obtained are tremendous, the initial quantity is usually sufficient for a number of years.

## 6.1.2 The DNA Fragment

The most difficult step is to find a suitable cutting site for cloning. After this problem is solved, the DNA is digested, the fragments are separated by means of an agar gel, and the desired DNA fragment is isolated from the gel (see Chapter 3, Section 3.2.2). The fragment is subsequently ligated in a vector DNA, which had been cut with the same restriction enzyme.

Occasionally, an experimenter wants to use a specific vector for the cloning but then discovers that it has no suitable cutting site. Before giving up, the researcher should briefly examine whether at least a couple of cutting sites can be found that produce compatible overhangs (Table 6-1). These can no longer be cut, but the cloning process can still function. Fragments with smooth ends can be ligated arbitrarily, although with great difficulty.

Table 6-1 shows restriction enzymes that result in compatible ends after digestion. A review can be found in the appendix of the New England Biolab catalog (see Appendix 2). The 4-cutters are marked with a superscript 4, 5-cutters with a superscript 5, and 8-cutters with a superscript 8. Restriction enzymes that cut outside of their recognition sites (labeled NNNN) can be used in cloning PCR fragments. With their aid, arbitrary overlaps can be inserted as desired.

Finding compatible cutting sites, however, is more of an exception, but in this case, there are other possibilities. The experimenter can smooth overlapping fragment ends and then ligate the fragment in a vector with smooth, dephosphorylated ends.

**Table 6-1.** Enzymes with Compatible Ends

| Ends | Four Bases | Enzyme* | Two Bases | Enzyme* |
|---|---|---|---|---|
| 5′ Overhang | AATT | *Eco*RI, *Mfe*I, *Tsp*509I[4] | CG | *Bsp*D1/*Cla*I, *Bst*BI, *Hin*PI[4], *Mae*II[4], *Nar*I, *Psp*1406I, *Taq*I[4] *Ase*I, *Mse*I[4] *Ace*I, *Mse*I[4] |
| | CATG | *Bsp*HI, *Bsp*LU11I, *Nco*I | | |
| | CCGG | *Age*I, *Bsp*EI, *Ngo*MI, *Xma*I | | |
| | CGCG | *Asc*I[8], *Bss*HII, *Mlu*I | | |
| | CTAG | *Avr*II, *Nhe*I, *Spe*I, *Xba*I | | |
| | GATC | *Bam*HI, *Bcl*I, *Bgl*II, *Dpn*II[4] /*Mbo*I[4]/*Sau*3AI[4] | TA | |
| | GGCC | *Bsp*120I, *Eag*I, *Not*I[8] | | |
| | GTAC | *Acc*65I, *Bsi*WI, *Bsr*GI | | |
| | TCGA | *Sal*I, *Xho*I | | |
| | TGCA | *Apa*LI, *Ppu*10I | | |
| | NNNN | *Alw*26I[5]/*Bsm*AI[5], *Bbs*I/*Bbv*16II/*Bpi*I/*Bpu*AI, *Bbv*I[5]/*Bst*71I[5], *Bsa*I/*Eco*31I, *Bsm*BI/*Esp*3I, *Bst*2BI, *Fok*I[5], *Sfa*NI[5] | | |
| 3′ Overhang | ACGT | *Aat*II, *Tai*I[4] | NN | *Bpm*I/*Gsu*I, *Bse*RI, *Bsg*I, *Bsr*DI, *Bst*F5I[5], *Eco*57I |
| | CATG | *Nla*III[4], *Sph*I | | |
| | TGCA | *Nsi*I, *Pst*I, *Sse*8387I[8] | | |
| Smooth ends | *Alu*I[4], *Bst*1107I, *Bsi*UI[4], *Dpn*I[4], *Dra*I, *Ecl*136II, *Eco*47III, *Eco*RV, *Ehe*I, *Fsp*I, *Hae*III[4], *Hpa*I, *Msc*I, *Nae*I, *Nru*I, *Pme*I[8], *Pml*I, *Pvu*II, *Rsa*I[4], *Sca*I, *Sma*I, *Sna*BI, *Srf* I[8], *Ssp*I, *Stu*I, *Swa*I[8] | | | |

* The 4-cutters are marked with a superscript 4, 5-cutters with a superscript 5, and 8-cutters with a superscript 8. Restriction enzymes that cut outside of their recognition sites (labeled NNNN) can be used in cloning polymerase chain reaction fragments. With their aid, arbitrary overlaps can be inserted as desired.

# 6.1.3 Fill-in Reaction

Ends with a 5′ overhang can be filled without any problem by carrying out a normal polymerase reaction. You then add DNA, T4 DNA polymerase or Klenow fragments (i.e., the large fragment of DNA polymerase I from *Escherichia coli*), nucleotides (final concentration of $100\,\mu M$ per dNTP), and buffers and incubate for 20 minutes at room temperature (Klenow) or 5 minutes at 37°C (99°F) (T4 DNA polymerase). The polymerase is inactivated by heating to 75°C (167°F) for 10 minutes. The 3′ overhangs can be smoothed, because both enzymes, in addition to polymerase activity, demonstrate a 3′-5′ exonuclease activity (although the Klenow polymerase is rarely employed for this purpose because of its lower exonuclease activity). In this case, there are no insertions, but rather disassemblies, the result being a fragment that is shorter by about 2 to 4 nucleotides. The nucleotides cannot be left out. They are required to stop further disassembly. In double-stranded DNA, the degradation and neosynthesis are in balance, and the result is a smooth end. However, as soon as the nucleotides are exhausted, the degradation continues, and it is important not to let the preparation stand around for too long.

You can use the variant of a **partial fill-in** (Figure 6-3). Many enzymes form overhangs that are at least compatible at the 5′ end, such as *Xba*I (5′ c̲tagA . . . ) and *Hind*III (5′ a̲gctT . . . ; the overhanging bases are written as lower-case letters, and the underlined letter represents the complementary base). If the respective unfitting bases at the 3′ end of the overhang are filled, a 2-base overhang remains. This is small, but it can be ligated better than the smooth ends. The procedure is the same as that

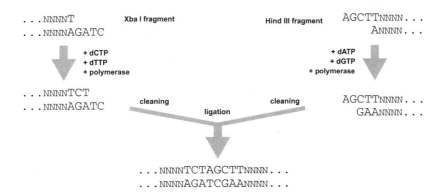

**Figure 6-3. Partial fill-in method shown by the example of an *Xba*I and a *Hind*III cut fragment.** The DNA fragments are initially filled with the suitable nucleotides until compatible ends are found, purified, and ligated with one another.

involving the smoothing of 5′ overhangs using Klenow fragments, but only the two nucleotides are added that should be used for the filling.

The standard enzyme used for ligations is the T4 DNA **ligase**, because it is extremely fast and adequate for these buffer conditions. Ligations function in the fitting ligase buffer and in almost every restriction buffer, as long as some ATP is added (final concentration of 1 to 5 mM; do not use dATP); without that, nothing happens. Between 0.5 and 1 Weiss unit (a detailed definition of the Weiss unit can be found in the Promega catalog) is used per ligation preparation (20 to 50 μL). Unfortunately, a certain wild growth proceeds in this area, and a large number of manufacturers choose to define their enzyme activity differently. You can occasionally recognize them because the number of units per microliter is astronomically high. It usually is safe to employ 0.5 to 1 μL of enzymes, or you can follow the recommendations of the manufacturer instead.

## 6.1.4 DNA Quantities

It is recommended that you use about five times more DNA fragments than the quantity of the vector. Quantity refers to calculation of the number of molecules, not the amount in nanograms.

You can calculate everything in moles. This is the more scientific method, but working with very small amounts, such as femtomoles and picomoles, frightens many away. It is not as difficult as it seems, and a small excursion into the world of moles follows.

A **mole** corresponds to $6.02 \times 10^{23}$ molecules. Perhaps we could have found something more practical for our purposes, but the mole is the unit of volume for chemists. The **molar mass** is the mass that a mole of a certain molecule has.

$$\text{Molar mass (of a DNA fragment) [g/mol]} = \text{DNA mass [g]/DNA quantity [mol]} \qquad (6\text{-}1)$$

The molar mass is numerically identical to the relative molecular mass, previously known as the *molecular weight* (MW) in the technical literature published in the English language. The molecular weight is also found on all chemical containers. It is unfortunately not found on containers with DNA, but it can be calculated. A base has an average molecular weight of 330 g/mol, and the molecular weight of a base pair is consequently 660 g/mol. The molecular weight of a DNA fragment can be calculated from the amount of base pairs multiplied by 660:

$$\text{Molar mass (of DNA)} = \text{molar mass (of a base pair)} \times \text{number of base pairs} \qquad (6\text{-}2)$$

The molecular weight of a 1-kb fragment is 660 g/mol. A simple transformation produces an equation based on Equation 6-11 that can be used to determine the DNA quantity:

$$\text{DNA quantity [mol]} = \text{DNA mass [g]/molar mass (of the DNA) [g/mol]} \qquad (6\text{-}3)$$

A new formula is produced through the application of Equation 6-2:

$$\text{DNA quantity [mol]} = \text{mass [g]/(molar mass of a base pair [g/mol]} \times \text{number of base pairs)} \quad (6\text{-}4)$$

or

$$\text{DNA quantity [pmol]} = \text{mass [pg]/(660 [g/mol]} \times \text{number of base pairs)} \qquad (6\text{-}5)$$

For example, 20 ng of DNA that is 1 kb long corresponds to 0.03 pmol, 30 fmol, or $3 \times 10^{-14}$ mol. In the previous example, it is $1.8 \times 10^{10}$ molecules. It is beneficial to perform such calculations more frequently to get a feeling for what you are handling and to understand what yields you can expect from a particular attempt. Calculate how much of a 1-kb DNA fragment you can obtain from 10 pmol of primer through the use of PCR and how much nucleotide is required for this, and then compare the result with the quantity of DNA you obtain from a PCR reaction—and you will be astonished!

The other possible method for determining the correct quantity of DNA fragments is simpler. The recommendation to use five times more fragment than vector leads to the following formula:

$$\text{DNA quantity}_{\text{Fragment}} = 5 \times \text{DNA quantity}_{\text{Vector}} \qquad (6\text{-}6)$$

By making use of Equation 6-5 and converting some terms, a new formula is obtained:

$$\text{Mass}_{\text{Fragment}} \text{ [ng]} = 5 \times \text{Mass}_{\text{Vector}} \text{ [ng]} \times \text{Length}_{\text{Fragment}} \text{ [bp]/Length}_{\text{Vector}} \text{ [bp]} \qquad (6\text{-}7)$$

If a standard DNA vector of 25 ng is inserted for the ligation, the entire equation can be simplified to the following form:

$$\text{Mass}_{\text{Fragment}} \text{ [ng]} = 125 \text{ [ng]} \times \text{Length}_{\text{Fragment}} \text{ [bp]/Length}_{\text{Vector}} \text{[bp]} \qquad (6\text{-}8)$$

## 6.1.5 The Ligation

According to the textbooks, ligation is carried out at 14°C to 16°C (57°F to 61°F) for 1 to several hours. It can function simply because fragments with overlapping ends generally only have to be incubated for 1 hour at room temperature. In cases of simple cloning (i.e., a fair amount of fragment DNA with ends having a four-base overhang), even 15 to 30 minutes is sufficient. The ligation of smooth ends is substantially more difficult because the reaction is less effective. In this case, incubation is carried out at 16°C (61°F) for 4 to 18 hours or, even better, at 4°C (39°F) overnight. Application of 15% polyethylene glycol (PEG) and reduced ATP concentrations should help to increase the yield.

An amount between 1 and 2 μL of a 20 μL preparation is sufficient for accomplishing a successful bacterial transformation. This procedure is described in Section 6.4.

A warning is in order. Many of you will have stumbled across quick-and-easy ligation preparations that have been available on the market for several years. These tubes contain buffers, ATP, and ligase in a lyophilized form, and you need only to add your DNA and water to it, and off you go. As tempting as it may appear to be able to dispense with a freezer, you should be aware of the possible problems. The efficiency, for instance, may be substantially smaller than for the classic procedure, or the kit may prove to be unsuitable for electroporation.

## 6.1.6 Cloning Polymerase Chain Reaction Products

A particular challenge is presented in the cloning of PCR products. The classic procedure, described by Sharf and colleagues (1986), consists of inserting a restriction cleavage site on the 5' ends of both of the primers used. After the purification, the PCR product is cut and cloned into a fitting vector. The difficulty is that some restriction enzymes can be cut only poorly at the ends of a fragment (Kaufman and Evans, 1990). Because the success of such an operation is difficult to evaluate, the process always represents a bit of a gamble. Another disadvantage to this method is that you do not always have two primers available with restriction cleavage sites.

Resourceful spirits have worked for some time on finding a simpler method that can be applied more generally. A classic solution consists of cloning a smooth PCR fragment, such as using *Eco*RV vectors (Sharply et al., 1986). Amplification with Pfu or Pwo functions well, although with the use of Taq polymerase, as is frequently employed, the yield is only mediocre, because it has a tendency to add an additional base to the 3' end of the synthesized DNA (discussed later). If the base at the 5' end of the primer is a T, there is a good chance of receiving a smooth end. Otherwise, you can smooth the ends with T4 DNA polymerase to increase the effectiveness of cloning. Stratagene (Stratagene Cloning Systems, La Jolla, CA) offers a similar kit for this purpose, which functions with Pfu polymerase.

To increase the proportions of clones with an insert during cloning, a number of clever methods are available. One is offered by Stratagene, in the form of a kit (pCR-Script), which can be adapted for everyday use. Surprisingly, this method appears to have no name of its own, and I would therefore like to take this chance to identify it as a **cut ligation**. The fragment is ligated with a vector cut with *Srf* I, and during the ligation, it is simultaneously digested with *Srf* I (Figure 6-4). Vector DNA, which can ligate with itself, is immediately cut again, the cutting site is destroyed through the insertion of the fragment, and the desired ligation products remain protected from restriction digestion—and the experimenter is spared dephosphorylation of the vector. You can also perform this method with

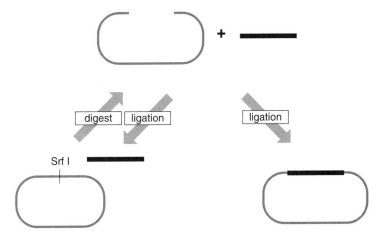

**Figure 6-4. Techniques for pCR script kits (Stratagene, Stratagene Cloning Systems, La Jolla, CA).** The cut vector and the fragment are ligated in the presence of *Srf*I with a T4 DNA ligase. The vector preferentially ligates with itself, but because the restriction site is restored in this manner, it is severed again by the restriction enzyme. Ligation with the fragment destroys the restriction cleavage site. That is principally the case with each restriction enzyme, in so far as it does not cut the fragment and as long as the ends of the fragment have not been cleaved by an enzyme other than that of the vector. The system employed by Stratagene is especially practical, because *Srf* I is an 8-cutter and is therefore less likely to cut within the insert. Additional restriction enzymes that leave smooth ends behind are listed in Table 6-1.

*Eco*RV or *Sma*I, together with a run-of-the-mill vector. The advantage of *Srf* I is that the enzyme is an 8-cutter. The probability that the fragment itself has an *Srf* I cutting site, which would then be cut as well, is relatively low.

The other method consists of cloning the fragment into the **pZErO** vector from Invitrogen. Its *multiple cloning site* is found in a killer gene; if a fragment is ligated therein, the expression of the gene is interrupted, and the bacteria survive. If the vector ligates with itself, the bacteria depart from this life and do not emerge as colonies on the plate, so that the experimenter is saved another selection among the thriving colonies.

Another preparation is based on the discovery that the Taq polymerase (as well as the Tfl and the Tth polymerases) produces no fragments with smooth ends, but it generally results in a nonspecific overhang of one base (Clark, 1988). In most cases, this base is an adenosine (Costa and Weiner, 1994; Hu, 1993). A single base overhang is not much, but it is better than nothing. Through the construction of a vector with a thymine overhang (Marchuck et al., 1991), PCR fragments can be cloned more easily than in vectors with smooth ends. The method is known as **TA cloning**, and there are kits with completed T vectors, such as the one from Invitrogen. You can manufacture the vector by incubating the smoothly cut vector with Taq polymerase and dTTP in a fitting buffer for 2 hours at 70°C (158°F) and then purifying the solution.

Because you have your TA vector in a drawer, you may wish to clone a PCR fragment that has been amplified through a polymerase with corrective activity into this. A short protocol can enable you to obtain your fragment with the necessary TA overhang with little effort. Perform a PCR, place the tube on ice, add 0.5 to 1 unit of Taq polymerase (another addition of buffers or nucleotides is unnecessary), and incubate for 10 minutes at 72°C (162°F). Afterward, place it on ice, and continue to work with it directly.

If you wish to clone with a TA overhang, choose the primer accordingly, because the base at the 5′ end of the primer has a substantial influence on the specific kind of overhang. It is best to select an adenosine or a guanine for the 5′ end of the primers (Table 6-2).

A variant of this is the **TOPO TA Cloning Kit** from Invitrogen. Instead of a ligase, a **topoisomerase I–activated** TA vector from a vaccinia virus is used. The topoisomerase recognizes the (C/T)CCTT cutting site, severs the DNA, and remains covalently bonded by means of one 3′ phosphotyrosyl bond. Because the TOPO vector contains two adjacent cleavage sites so that both DNA ends are blocked by topoisomerase, self-ligation is prevented. According to the manufacturer, the ligation reaction is completed after 5 minutes, making this much quicker than the usual T4 ligase reaction. That is amazingly fast, but you must always purchase the activated vector first. Only fragments with free 5′OH ends can become ligated (because of the 3′ phosphotyrosyl bonds), because the PCR primer cannot be phosphorylated. Shuman (1994) explains how the entire procedure functions in detail.

A similar kit (Zero Blunt TOPO Cloning Kit) is available for cloning PCR fragments, which are amplified using a polymerase with corrective activity and therefore have no overhanging adenosine. Other DNA fragments with smooth ends can be cloned in this way, although it is also imperative that the ends be dephosphorylated.

**Table 6-2.** Modification at the 3′ End by Taq Polymerase

| 5′Nucleotide of the Primer | 3′ Modification of the Synthesized Strand |
| --- | --- |
| A | −T, +A |
| C | +G > +A > +C |
| G | +A > +C |
| T | (+A)* |

The Taq polymerase rarely produces fragments with smooth ends. The base, which is hung (+) or not hung (−) onto the 3′ end of the newly synthesized of the DNA strand, depends on the 5′ base of the complementary primer.
* If the 5′ base is thymidine (T), an adenosine might be added.

**Literature**

Clark JM. (1988) Novel non-templated nucleotide addition reactions catalyzed by procaryotic and eucaryotic DNA polymerases. Nucleic Acids Res 16:9677–9686.

Costa GL, Weiner MP. (1994) Increased cloning efficiency with the PCR Polishing Kit. Strategy 7/2:48.

Hu G. (1993) DNA polymerase-catalyzed addition of non-templated extra nucleotides to the $3'$ end of a DNA fragment. DNA Cell Biol 12:763–770.

Kaufman DL, Evans GA. (1990) Restriction endonuclease cleavage at the termini of PCR products. Biotechniques 9:304–305.

Marchuk D, Drumm M, Saulino A, Collins FS. (1991) Construction of T-vectors, a rapid and general system for direct cloning of unmodified PCR products. Nucleic Acids Res 19:1154.

Sharf SJ, Hirn GT, Ehrlich. (1986) Direct cloning and sequence analysis of enzymatically amplified genomic sequences. Science 233:1076–1078.

Shuman S. (1994) Novel approach to molecular cloning and polynucleotide synthesis using vaccinia DNA topoisomerase. J Biol Chem 269:32678–32684.

## 6.1.7 Cloning with Recombinase Systems

Cloning of the PCR fragment usually is the beginning of the experiment. If the fragment has been sequenced successfully, it is then recloned into a true determination vector (normally an expression vector).

For anyone who has recloned cDNAs, it maybe interesting to contemplate the application of a recombinase system, which has been available on the market for a short time. These systems facilitate the cloning work to a small extent and deliver a higher yield of positive clones, a feature that reduces the effort required. For inexperienced researchers, who have an almost unlimited time to complete their doctoral theses, this may prove to be less important, but anyone who must often perform recloning in the course of any more extensive project will be happy to simplify this step. Because of their high specificity, these systems offer the possibility of automating the cloning process for the most part.

There are three cloning systems available that function as a recombinase from Life Technologies (GATEWAY), Invitrogen (GATEWAY and Echo), and Clontech (Creator). GATEWAY is based on the recombinase system of bacteriophage λ, whereas Echo and Creator make use of the Cre recombinase of bacteriophage PI. Although the logic of the system is principally equivalent in all cases—a recombinase takes care of the recombination of two different DNA strands, while the recombination of specific recognition sites are available on both strands—the three systems differ technically from one another to a relatively large extent, which does not help to make the understanding any more beneficial.

The simplest is the Echo system (Figure 6-5A). The cDNA is cloned into a donor vector, which contains a loxP recognition sequence and which lies before the cDNA. The donor construct is recombined with a fitting acceptor vector, which also has an loxP sequence. The acceptor vector contains all elements that are required for the expression of the cDNA in a certain system (i.e., bacterial, yeast, insect, or mammalian cells); through "fusion" of both vectors, a new, completed expression construct is produced. The fusion of both vectors is possible only because the donor vector has an R6Kγ replication origin, whereas the acceptor vector has a normal ColE1 origin, and the origins can coexist in the same plasmid. The loxP sequence, which is recognized by the Cre recombinase, is composed of two *inverted repeats* (i.e., two identical sections with a counter orientation 13 bp long) that are separated from one another by a nonsymmetrical 8-bp sequence. This ensures that the recombination proceeds only in a defined orientation. More information concerning vector fusion can be found in the article by Liu and coworkers (1998), and details about the Cre-loxP system can be found in two other articles (Abremski et al., 1983; Hoess et al., 1982).

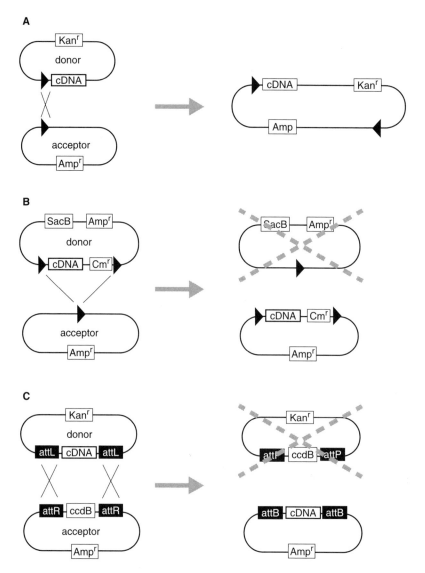

**Figure 6-5. Overview of the three recombinase systems. A,** Echo system. **B,** Creator system. **C,** Gateway system. Recombinase recognition sequences are presented in black. Amp$^r$, ampicillin resistance; Cm$^r$, chloramphenicol resistance; Kan$^r$, kanamycin resistance.

The Creator system (see Figure 6-5B) has two loxP sequences in the donor vector, between which the cDNA is cloned. For selection purposes, the donor also contains a chloramphenicol-resistance gene between the loxP sequences and a *sacB* gene (i.e., sucrase gene from *Bacillus subtilis*) in the vector *backbone* ("spine"), which inhibits the growth of gram-negative bacteria in the presence of 5% sucrose (Gay et al., 1985). Through recombination, the region between the two loxP sequences is inserted into the desired acceptor vector, and the undesirable products of the reaction are suppressed through the addition of chloramphenicol and sucrose. Instead of fusion, a DNA fragment is transferred from the donor to the acceptor vector.

Gateway (see Figure 6-5C) is based on the recombination system of bacteriophage λ, which makes use of a total of four recognition sequences that are recombined with each other in a specific manner (attB × attP ⇌ attL × attR). Each of these att recognition sequences is composed of two cassettes, which can be combined with one another according to the scheme AB × CD ⇌ AD × CB. Every vector has two (equivalent) att sequences, which means that the DNA segments lying between the att sequences are exchanged from the donor to the acceptor vector during recombination. The undesirable recombination products are selected through the resistance gene in the vector backbone (i.e., kanamycin in the one case and ampicillin in the other) and through the *CcdB* killer gene (see Section 6.2.1), which is first situated in the acceptor vector and then, after recombination, in the donor vector. The fact that the donor vector of this system is known as an *entry vector* and the acceptor vector is known as a *destination vector* may prove to be somewhat confusing when comparing these two systems.

In all systems, recombination proceeds along a defined orientation, and the acceptor vector is decisive for the possible applications of the expression construct. The Gateway system distinguishes elements by using two peculiarities. First, it permits a direct cloning of the PCR fragments in the donor vector, as long as the primer has been furnished with the necessary attB sequences (which have a length of at least 25 bp). Second, aside from a series of completed acceptor vectors, the kit for this system allows transformation of your favorite vector into a Gateway-compatible acceptor vector, which can be especially useful in some cases.

Depending on the palette of acceptor vectors (and on your finances), you can produce constructs for various expression systems quite rapidly with all of these systems. You can then distribute them to laboratories run by friends and wait for responses to determine with which systems the protein expression functions best. If the cDNA should prove to be toxic for the host cell in these pilot tests, you could reclone the fragment without taking too much time in a vector with a controllable promoter. However, if everything functions unexpectedly, you could quickly produce various fusion proteins to determine how best to purify your protein. That is approximately how the recombinase system works, although not everything functions so easily in practice. If you find such unusual projects to be interesting, you should learn about the various recombinase systems to decide which of these purposes may be suitable for you. Have fun in the process.

**Advantage:** With a great percentage of positive clones, you are not dependent on suitable restriction cleavage sites, and the orientation remains with the recloning.

**Disadvantage:** For all recombinase systems, the recognition sequences must be available, or nothing will function. You must be committed to a particular system at the beginning, because a subsequent selection or transfer frequently involves more work, rather than saving time, and all initial clones must be prepared anew.

Ultimately, you should not forget the conditions for licensing. Much of gene technology is protected from A to Z through patents. When considering leaving the path of pure, basic research, you are better off keeping things clear as far as the possible costs or proprietary questions are concerned. Many patents (especially those in the United States) lay claim to any later applications that had not been taken into consideration at the time of the initial application.

**Literature**

Abremski K, Hoess R, Sternberg N. (1983) Studies on the properties of P1 site-specific recombination: Evidence for topologically unlinked products following recombination. Cell 32:1301–1311.

Gay P. Le Coq D, Steinmetz M, et al. (1985) Positive selection procedure for entrapment of insertion sequence elements in gram-negative bacteria. J Bacteriol 164:918–921.

Hoess RH, Ziese M, Sternberg N. (1982) P1 site-specific recombination: Nucleotide sequence of the recombining sites. Proc Natl Acad Sci U S A 79:3398–3402.

Liu Q, Li MZ, Leibham D, et al. (1998) The univector plasmid-fusion system, a method for rapid construction of recombinant DNA without restriction enzymes. Curr Biol 8:1300–1309.

# 6.2 Choosing Vectors for Cloning

The decision about the right vector is one half of the battle in cloning, it depends on what you want to carry out with the clone and on what you want to avoid. This section provides an overview of the different kinds of cloning vectors.

## 6.2.1 Plasmids

Plasmids are circular, double-stranded DNA molecules that can multiply in bacteria independent of the bacterial genome. The minimal features of a plasmid consist of a **start of replication** (i.e., *origin of replication* [ori]), a **selection gene** (usually an antibiotic-resistant gene), and a *cloning site* to introduce foreign DNA into the plasmid. A quantity of other sequences can be included, such as a second selection marker or a promoter for the purpose of expressing the included gene.

Plasmid vectors are usually 2.5 to 5 kb long, depending on which characteristics they demonstrate (Figure 6-6). Because the size of a plasmid is not limited, in principle, you can clone any amount of DNA into them, which makes them wonderful tools in molecular biology. In practice, because the size of a plasmid is not limited, They are used primarily for cloning fragments that are almost 0 to 5 kb long. Longer fragments can be inserted into plasmids, although the difficulty of cloning usually increases with longer fragments.

The number of plasmid copies containing a bacterium depends on the origin of replication. If there are fewer than 20, it is known as a **low-copy plasmid**, whereas **high-copy plasmids** can exist in several hundred copies per bacterium. High-copy plasmids typically are used, because the yield is higher in plasmid preparations, although the dimensions of the copy numbers may occasionally be troublesome.

The cloning sites contain cleavage sites, which usually occur only once in the vector. This allows the experimenter to cut the vector for cloning without decomposing it into a thousand smaller pieces. Theoretically, such a cleavage site makes it possible to vary the vectors as much as possible. Most plasmid vectors contain 10 to 20 of these, and molecular biologists speak of a **multiple cloning site** (MCS).

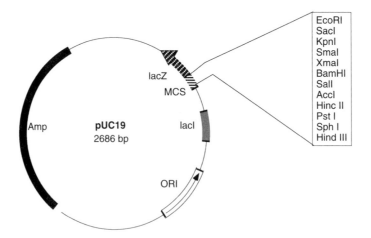

**Figure 6-6. Map of a plasmid.** Amp, β-lactamase; lacI, lac promoter; lacZ, α-fragment of the β-galactosidase; MCS, multiple cloning site with a series of singular cleavage sites; ORI, replication origin site.

The cloning site may lie in a gene, which becomes inactivated if a DNA fragment is cloned into the vector. This permits simple selection of the desired clone after cloning is completed—a useful characteristic because only a small part of the colonies found on the agar plate after cloning contains the desired clone. Most frequently, **blue-white selection** is used. It is based on an interruption of the *lacZ'* gene. *LacZ'* codes for the N-terminal α-fragment of β-galactosidase, which demonstrates no galactosidase activity. If it is combined with the inactive C-terminal ω-fragment, this activity is miraculously reestablished, a process known as α-**complementation**. More information on this subject has been provided by Ullmann and Perrin (1970).

Bacterial colonies in which the *LacZ* gene is destroyed through the insertion of a DNA fragment remain white after incubation with IPTG and X-Gal, whereas clones without any insertion express the α-fragment and turn blue. Instead of X-Gal, you can use Bluo-Gal (5-bromo-3-indolyl-β-D-galactopyranoside), which yields a somewhat darker blue. For the blue or white selection, you distribute 50 µL of IPTG solution (100 mM) and 50 to 100 µL of X-Gal solution (4% in dimethylformamide) onto an LB agar plate before applying the bacteria and incubating overnight. As an alternative, you can add IPTG and X-Gal directly to the bacterial solution before distributing it on the plate.

For those of you who trust the results only if they lie before you in black or white, Sigma offers a dye called S-Gal (3,4-cyclohexenoesculetin-β-D-galactopyranoside) with which lacZ-positive bacterial colonies can be stained black.

The whole process functions only if you also use a bacterial strain that expresses the ω-fragment. The bacteria have the corresponding gene; it is designated *LacZΔM15*. Another problem is that the blue dye first develops fully when the bacterial colonies have a diameter of at least 1 mm. The blue dye is frequently somewhat indefinable, and you may find pale blue colonies or colonies that are white with a blue center. A reason is that the dye is quite weak when the plates are removed from the incubator, and the colonies develop to their full splendor after 1 to 2 hours in the refrigerator. Weak color may also indicate that the insert is too small, and despite successful cloning, only small quantities of functional α-fragments are expressed, or there may be a delay until a fusion protein develops that can fulfill the function of the α-fragment. This phenomenon primarily occurs for inserts less than 1 kb long. It is therefore beneficial to select such colonies.

An interesting system involves the pZErO plasmids from Invitrogen. Their cloning site lies in a killer gene, *CcdB*, whose product poisons the bacterial DNA gyrase (i.e., topoisomerase II), a process that is deadly for the bacterium. The vector also contains some antibiotic resistance. With the antibiotic, the experimenter can select bacteria that contain a plasmid that can only survive in the event that the killer gene has been inactivated, usually through the cloning within a piece of DNA (Table 6-3). The beautiful part of this process is that most clones that develop also contain an insert, sparing the researcher much additional screening work. The mechanisms of action pf CcdB are described elsewhere (Bernard and Couturier, 1992; Bernard et al., 1993).

These are a few examples of the possibilities offered by plasmid vectors. The number of commercially and noncommercially obtainable plasmid vectors cannot be assessed, because there is a practical construct for every application, and many more are developed each year. Before you begin cloning, you should review what the market offers in plasmids for simple cloning, for PCR cloning,

**Table 6-3.** Antibiotics

| Antibiotic | Effect | Working Concentration | Stock Solution |
|---|---|---|---|
| Ampicillin (Amp) | Kills replicating cells | 50–100 µg/mL | 50 mg/mL in H$_2$O |
| Chloramphenicol (Cm) | Bacteriostatic | 20–170 µg/mL | 34 mg/mL in ethanol |
| Kanamycin (Kan) | Bactericidal | 30 µg/mL | 50 mg/mL in H$_2$O |
| Streptomycin (Sm) | Bactericidal | 30 µg/mL | 50 mg/mL in H$_2$O |
| Tetracycline (Tet) | Bacteriostatic | 10 µg/mL in liquid culture 12.5 g/mL in plates | 12.5 mg/mL in ethanol |

or for expression vectors. You may be able to occasionally save yourself work. Larger companies that offer supplies for molecular biologists also have many vectors available. Especially interesting is the catalog from Invitrogen, which has become a specialist in this field.

**Literature**

Bernard P, Couturier M. (1992) Cell killing by the F plasmid CcdB protein involves poisoning of DNA topoisomerase II complexes. J Mol Biol 226:735–745.

Bernard P, Kedzy KE, Van Melderen L, et al. (1993) The F plasmid CcdB protein induces efficient ATP-dependent DNA cleavage by gyrase. J Mol Biol 234:534–541.

Ullmann A, Perrin D. (1970). In: Beckwith J, Zipser D (eds): The Lactose Operon, pp 143–172. New York, Cold Spring Harbor.

## 6.2.2 Phages

Phages are viruses that specifically attack bacteria and are consequently harmless for humans (Figure 6-7). Many clever individuals have written many books about phages, because they are among the earliest "little laboratory animals" used by *Homo laboriensis*. They have also been abused for cloning purposes. However, they have lost some significance, because they are more difficult to deal with than plasmids or cosmids.

The most important phage for molecular biology is phage λ (**lambda**). In its wild-type form, it is a phage with a 48.5-kb, linear, double-stranded DNA genome, which is packaged in a protein envelope. Once constructed, it can automatically infect *E. coli* and replicate within this bacterium. Because there is still a little space within the phage envelope, a 12-kb-long fragment can be inserted into the λ genome (according to the vector used) without disturbing the phage. λ-Derived vectors that function according to this principle are known as **insertion vectors**. Because the λ genome is organized in blocks of functionally related genes and not all genes are necessary for the replication of the phages, it is possible to excise parts of the genome and replace them with foreign DNA. Vectors that function according to this principle are known as **replacement vectors**; they contain 9 to 24 kb of foreign DNA. All together, the λ genome must be 38 to 53 kb long to be able to be packaged. Especially during the early period of cloning, this characteristic was an important reason for the application of λ, because no other vector permitted the insertion of DNA fragments that were so large.

λ-DNA (or what the experimenter has made of it) can be packaged in a functioning phage in vitro. The effectiveness of this process is only about 10%; this means that a maximum of 1 in 10 DNA molecules can be packaged so that a bacterium can be infected with it. This rate is clearly above that for plasmids, for which only 1 in 100 or in 1000 arrives in a bacterium and is replicated there. The packaging is expensive, and experimenters use only λ for the manufacture of DNA banks, such as

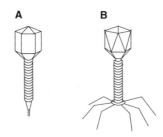

**Figure 6-7. The phage** λ (**A**) and, because of its attractivity, phage T4 (**B**) as they are typically portrayed. Construction of the phages is very complex and demonstrates little variability. The capsid offers space for a restricted quantity of DNA, so that the length of the clonable fragments is highly limited.

for cDNA or whole genomes or for expression banks. The manufacture of such a bank is not entirely trivial, and for someone who is not experienced in this area, it may require several months, and some even give it up, unnerved at the end.

Classic, frequently used λ-vectors are the EMBL3, λ2001, and Charon 4 replacement vectors and the λgt10 and λgt11 insertion vectors. Technically interesting is the λZAP (Stratagene), which has a Bluescript SK⁻ plasmid (or a phagemid) vector in its genome, in whose *multiple cloning site* the foreign DNA is cloned. If you have found the phage of your desire, you isolate the plasma portion of the phage with the aid of helper phages and transfer this into *E. coli*, where it replicates as a normal plasmid. Because plasmids can far more easily be replicated and purified than phages, the experimenter is then out of the woods.

All phage vectors have RNA polymerase promoters and permit in vitro transcription of foreign DNA. Others have a cloning site in an inducible phage gene and make it possible to manufacture a bank of fusion proteins; λ cloning systems can be purchased from manufacturers such as Stratagene, Clontech, and Promega.

**Advantage:** The screening of phage banks has been a standard method for many years and is relatively simple. The procedure is described in Section 7.4.

**Disadvantage:** If you have isolated a clone, it is frequently quite difficult to isolate large quantities of DNA. In practice, many problems are encountered that do not occur with plasmids. There is still no truly rapid, reliable protocol for the production of very clean λ-DNA. The most successful method is to use anion exchange columns. The most frequent problem is that the preparation contains dirt that makes further processing, such as a restriction digestion, difficult or impossible. Even in the replacement vectors, almost two thirds of the DNA is made up of vector sequences. If possible, you should clone the sections that are of interest using plasmids. λZAP banks can save work, because the plasmid portions are cut out in vivo, along with inserted DNA. That process is highly efficient, requires only a relatively few work steps, and lasts only 1 to 2 days.

**M13** is another significant phage, a filamentous phage whose genome can be found as double-stranded or single-stranded DNA. Single-stranded DNA can be sequenced better than double-stranded DNA.

At the end of the 1970s and beginning of the 1980s, a series of M13mp vectors was constructed based on M13, which contained a polylinker for the introduction of DNA fragments and a *lacZ* gene for the blue-white selection of clones with DNA inserts. The double-stranded form can be isolated from infected bacteria and used like plasmids for cloning. Like plasmids in bacteria, the constructs are transformed; instead of bacterial colonies, the experimenter has phage plaques. From the plaques, infected bacteria can be obtained that express phage particles with single-stranded DNA into the media.

Work with M13 phages is relatively troublesome because double-stranded DNA frequently cannot be isolated in large quantities and because portions of the laboriously cloned DNA in the M13 vectors are simply thrown out. A more elegant manner consists of the construction of plasmid vectors with an M13 or an f1 replication origin (**phagemids**), which can be increased in bacteria like normal plasmids. Only after these bacteria are infected through the addition of M13 phages is single-stranded phagemid DNA packaged beside the M13 and then secreted into the medium. The pUC and pBluescript are examples of such phagemids.

## 6.2.3 Cosmids

An interesting development in phage vectors is the cosmids. They are plasmids with typical plasmid characteristics: circular, double-stranded DNA with a replication origin, a selection marker, and a cloning site, which also contain a typical sequence for the λ **cos site**. The λ genome, a linear DNA molecule, has a 12-base-long, single-stranded, complementary overhang at both ends, which emerges during the packaging process through splitting of the cos site. Once inserted into the bacteria, both ends find one another and are ligated by the bacterial enzymes. The circulating λ genome is replicated in the bacterium, and because the responsible DNA polymerase continues in a circle, a very long copy emerges that consists of many λ genomes that have been strung together. λ-Specific enzymes cleave this concatemer at the cos sites and generate handy, linear λ genomes with single-stranded overhangs that can be packaged in phage particles.

This process functions somewhat differently with cosmids. The cosmid vector is digested and ligated with the desired DNA. The wild mixture that emerges is cut at the cos sites with the help of packaging extracts and packaged into phage envelopes; everything between 40 and 50 kb long that contains two *cos* ends is packaged. These particles are as infectious as phages, although as long as the DNA is inserted and ligated into a bacterium, it behaves like a plasmid. Based on the selection according to size that proceeds with the packaging process and considering the small size of the vector portions (5 to 6 kb), the DNA fragments cloned in this way are larger than everything that can be achieved using phage vectors. Cosmids are suitable for constructing genome banks because fewer clones are required to cover an entire genome.

**Disadvantage:** With cosmids, the dimensions are such that the standard molecular biologic protocols are no longer valid. The manufacture of cosmid banks is therefore not trivial. If you wish to try it, you should consult people who have experience. Another problem with cosmids is that clones of this size can lead to a recombination within the sequences that have been cloned.

**Advantage:** The DNA is not increased in bacteria as with phages. After the desired clone is identified, production of DNA and storage of the clone are rather simple. The size of the cloned DNA fragments makes an analysis possible without too many difficulties.

**Literature**

Gibson TJ, Coulson AR, Sulston JE, Little PF. (1987) Lorist 2, a cosmid with transcriptional terminators insulating vector genes from interference by promoters within the insert: Effect on DNA yield and cloned insert frequency. Genes 53:275–281.

Ish-Horowitz D, Burke JF. (1981) Rapidly and efficient cosmid cloning. Nucleic Acids Res 9:2989–2998.

## 6.2.4 P1 Artificial Chromosomes and Bacterial Artificial Chromosomes

The development of P1 artificial chromosomes (PACs) and bacterial artificial chromosomes (BACs) permitted cloning of far larger DNA fragments. BACs consist of vectors that are made up of fragments up to 300 kb long. They are based on F factors, the naturally occurring sex factor plasmids of *E. coli*. The basic elements of the PAC vectors are derived from P1, a temperate phage with a genome that is approximately 100 kb long. Both cloning systems are relatively new, experience is limited, and the problems have not yet been solved completely.

**Advantage:** BACs and PACs are circular molecules that are replicated in bacteria, and they are as easy to prepare as plasmids. BAC and PAC banks can be analyzed just about as easily as cosmid banks, because the bacteria including BAC and PAC bred on agar plates can be screened by means of colony lift-hybridization. The transformation is simpler than that seen with YACs (discussed later).

**Disadvantage:** The production of corresponding banks is not a trivial feature.

**Literature**

Ashworth LK, Algeria-Hartman M, Burgin M, et al. (1995) Assembly of high-resolution bacterial artificial chromosome, Pl-derived artificial chromosome, and cosmid contigs. Anal Biochem 224:564–571.

Ota T, Amemiya CT. (1996) A nonradioactive method for improved restriction analysis and fingerprinting of large P1 artificial chromosome clones. Genet Anal 12:173–178.

Shizuya H, Biren B, Kim UJ, et al. (1992) Cloning and stable maintenance of 300-kb-pair fragments of human DNA in *Escherichia coli* using an F factor-based vector. Proc Natl Acad Sci U S A 89:8794.

## 6.2.5 Yeast Artificial Chromosomes

Among the mega-cloning vectors, the yeast artificial chromosomes (YACs) are the oldest. Their greatest advantage is that they allow the cloning of extremely long DNA fragments (i.e., length of several hundred kilobases to more than 1 Mb), producing the largest clones achieved so far. For this reason, YACs are extremely popular for those trying to analyze entire genomes.

The principle is similar to that for plasmids or cosmids. The experimenter introduces some typical elements that are necessary for correct replication. In the case of YACs, the replication origins are the centromeres and telomeres of the yeast chromosomes, which must be inserted into the DNA being cloned. The constructs can be transformed in yeast spheroblasts and are then replicated there. In contrast to the vectors described earlier, YACs are not circular; they are made of linear DNA.

**Disadvantages:** The cloning of YACs is too complicated to be carried out by a lone researcher. Anyone who requires them should take advantage of a laboratory that specializes in these procedures, but this should not be necessary, because many YAC banks are available and accessible to researchers. Entire chromosomes have been cloned in YACs and mapped there.

A large problem is the modification of DNA in the yeast cells. This problem is generally observed with especially large DNA clones, but is more distinctive in YACs. Two different DNA fragments are frequently seen in the same YAC (i.e., chimeric YAC clones), which can complicate the mapping procedure considerably. Early YAC banks contain 40% to 50% chimeric clones. Newer banks are much better, although only through fluorescence in situ hybridization (FISH) of chromosomes (see Chapter 9, Section 9.1.4), in which the YAC to be examined is used as a probe, can you ascertain whether this clone involves a chimera or several different chromosomes are labeled. Moreover, recombinations can occur within a YAC, during which parts are exchanged or eliminated.

The use of YACs is not completely free of problems. Even the preparation is not simple, because the chromosomes of the yeast must be separated from the artificial chromosomes. This is based on the different sizes of the chromosome, and the separation requires the use of pulse-field gel electrophoresis (see Chapter 3, Section 3.2.5) if you require a YAC. If the YAC chromosome is by chance equal in size to that of the yeast chromosome, it can then be identified only with the aid of hybridization. The YAC must be isolated from the agarose, and the quantity of DNA that can be won through purification is therefore far smaller than that with other vectors. Because the YACs are so large, special protocols are needed.

The screening of YAC banks demonstrates its own difficulties. Because yeast colonies cannot be transferred from the agar plate to a membrane like bacterial colonies, to then undergo a screening, the yeast clone must then be isolated and analyzed piece by piece–a process that cannot be carried out by a single individual. The experimenter is therefore well advised to leave the preparatory work to others and to work only with defined YAC clones. The most clever method involves decomposing the clone into smaller fragments and then cloning these subsequently in another vector that can then be handled more easily.

**Advantage:** The work with YACs has been carried out for some time, and experimenters can take advantage of more values from past experiences than are available for other mega-cloning vectors.

### Literature

Burke DT, Carle GF, Olson MV. (1987) Cloning of large segments of exogenous DNA into yeast by means of artificial chromosome vectors. Science 236:806–812.

Imai T, Olson MV. (1990) Second-generation approach to the construction of yeast artificial-chromosome libraries. Genomics 8:297–903.

# 6.3 Choosing the Bacteria

Not all bacteria are equivalent. Vast numbers of different strains have been investigated, with well-described characteristics. The vectors and bacterial strains form a team, which gets along more or less well with each of the parts. The one plasmid appears to prefer the one strain, whereas the other prefers another. The companies that manufacture and sell the plasmids also provide information concerning the strains that can be used. This does not mean that you can use no other bacteria, but rather that those indicated are more likely to function.

For work with phages, you need a bacterial strain that permits the use of specific phages. **M13**, for example, requires a strain with an $F'$ episome. Other phages require other bacterial strains. You should direct inquiries concerning phages to those from whom you have acquired them.

To make use of the **blue or white selection** in the course of cloning (see Section 6.2.1), you need a strain that complements the *LacZ* gene in the vector, or it will not result in a colony that stains blue. Corresponding strains have a ω-fragment of the β-galactosidase and are known as ZΔ M15 in the genotype description.

The bacteria can be differentiated in their **methylation** and **restriction systems**. The former can cause difficulties when cutting plasmid DNA preparations using methylation-sensitive restriction enzymes. Most bacterial strains are Dam and Dcm positive, although there are strains that are deficient in these methylases. If necessary, you can transform a DNA clone in such a bacterial strain and repeat the plasmid preparation. The bacterial restriction system can lead to difficulties in cloning, because DNA that has become amplified by means of PCR, for instance, is not methylated and may be recognized and disassembled by the bacteria as being a foreign substance. Mammalian DNA demonstrates a different methylation pattern than the bacterial DNA and can be recognized as being foreign. The most favorable strains are deficient for the corresponding restriction systems, such as those that have an $McrA^-$ $McrBC^-$ $Mrr^-$ $HsdR$ ($r_K^-$ $m_K^-$) phenotype.

A list of bacterial strains and their characteristics is found in the *Current Protocols in Molecular Biology*, and a review of the restriction systems of these strains is printed in the appendix of the New England Biolabs catalog.

# 6.4 Construction of Competent Cells and Their Transformation

Inserting DNA into bacteria is an everyday job for the experienced cloner. If the method functions for the first time, it is a trivial affair, which becomes exciting if the competent bacteria are used up and the whole laboratory team must pray that the new ones will be just as good as the old. If the transformations then work only moderately, all hell breaks loose. After all, this step is just as important for the success of the entire cloning process as the previous ligation. In the event of a difficult ligation, an efficient transformation is the only way to save the situation and to at least obtain a handful of positive clones.

The classic method of making a bacteria competent to transformation functions with the aid of calcium chloride. There is an entire series of additional protocols available for making bacteria competent with the aid of specific chemicals and many more variants that frequently result in a higher **competency** (i.e., produce more transformed bacteria). A quantum leap in the transformation of bacteria is achieved with electroporation. If the concentrations of chemically competent bacteria are in the area of $10^6$ to $10^8$ colonies per 1 μg of test DNA (the latter figure is quite a good value), you can by means of electroporation, even with the worst of bacterial preparations, obtain a competency of $10^7$ and even values of $10^{10}$ with good preparations.

If possible, you should purchase an electroporation device. The cost depends on the particular equipment associated with the individual device. An instrument that is suited for the electroporation of bacteria can be purchased for about $2000, and an instrument that is suitable for bacteria and for eukaryotic cells costs about $8000.

The protocols for the manufacture of competent bacteria are always based on cultures in LB medium. The amount of bacteria attained can be increased through the use of other, richer media, although they should first be tested carefully, because not every medium functions equally well with each method.

With all protocols, you can preserve the competent bacteria by aliquoting them (aliquots of 100 to 400 µL are usually good), shock-freezing them in liquid nitrogen, and storing them at $-70°C$ ($-94°F$). Afterward, however, the competency is usually poorer by a factor of 10.

## 6.4.1 Calcium Chloride Method

The calcium chloride method is the forbearer of transformation protocols. You inoculate 400 mL of an LB medium with 1 mL of an overnight culture and incubate it at 37°C (99°F) on a platform shaker at 200 to 300 rpm. The volume of the Erlenmeyer flask should be at least fivefold the volume of the medium, and it should have a spout extending from within to provide good ventilation of the culture. The culture is allowed to grow until it has an $OD_{595nm}$ of 0.6. That takes about 4 hours, although the times can vary extensively depending on the particular test conditions. Higher OD values lead to lower competency. All of the following steps are carried out on ice or at 4°C (39°F). The culture is cooled on ice and then centrifuged at 3000 $g$ (the time of the centrifugation is lacking in the original article on this topic), and the pellet is resuspended in 100 mL of $CaCl_2$ solution (60 mM $CaCl_2$/15% glycerin/10 mM PIPES, pH 7.0; sterilized by autoclaving or with a filter). After that, it is centrifuged, resuspended, and centrifuged again. The pellet is resuspended in 10 mL of $CaCl_2$ solution and can then be used. The suspension can be kept on ice for several days, and the competency reaches its maximum after approximately 12 to 24 hours, depending on the particular bacterial strain used.

The transformation proceeds as follows: 100 µL of competent bacteria are mixed with the DNA to be transformed (usually with 5 µL of a 20-µL ligation preparation or 1 ng of plasmid DNA) and then placed it on ice for 10 to 30 minutes. After that, the bacteria are heated to 42°C (108°F) for 2 minutes (i.e., heat shocked), 1 mL of LB medium is added, and it is incubated for 30 to 60 minutes at 37°C (99°F). The bacteria are cultured on a selective medium (e.g., LB-Amp plates) and incubated overnight at 37°C (99°F).

**Suggestions:** Heat shocking is a mysterious matter. No one knows why it functions, but everyone confirms that it must be carried out for exactly 2 minutes, exactly 60 seconds, or exactly 45 seconds at 37°C (99°F). Otherwise, it will not function. To achieve the best results, you need to determine the optimal conditions, because success depends on the medium, the bacterial strain, the DNA, and the air pressure. If you wish to perform this more easily and do not need to have excessively high competency of the bacteria, you can dispense heat shocking and culture the bacteria directly using cooling with ice after 30 minutes of incubation.

Chemically competent bacteria are sensitive to quantities of DNA that are too large. They are typical examples that a lot does not always help a lot. If you find smaller yields than expected, make another attempt with less DNA.

It has been reported that the competency of the bacteria becomes better if they reproduce at lower temperatures. They require about 24 hours at 18°C (64°F).

## 6.4.2 Rubidium Chloride Method

The rubidium chloride method is a variant of the calcium chloride method that offers somewhat higher competency. Inoculate and incubate 500 mL of LB medium as with the calcium chloride method, and pellet the bacteria as soon as the culture demonstrates an $OD_{595nm}$ of 0.4 to 0.7. The pellet is resuspended in 150 mL of TFB I (10 mM $CaCl_2$/15% [v/v] glycerin/30 mM potassium acetate, pH 5.8/100 mM rubidium chloride/50 mM manganese chloride), put on ice for 15 minutes, centrifuged again, and then resuspended in 20 mL of TFB II (10 mM MOPS, pH 7.0/10 mM $RbCl_2$/75 mM $CaCl_2$/15% glycerin). The bacteria are fresh or shock-frozen in liquid nitrogen in 400 μL aliquots, which are stored at −70°C (−94°F). The transformation is carried out as for the $CaCl_2$ cells.

## 6.4.3 TSS Method

Because the true experimenter is actually a lazy individual, someone has found a single-step method for the generation of competent cells. To carry out this simple method, inoculate and incubate 100 mL of LB medium as described in the calcium chloride method. As soon as the culture demonstrates an $OD_{595nm}$ of 0.4 to 0.7, the bacteria are pelleted and resuspended in 10 mL of 1× TSS (LB medium, pH 6.5/10% [w/v] PEG 8000/5% DMSO/20 to 50 mM $MgSO_4$).

There is a method that is even easier. Mix the culture with an equal volume of ice cold 2× TSS (LB medium, pH 6.5 with 20% [w/v] PEG 8000/10% DMSO/40 to 100 mM $MgSO_4$) and put it on ice. In both cases, the bacteria can be used directly for the transformation or be aliquoted and stored in a frozen state.

For the transformation, mix 100 μL of competent bacteria with the DNA (0.1 to 10 ng), incubate the mixture for 5 to 60 minutes at 4°C (39°F), add 900 μL of LB medium with 20 mM glucose, and incubate for 30 to 60 minutes at 37°C (99°F). After that, it is cultured and then incubated overnight.

The method should make a competency of up to $5 \times 10^7$ colonies per μg DNA possible. I have not been able to achieve this quantity successfully, but I would be very happy to receive any reports from others who might have been successful.

**Literature**
Chung CT, Niemela SL, Miller RH. (1989) One-step preparation of competent *Escherichia coli*: Transformation and storage of bacterial cells in the same solution. Proc Natl Acad Sci U S A 86:2172–2175.

## 6.4.4 Electroporation

Electroporation is the path taken by royalty. Preparation of the bacteria is very simple, and the competency achieved is very high. For ordinary transformations, many individuals struggle more with the problem that their yields are too high and that they develop no true colonies, but actual bacterial lawns. Another positive aspect is the fact that the bacteria are substantially more tolerant to large DNA quantities with this method than, for instance, with the calcium chloride method, which delivers almost no transformed cells from 1 μg of DNA. Electroporation devices can be purchased from several manufacturers, including Stratagene, BTX, and BioRad.

As in the calcium chloride method, you inoculate and incubate 500 mL of LB medium, and pellet the bacteria as soon as the culture demonstrates an $OD_{595nm}$ of 0.4 to 0.7. The pellets are resuspended carefully in 100 mL of cold $H_2O$, and the bacteria are then centrifuged immediately at a temperature of 4°C (39°F). This is repeated three times, and the pellets are then resuspended in 2–5 mL of a cold, 10% (v/v) glycerin solution. The bacteria can be preserved in this manner for a number of hours on ice, or you can freeze 100- to 200-μL aliquots in liquid nitrogen and store them at −70° C (−94°F).

For the transformation, you mix 50 μL of competent bacteria with DNA, transfer the mixture into a 2-mm electroporation cuvette with a 2-mm interelectrode distance, and proceed according to the directions provided by the manufacturer of the device. The settings are usually 2.5 kV, 25 μF, and 200 to 400 Ω; 200 to 1000 μL of SOC medium are added to the deeply shocked bacteria in the cuvette directly after the pulse (see Appendix 1, Standard Solutions). Everything is then transferred to a 1.5 mL vessel and incubated for 30 minutes at 37°C (99°F). After that, a part is cultured on selective LB agar plates and is incubated overnight. The quantity of bacteria to be cultured is largely a matter of experience and depends on the DNA used and the local influences. Usually, 10% of the preparation will provide far more colonies than you desire.

Because fresh cells clearly provide better results than the deep-frozen cells, I provide a fast **variety for bacterial purification** that only takes 30 minutes. With the method, 100 mL of TB medium are inoculated as described for the calcium chloride method and incubated for 5 to 6 hours until the culture demonstrates an $OD_{595nm}$ of 2. Because the bacteria essentially feel better in TB than in LB, you can allow them to grow much more densely without losing any competency. The culture is then transferred into two normal 50-mL polypropylene vessels with screw-on lids (e.g., from Falcon) and centrifuged in a refrigerated centrifuge with a swing-out rotor at 6000 $g$ for 5 minutes and at 4°C (39°F). The pellet is resuspended carefully on the vortexing device in a small quantity of fluid (approximately 1 mL), cold water is added to reach a quantity of 50 mL, and this solution is then centrifuged again. After washing the bacteria three to five times in this manner, resuspend them in 2 mL of 10% (v/v) glycerin and store them on ice until they are to be used.

**Suggestions:** Decisive for success in this procedure is the **salt content of the bacteria**. If it is too high, the transformation preparation in the cuvette will explode. In case of doubt, increase the number of washing steps for manufacturing competent cells. Decreased salt concentration in the bacteria means that less firm pellets precipitate. Although this facilitates resuspension, it is somewhat troublesome when removing the supernatant. It is better to plan for an additional washing step than lose one half of the pellet each time with the supernatant.

If the preparation contains too much salt, it will explode during the electroporation. This is a normal procedure in the life of a researcher, but it is not desirable. If it occurs frequently despite properly washing the bacteria, the DNA solution probably has too high a concentration of salt; this problem will most likely affect the ligation preparation. You can reduce the quantity of the DNA solution in the preparation with 2 μL of a ligation preparation, or reduce the salt concentration by precipitating the DNA with ethanol and washing well with 70% ethanol.

An unusual, but easily reproducible method is **precipitation with butanol**. Add a 5- to 10-fold quantity of *n*-butanol to the DNA solution, mix it until the watery phase has disappeared, and centrifuge the preparation at room temperature. The supernatant is removed, and the pellet, which is frequently not visible, is carefully washed with 70% ethanol. The DNA is then dried and dissolved in $H_2O$.

The density of the bacteria when you begin with the washing is primarily based on the bacterial medium used. LB, although it has become the standard medium throughout the world, is a rather mediocre medium in which the bacteria feel uncomfortable at an early point. The duplication time of the bacteria in an LB culture at an OD of 0.2 is 55 minutes, whereas it is about 32 minutes (see Fig. 2.7) in a *Terrific Broth* (TB) culture, because the bacteria seem to feel substantially more comfortable in the TB medium. Generally, bacteria can be cultured much more densely in a medium that is richer than LB medium without any loss in competency.

**Electroporation cuvettes** can be reused, news that will make those in the poorer laboratories very happy. At an individual price of more than $3.25, quite a bit can be saved over the course of an entire year. Rinse the cuvette carefully a number of times with distilled water after each use and tap out the remaining fluid. The cuvette can be reused immediately or stored in a dry condition. The emphasis is on the term *dry*, because the electrodes are made of aluminum, which corrodes easily when stored under damp conditions, a situation that would then make the cuvette useless. Well cared for, such a cuvette can perform over the course of several months, until an impure ligation preparation puts an end to it. The problem of such a **carry-over** is relatively small, although as a precaution when performing difficult cloning procedures, you should use a fresh cuvette, or you may discover after 3 days of work that the three clones on the plate are from a previous transformation.

All protocols are in agreement that the survival time of freshly transformed bacteria is poorer, the longer you must provide it with medium. After 1 minute, more than 90% of the bacteria are dead. Instead of SOC, you can use LB or other media, although the survival time is generally even worse.

**Literature**

Dower WJ, Miller JF, Ragsdale CW. (1988) High efficiency transformation of *E. coli* by high voltage electroporation. Nucleic Acids Res 16:6127–6145.

# 6.4.5 Testing the Competency of Bacteria

The quality of the bacterial preparations can be quite varied as related to the use of the various methods. It is advisable to test the competency of the bacteria. This can save you the usual doubts when you find an empty plate and are unaware of whether the ligation did not function or the bacteria did not grow.

The competency of a bacterial stock is always indicated in number of colonies per μg of DNA. Although there are no conventions, a vector DNA such as pBR322 or pUCl8 typically is used for this test. It allows you to make comparisons between different laboratories. However, transformation efficiencies for large DNA pieces (>10kb) are worse than for smaller fragments and worse for litigation preparations than for purified plasmid DNA.

Stratagene reports of large differences in the effectiveness of different strains (see their company newspaper, *Strategies* 10:37). Accordingly, the effectiveness of XL2-Blue (in this case, a 25-kb plasmid) should be eight times larger than that for DH10B, and that of XL10-Gold should be 80 times more efficient. This is especially interesting for the cloning of large fragments, such as for DNA banks.

Because no one is able to count out $10^8$ colonies on a single plate and because the yield decreases with many methods if too much DNA is inserted, only small quantities of DNA are used for the test, in respect to the competency expected. For an expected competency of $10^7$ colonies/μg of DNA, 10 pg of DNA can supply about 100 colonies, an amount that can still be counted out on a Petri plate. For the test, you should use freshly thinned DNA, because DNA is not particularly stabile at low concentrations (<10 ng/μL) and appears to have a smaller competency. The DNA quantity used should not be too small either; for electroporation, the effectivity decreases substantially for quantities less than 5 pg.

If you wish to freeze and store the bacteria, it is better to carry out the test on a deeply frozen aliquot to avoid later disappointments. The competency of fresh bacteria is higher by about 10-fold. Anyone requiring bacteria with an especially high competency can obtain them by purchasing them, such as from Stratagene or Promega. The time required to produce extremely good cells is too long for the average experimenter.

**Literature**

Hanahan D. (1983) Studies on transformation of *Escherichia coli* with plasmids. J Mol Biol 166:557–580.

# 6.5 Problems Associated with Cloning

*Und hat mit diesem kindisch-tollen Ding*
*Der Klugerfahrne sich beschäftigt,*
*So ist fürwahr die Torheit nicht gering,*
*Die seiner sich am Schluß bemächtigt.*

*And if the old campaigner thus could wholly*
*Bemuse himself with such a childish whim,*
*It clearly was no common fit of folly*
*Which in the end o'ermastered him.*

Anything can go wrong in cloning. No colonies may be obtained, or none of the clones contains the desired insert. Sometimes, the experimenter even finds out why the process has not functioned:

- The **ligase**, for instance, can give up the ghost. This possibility can be examined with the aid of a simple test by ligating 1 μg of digested DNA (e.g., molecular-weight marker DNA without any blue marker) for 30 minutes. The largest part of the DNA should appear in the agarose gel as a high-molecular-mass smear.
- Less evident is the possibility that **ATP** may be lacking. Some manufacturers have the necessary ATP in the 10× ligation buffer, but others do not, so you should pay attention to the buffer you use. If the buffer is old or has been stored at 4°C (39°F) for a longer period, the ATP in the buffer might have been degraded. In this case, the addition of some fresh ATP can help (final concentration of 5 mM).
- The problem may be the **DNA**. Some purification methods, such as different glass milk kits, can disturb the ligation. Dispense with the purification and use the digested, only heat-inactivated DNA directly for the ligation. That always works, although it does demonstrate a small flaw—a large proportion of the clones does not contain the desired insert. You must then solve this problem with a suitable screening method or become industrious and perform a very great number of mini-preparations.
- **Large fragments** (>3 to 5 kb) can fundamentally be ligated more poorly in vectors than the smaller fragments. They also are taken up by the bacteria less efficiently, although this depends on the bacterial strain employed, because some accept large fragments better than others (see Section 6.4).
- Sometimes the fragment you wish to clone contains a sequence that is recognized by the **bacterial restriction system** and then cleaved. This is unusual, so you may be happy to forget this possibility. Characteristically, the clone reproduces splendidly if you have, after many futile transformations, finally been able to "force" this development. You can frequently overcome the problem by using a Dam⁻ Dcm⁻ bacterial strain with a defective methylation system for cloning (see Chapter 3, Section 3.1.4).
- The cloned DNA may code for a product that is toxic to the bacterium. According to the vector used, it may have more or less extensive expression, which can strongly hinder the growth of the bacteria. In this case, it helps to consider the cloning strategy anew.

Experience shows that cloning functions or it does not. When it functions, you can drown in clones and suffer more from the problem that this amount can be reduced to a sensible degree. If you obtain only 10 clones, there is a good chance that you will not find the desired insert. If you have a large number of clones and none of them has the desired insert, it may be better to rethink your cloning concept.

# 6.6 Storage of Clones

After identifying your clone, you would naturally like to keep it. In **liquid culture**, the bacteria unfortunately do not survive longer than a week, a situation that is unsatisfactory for storage. Yeast frequently develops after a few days and then spreads throughout the culture. On an **agar plate**, the bacteria can still survive for up to 4 weeks, and this works best when you seal the plate with parafilm to prevent drying. This method is quite practical if you want to wait until you have found the desired clone and characterized it, such as after a transformation and before throwing the plate away. For longer-term storage, however, this is not of any use.

Storage at room temperature in so-called agar vials is better. This approach deals with agar stabs (1.5 to 5 mL), which are made of glass or plastic with a screw-on cap (e.g., Cryotube vials, Nunc), which should be filled to two-thirds full with sterile stab agar (LB medium with 0.6 [w/v] agar; add 10 mg of cysteine per liter to increase the survival time of the bacteria). The vessels can be laid out in reserve, and if required, you can inoculate one with bacteria by pulling an oblong, sterile object (e.g., toothpick, inoculating loop) through the culture or the agar plate containing the desired bacteria and by pushing it repeatedly and heartily into the stab agar, without causing it to be plowed too extensively. The *stab* is incubated for about 8 hours at about 37°C (99°F) and stored tightly closed in a dark, dry place at room temperature. The bacteria should be able to survive in this manner for years, although 3 to 6 months is more realistic. To revive the bacteria, stick a sterile inoculating loop into the stab agar, smear it onto an LB agar plate, and incubate this overnight at 37°C (99°F).

To preserve bacteria for a very long time, make **glycerin** or **DMSO stocks** and store these at −70°C (−94°F). To this, add 1 mL of a fresh culture with 1 mL of glycerin solution (65% [v/v] glycerin/0.1 M $MgSO_4$/25 mM Tris HCl, pH 8) or 1 mL of DMSO solution (7% [v/v] dimethyl sulfoxide), transfer the mixture into a suitable 2-mL vessel (preferably with a screw cap, because normal tubes occasionally pop open and distribute their contents when they become warmer), and store this at −20°C (−4°F) or −70°C (−94°F); the latter temperature increases the survival time. The most important advantage of the DMSO solution is that, in contrast to the glycerin, it can easily be pipetted. However, it stinks a bit, at least for that part of the human population who can smell DMSO. To rouse the bacteria to life again, scratch the surface with a sterile but stable object (e.g., a toothpick), and use this to inoculate an LB agar plate. The bacteria then grow overnight at 37°C (99°F).

**Freeze-drying** and its low-cost version, **vacuum-drying**, methods at best demonstrate a shadowy existence in normal cloning laboratories because of the great expense of the apparatus and its only slight flexibility. For this procedure, the bacteria are frozen, the water is removed by means of a vacuum, and all of the remnants are collected in glass tubules in an anaerobic state. Depending on the mode of storage and the specific microorganisms, the survival rate can be up to 30 years. The opening of the ampule, however, demands the breeding of the bacteria, because subsets cannot be removed. More information can be found in the technical literature, such as a book by Bast (2001).

Probably the most reliable method is storage of a **DNA stock** at −20°C (−4°F). As long as the DNA is free from nucleases, it will not degrade, even if a laboratory MCA (maximum credible accident) were to set the refrigerator into checkmate. The concentration should, as far as it is possible, be more than 0.1μg/μL. At lower concentrations, the DNA can break down, although the reasons for this are not clear. Revival is somewhat more laborious than with bacterial stocks, because you must first transform the DNA into bacteria (see Section 6.4).

You can also make a attempt with a different method that I have been testing successfully for a number of months and would like to hereby call **pickled DNA**. A stately quantity of DNA (e.g., 10 μg) is mixed with a fivefold quantity of a buffer containing chaotropic salts (6 M guanidium thiocyanate or sodium perchlorate), as is used in glass milk purification. The mixture is stabile at 4°C (39°F) and presumably also at room temperature for a number of months. If you need the DNA, you must carry out a glass milk purification (see Chapter 3, Section 3.2.2). The disadvantage of a larger amount of work is also an advantage, because the temptation to reach for the DNA stock when you need some DNA for a digestion on the "spur of the moment" is very small. Besides, you can save some precious space in the refrigerator.

The most modern mini-preparation kits are based on an alkaline lysis of bacteria followed by a glass milk purification, in which the glass milk, for the sake of easier operation, has been replaced by a silica membrane in a minifilter. Because about 90% of the DNA can be removed from the membrane during elution, you can save the minifilter (dry and at room temperature), and in the event of an urgent situation, perform a second elution. The quantity of DNA obtained in this way should be sufficient for a renewed transformation.

Another interesting idea is to use this method for the storage of DNA (labeled the "Lewinsky method" by an American cynic). One drop of the DNA solution is placed on a clean leaf of filter paper, the site of application is marked with a circle, and it is left to dry and then stored. To reactivate it, you must cut out the site, eluate in $H_2O$, and then perform a transformation with the solution.

The more "noble" variant is offered by Whatman (CloneSaver Card) at a more or less reasonable price. It consists of a small "booklet" with membranes that contains the correct leaflet for 96 individual tests. The pink colored membrane demonstrates a change in color at the site that has come into contact with liquid, a great aid for those who cannot remember where they have applied the last test. The special feature of this membrane is that it is coated with a lysogenic substance; if you apply bacteria or other cells to this surface, according to the manufacturer, the cells are degraded, the proteins are inactivated, and the DNA becomes bound to the membrane. This also functions with simple DNA-containing solutions. The DNA is reactivated by punching out a small bit of membrane, which can then be used for a transformation or for PCRs.

**Literature**

Bast E. (2001) Mikrobiologische Methoden: Eine Einführung in grundlegende Arbeitstechniken, 2nd ed. Heidelberg, Spektrum Akademischer Verlag.

# 7 Hybridization: How to Track Down DNA

Tracking down DNA is primarily known as *hybridization*. The methods for transferring DNA and RNA to a membrane have been explained in Chapter 3, Section 3.3. This chapter covers what can be done with these substances. There are essentially six different areas for the use of hybridization:

- With a **Southern blot**, you have separated restriction-digested DNA in the gel and want to show which of the many fragments contains the sequence you are looking for.
- The **Northern blot** contains RNA from different tissues or cells that have been differentiated according to size and reveals how large the corresponding mRNA is if it is all expressed in the tissue.
- With the **dot blot**, you can omit the separation according to the size of the fragments, because you are interested in determining only how much of the fragment being sought is available in the DNA or RNA probes.
- **DNA banks** are made up of a mixture of different clones, from which the desired clone must be differentiated. The only way to do so is to plate out each of the clones, transfer them to membranes, perform a hybridization, isolate them, and plate them out again. After many such cycles, one or several specific clones are obtained.
- Other methods include **in situ hybridization** and **fluorescence in situ hybridization**, which are described in more detail **in** Chapter 9, Section 9.1. In one case, you can identify in the tissue whether and where a certain mRNA has been expressed, whereas in the other case, you can determine on which chromosome and at which site of the gene an alteration is hidden.

Until about 10 years ago, hybridization was the most sensitive method to provide evidence of DNA or RNA. Since then, polymerase chain reaction (PCR) has taken over in all areas in which it is to be determined *whether* a certain nucleic acid is present. The intensity of the hybridization provides additional information concerning *where* the nucleic acid is located.

The procedure is the same for all types of hybridization.: You produce a labeled probe, hybridize a membrane with this (or whatever else you would like to hybridize), wash, and then demonstrate what has remained hanging on the probe.

## 7.1 Production of Probes

In the manufacture of probes, you must be fully aware of the purpose for which the probe is to be employed. You can choose among **oligonucleotide, DNA**, and **RNA probes**. The RNA probes are hybridized best, because the stability of RNA-RNA hybrids is highest, followed by RNA-DNA hybrids. The binding of DNA-DNA hybrids is weakest, and they are the most difficult to produce, because they can be obtained only by means of in vitro transcription (see Chapter 5, Section 5.4). Consequently, most molecular biologists use DNA probes.

The second decisive question is about the type of labeling. **Radioactively labeled probes** are the classic method, and much experience has been collected, facilitating the matter, because almost every problem has been encountered somewhere by somebody, who can offer assistance. The sensitivity of radioactive evidence is extremely high, and its verification is unusually direct. After hybridization and washing, you must apply a film or place the preparation into a counter for measurement. That makes quantification very simple and quite reliable. Detrimental factors include the radiation (although the danger of the quantities of radioactivity that you generally encounter in the laboratory are usually overestimated, because the monitors for contamination always clatter and whimper in such a manner as to appear quite fearful), the short half-lives (of $^{32}$P or $^{35}$S), and the longer half-lives (e.g., of $^3$H or $^{14}$C). Although that information may sound contradictory, the probes are quite short-lived in one case, but you have problems with the disposal of the wastes in the other. You also require a permit from the authorities for carrying out procedures with radioactive substances, an administrative action that can normally be done only by an entire institute.

For a number of years, **nonradioactive methods of verification** have increasingly been observed on the market, and they offer a good alternative to radioactive probes. Working with these substances is not very dangerous, even if the makers warn about the possible dangers in the event that you should consider eating this stuff, and labeling can be carried out on the job without difficulties. If you have worked in a correctly organized isotope laboratory and are familiar with the work required to cleanly make use of such radioactive agents, you will presumably consider this to be the most important positive point. The other, enormous, advantage is the long half-life of nonradioactive probes. Whereas $^{32}$P has a half-life of 14 days, and the $^{32}$P-labeled probes cannot be used after 2 to 4 weeks, nonradioactive-labeled probes can be preserved in the refrigerator for years. The probes can also be used repeatedly in some cases. Consequently, anyone who performs the same hybridization procedure repeatedly needs only to produce a larger quantity of the labeled preparation every leap year. Some time is needed to establish the system so that the sensitivity remains high while the radioactive methods and the background remain low. Because most methods used are indirect and require incubation with antibodies and enzymatic evidence, quantification is more problematic than with the use of radioactivity.

Ultimately, you must select a particular **labeling method** (Table 7-1). Not every method is suitable for every kind of probe, and the application of the probe at a later point in time also plays an important role. The following section offers an overview of available labeling methods.

## 7.1.1 Methods for the Production of Labeled Probes

### Nick Translation

*Nick translation* is the oldest of the methods presented, although it still continues to be employed quite frequently. The template DNA is first digested with DNase I in the presence of $Mg^{2+}$ ions. Under these conditions, the enzyme cuts only one of the two strands, so that the double strand

**Table 7-1.** Labeling Methods for Particular Probes

| Method | Oligonucleotides | dsDNA | ssDNA | RNA |
|---|---|---|---|---|
| 5′ End labeling | +*† | +* | +* | +* |
| 3′ End labeling | + | (+) | + | − |
| Nick translation | − | + | − | − |
| Random labeling | − | + | + | − |
| Fill-in | − | + | − | − |
| Polymerase chain reaction | − | + | (+) | − |
| Photo biotinylation | − | +† | +† | +† |

* Most methods are suitable for radioactive and nonradioactive labeling. The exceptions are marked with an asterisk (only radioactive) or single dagger (only nonradioactive).
† Oligonucleotides that can be ordered with 5′ end labeling.

maintains its structure. In a second step, the template is processed with DNA polymerase I in the presence of nucleotides, one of which should be labeled. The polymerase recognizes the *nick*, which has been brought about by the DNase I, and prolongs the free 3' end, while the 5' end is degraded simultaneously. The old strand is replaced by a new one that is labeled. The method functions with template quantities of 20 ng to 1 μg.

The trick involves choosing the correct DNase I concentration. If it is too high, the template DNA is degraded into small fragments, and if it is low, places are cut here and there, and the specific activity is quite low. A good orientation value is approximately 20 pg of DNase I in 0.5 μg of DNA. After reaching the optimal concentration, fragments of 400 to 800 bp can be obtained with high specific activity. Through subsequent ligation, the fragment length can be increased, and the specific activity can be increased even further by including several labeled nucleotides. Nonradioactive labels are possible by using biotin- or digoxigenin-labeled dNTPs.

DNase I can be obtained from any company that has enzymes in their list of products. Some of them even offer the product as a kit, such as that by Promega.

### Literature

Kelly RB, Cozzarelli NR, Deutscher MP, et al. (1970) Enzymatic synthesis of deoxyribonucleic acid. XXXII. Replication of duplex deoxyribonucleic acid by polymerase at a single strand break. J Biol Chem 245:39–45.

Rigby PWJ, Dieckmann M, Rhodes C, Berg P. (1977) Labeling deoxyribonucleic acid to high specific activity in vitro by nick translation with DNA polymerase I. J Mol Biol 113:237–251.

## Random Priming

With this method, the double-stranded template DNA is denatured and hybridized with **random primers** (i.e., *random hexamers*), which serve as primers for DNA polymerase, primarily for T7 polymerase or Klenow fragments. The labeling occurs by the installation of radioactively or nonradioactively labeled nucleotides. If you label radioactively, you require [α-$^{32}$P]-dNTPs. With small amounts of template ($\sim$ 25 ng), specific activities of $5 \times 10^8$ to $4 \times 10^9$ cpm/μg DNA can be achieved. The lengths of the synthesized fragments, depending on the specific conditions, can be quite different. Stratagene assumes that a length of 500 to 1000 nucleotides is typical, although it may be shorter. Promega assumes a length of 250 to 300 nucleotides.

The advantage of this type of labeling is that all regions of the template are found in the fragment mix. This method is less problematic than *nick translation*, because the primer concentration, unlike the DNase concentration, does not represent a critical element.

Random priming is used for radioactive labeling, because specific amplification is obtained in this way. The first synthesized strands are displaced by the next bordering polymerases, and the DNA can be repeatedly read so that high specific activities can be achieved. Care must be taken, however, because probes with a high specific activity degrade more quickly and must therefore be used quickly.

Nonradioactively labeled nucleotides do not require such a high rate of incorporation; 10 to 30 labeled nucleotides per 1 kb of DNA are adequate because a higher rate of labeling leads to the antibodies, which are used for hybridization, getting in one another's way. The evidence at the end is produced by means of an alkaline phosphatase or a horseradish peroxidase reaction, which further intensifies the signal. Commercial kits can be obtained from most companies that produce materials required in the biosciences.

**Self-made random primers** can be produced. DNA (from calf thymus or herring sperm) is digested by DNase I in the presence of $Mn^{2+}$ (under these conditions, the enzyme severs both strands) and then denatured. In this way, you receive a colorful mixture of single-stranded oligonucleotides that are 6 to 12 bases long.

**Literature**

Feinberg AP, Vogelstein B. (1983) A technique for radiolabeling DNA restriction endonuclease fragments to high specific activity. Anal Biochem 132:6–13.

## 5′ Labeling with Polynucleotide Kinase

With this method, labeling is carried out through the exchange of the phosphate remnants at the 5′ end with a radioactive phosphate. The DNA to be labeled is dephosphorylated on the 5′ end and rephosphorylated with radioactively labeled phosphate.

The 5′ labeling functions with all nucleic acids, independent of their length, although it has only one label built in per molecule, and the intensity of the signal is relatively small. The 5′ label is preferably used for oligonucleotides, because they offer few alternatives. The 5′ label does not change the characteristics of the oligonucleotide, and it can be used as a primer for PCR amplifications.

For **dephosphorylation**, the DNA is incubated with alkaline phosphatase (see Chapter 6, Section 6.1). With synthetic oligonucleotides, however, you can skip this step because they have no phosphate residue on their 5′ end. The labeling is carried out with the T4 polynucleotide kinase (PNK) and radioactively labeled nucleotides. Use caution, because unlike all other radioactive labels, $[\gamma\text{-}^{32}P]$-NTPs is required in this case, because the last of the three phosphates is thereby transferred to the DNA. As experience has shown, this point frequently causes confusion for beginners.

In a typical labeling protocol, 10 pmol of template in 29 μL volumes is incubated with 15 μL of $[\gamma\text{-}^{32}P]$-ATP (3000 Ci/mmol, 10 mCi/mL), 5 μL of 10 × kinase buffer, and 1 μL of T4 polynucleotide kinase (10 u/μL) for 10 to 30 minutes at 37°C (99°F). The reaction is terminated by addition of 2 μL of 0.5 M EDTA.

## 3′ Labeling with Terminal Transferase

The 3′ ends can be labeled, although the procedure is based on a different principle and uses another enzyme. Unlike ordinary polymerases, the **terminal deoxynucleotidyl transferase (TdT)** can hang deoxynucleotides nonspecifically on the 3′ end of single-stranded and double-stranded DNA. This means that no complementary strand is required. The result is a homopolymer tail from a number of the same bases with a variable length (as long as you have employed a single dNTP). This presents a dilemma. You obtain more than one labeled product per molecule, which is good. However, the proportion of nonspecific sequences from all of the fragments becomes larger with a greater number of individual labels, and the danger increases that such probes may bind nonspecifically. You can evade the problem by using labeled dideoxynucleotide, in which only one single nucleotide is built in per molecule, a situation that is similar to that observed for 5′ labeling. Nevertheless, a difference does remain. With the terminal transferase, nonradioactively labeled nucleotides can be installed. Single-stranded DNA evidently prefers enzyme, so that double-stranded, labeled DNA with 3′ overhangs can be labeled better than DNA with smooth ends.

The 3′-labeled oligonucleotides are used, for instance, for in situ and Northern blot hybridizations. Terminal transferase and a protocol for how to employ this agent can be obtained from Roche, AGS, Promega, Stratagene, and other manufacturers.

## Fill-in from the Ends

The procedure of fill-in is equivalent to that used in smoothing DNA ends (see Chapter 6, Section 6.1), whereby single-stranded 5′ overhangs of a double-stranded DNA are filled-in with the help of Klenow or T4 DNA polymerase. The difference is that you insert at least a single (radioactively) labeled nucleotide for the full reaction. As for random priming, you require $[\alpha\text{-}^{32}P]$-dNTPs.Nonradioactively labeled nucleotides can also be inserted.

**Disadvantage:** Similar to the results seen with 5′ labeling, the number of labels per molecule is small, and the strength of the signals is therefore quite low.

## Polymerase Chain Reaction

PCR is an option. Labeling results by means of a normal PCR amplification, to which a labeled nucleotide is added.

The yield of probes is tremendous, although the specific activity is not particularly high, so that this preparation is less suitable for radioactive labeling. With nonradioactive labeling, the number of the labeled molecules is less critical, and labeling by means of PCR is particularly good in such cases. Roche offers a nucleotide mixture that contains nucleotides that are labeled or not labeled with digoxigenin in the fitting relationship and that can be inserted as an alternative to the normal nucleotide mixture. Labeling with this substance is child's play.

Single-stranded probes can be produced in this manner by using a single primer in the amplification. Replication then proceeds linearly, and the yield is therefore small.

## Photobiotinylation

Photobiotinylation is somewhat exotic and has therefore been banished to the end of the list. Instead of labeled nucleotides being built in by means of boring enzymatic pathways, you work with **photobiotin** [$N$-(4-azido-2-nitrophenyl)-$N'$-($N$-D-biotinyl-3-aminopropyl)-$N'$-methyl-1,3-propanediamine], a photo-activatible analogue of biotin that is bound covalently in DNA through a light-induced reaction. The sensitivity of probes labeled in this manner, however, should be less than that observed with enzymatically labeled probes.

### Literature

Forster AC, McInnes JL, Skingle DC, Symons RH. (1985) Non-radioactive hybridization probes prepared by the chemical labeling of DNA and RNA with a novel reagent, photobiotin. Nucleic Acids Res 13:745–761.

## Purification of Labeled Probes

After labeling, you often want to separate the labeled DNA from the labeled nucleotides that have been built in and that remain in most cases even after the most successful labeling reactions. They can increase the background signal during a subsequent hybridization, because even free nucleotides can bind to membranes or tissues. However, you cannot otherwise determine how well the labeling has functioned.

The best approach is through **ethanol precipitation with ammonium acetate**. The labeling preparation is brought to a 3 M ammonium acetate concentration, 2.5 volumes of ethanol are added, the solution is precipitated for 30 minutes at $-20°C$ ($-4°F$), and centrifuged. The pellets are resuspended in 2 M ammonium acetate, and the precipitation is repeated. In this way, you obtain DNA fragments with a length of 15 nucleotides or longer. The losses in material, however, can amount to up to 50%.

The alternative is **gel chromatography** (i.e., *size exclusion chromatography*) with custom-made columns. Among the purification methods, this is the clearest, because you can see what you have. You stuff a bit of silanized glass wool in a Pasteur pipette and pour in approximately 2 mL of Sephadex G-50 soaked previously in TE buffer (available from Pharmacia or Sigma). The labeled preparation is mixed with 5 μL of Dextran Blue and Orange G solution (0.01% [w/v] each) and then added directly to the Sephadex. Several milliliters of TE buffer are pipetted on top. You can

$$\text{specific activity} = \frac{\text{total incorporated activity}}{\text{quantity of DNA used for labelling}}$$

$$= \frac{\text{measured activity} \times \text{dilution} \times \dfrac{\text{original reaction volume}}{\text{volume after purification}}}{\text{quantity of DNA used for labelling}}$$

**Figure 7-1. Specific activity.**

then follow how the blue color separates very quickly from the orange. Dextran Blue is a very large molecule and behaves like the DNA fragments in gel chromatography; Orange G is small and migrates like a nucleotide. The blue fraction is collected and measured, and the orange-colored fraction is left in the column. If the blue and the orange fractions do not separate properly, the migration range is then too short, and you must use more Sephadex. Instead of Orange G, you can use bromophenol blue, which is available in every laboratory.

Many manufacturers produce **spin columns** that function according to the same principle, but they are easier to use, because the columns are prepared and the separation occurs during a short period of centrifugation. The purification is very rapid, although the columns are relatively expensive.

## Determination of the Specific Activity

To measure the success of radioactive labeling, you must know how much DNA template has been inserted and then purify this probe. You measure an aliquot in the scintillation counter to calculate the total activity of the purified probe based on the amount of radioactive decay per minute (counts per minute [cpm]) and divide this by the quantity of DNA employed to obtain the specific activity of the probe (Figure 7-1). The higher the specific activity, the higher is the sensitivity of the probe. How high the specific activity must be for it to make sense to continue working with the probe depends on the specific experiment, although you should normally have a specific activity of $10^8$ cpm/µg of DNA or more. If you find it too difficult to determine the specific activity, you can rely on the values obtained by colleagues and later extrapolate these findings to your own experiences.

Nonradioactive probes are far more difficult to evaluate. Because this labeling cannot be measured directly, test hybridization is used. A dilution series is performed on a membrane using the DNA template, and it is hybridized as in an actual case. The smallest, still just demonstrable DNA quantity gives you an idea of the sensitivity of the probe.

## 7.2 Hybridization

Hybridization is a science in itself. The conditions differ considerably depending on whether one performs this using a membrane with DNA or one hybridizes a tissue section, and also depending on the stringency with which one carries this procedure out.

Most frequently, Southern blot hybridizations are performed. This method is used to provide evidence for any DNA that has been transported to a particular membrane. This is assumed to be the normal case in the following discussion.

### 7.2.1 The Hybridization Buffer

The membrane on which the DNA is bound also binds everything else like the devil: fingerprints, dirt from the surface of the workplace, and the probe itself. You must therefore first form a block to saturate the remaining free binding sites, a process that is known as **prehybridizing**. Usually, the same prehybridization buffer is used as that used for the hybridization, although without the probe. The most important component is SDS; a 10% (w/v) SDS solution works just as well, and it is easy to manufacture. After 30 minutes of shaking softly, the membrane in the prehybridization solution is ultimately blocked.

For a hybridization buffer, there are almost as many solutions as there are researchers. A typical composition is $5 \times$ SSC or $5 \times$ Denhardt's ($25\,mM$ $NaH_2O_4$, pH 6.5/0.1% SDS/$100\,\mu g/mL$ of single-stranded herring sperm DNA) (see Appendix 1, Solutions). If you prefer lower hybridization temperatures, you should add 50% (v/v) formamide. Boehringer Mannheim (now Roche) previously offered a minimized hybridization buffer ($0.5\,M$ $NaH_2PO_4$, pH 7.2/7% [w/v] SDS/$1\,mM$ EDTA). You can see how extensive the range of possibilities is.

## 7.2.2 The Hybridization Vessels

Hybridization usually is carried out in **plastic containers** (with covers), which can be obtained in a variety of different sizes in every supermarket. The incubation is best performed in a heatable shaker with a water bath. You can carry out this hybridization with any amount of membranes, although you should not put more than 10 in a single bowl, because the membranes lying underneath will be pressed together from the weight of those above, and the hybridization solution will no longer get to them. The disadvantage of using such bowls is the large amount of liquid required for hybridization and washing.

A liquid-sparing alternative is to **hybridize in plastic bags**. The membranes are sealed in a suitable plastic wrap, which is a process requiring a foil seam-welding device and some talent, because this procedure is quite difficult, and the result usually is a considerable mess. Roche offers *hybridization bags* with a spigot that can be screwed shut, allowing them to be filled and emptied in a clean manner. You should grant yourself this luxury.

Another possibility is to use **hybridization tubes**. They can be handled easily, are easy to clean, and require only a small amount of hybridization solution. For a membrane with dimensions of $10 \times 13\,cm$, 5 to $10\,mL$ is sufficient. This is a practical solution. Unfortunately, a rotating hybridization oven is required, and it may not be considered part of the basic equipment in every laboratory. The number of the membranes that can fit in a hybridization tube is also quite limited. A trick is to place a strip of gauze between the membranes; in this way, two to three membranes can be layered over one another without sticking together. The side of the membrane with the DNA must always face inward. Hybridization ovens and bottles can be obtained from Techne, Bachofer, Hybaid, Stratagene, Hoefer, and Biometra.

## 7.2.3 The Hybridization Temperature

Temperature is one of the most decisive points in hybridization. The specificity of the whole procedure can be guided by temperature. However, which temperature is correct? In principle, there is an equation to determine the answer:

$$T_m = 81.5°C\,(179°F) - 16.6\,(\log_{10}[Na^+]) + 0.41\,(\%\ G+C) - 0.63\,(\%\ formamide) - 600/L$$

The sodium concentration $[Na^+]$ is given as mol/L, % $G+C$ is the share of guanosine and cytosine, and L is the length of the probe in bases.

The melting temperature, $T_m$, is the temperature in which one half of the nucleotides are dissociated from a double helix. The optimal hybridization temperature is officially 25°C (77°F) below the $T_m$, although this is not completely true. Other factors play a role in the success of this procedure, such as the sequence of the probe. The more unique it is, the simpler the entire affair is, although you must use a higher hybridization temperature if it displays a large homology to all kinds of other sequences to avoid faulty hybridization. In such cases, the signal is weaker. If the probe develops stabile secondary structures, it may not function at all.

The correct temperature remains a question of experience. In practice, it is possible to work fairly well with a temperature of 65°C (149°F) (42°C [108°F] using 50% formamide) for a specific hybridization. If you want to carry out a hybridization with a reduced specificity to include homologous sequences, you must experimentally determine how low to go with the temperature.

The duration of the hybridization is a matter of experience. Four hours enables you to hybridize most samples quite well, although you are better off hybridizing overnight for signals that are difficult to detect. The time required for hybridization can frequently be derived from the timetable for the entire procedure. If you are busy all day with the purification and labeling of the fragments, you are better off carrying out the hybridization overnight.

## 7.2.4 Washing

Another decisive point is washing correctly. It contributes to the specificity as much as the hybridization itself; some consider it to be even more important. If you have tried to hybridize as many probes as possible, you want to wash away everything that is not specific.

Characteristically, you purify for $2 \times 15$ minutes with $2 \times SSC/0.1\%$ (w/v) SDS at the same temperature with which you have previously hybridized. You can increase the specificity by decreasing the ionic strength of the washing solution (e.g., $0.1 \times SSC/0.1\%$ SDS) or by increasing the washing temperature. If you hybridize with a lower specificity, it helps to try shorter washing times, lower temperatures, or higher ionic strengths. It is easier with radioactive probes, because they can be examined intermittently and repeatedly with the contamination monitor to see whether washing has been sufficient.

# 7.3 Verification of Labeled DNA

## 7.3.1 Autoradiography

Probes that are labeled radioactively can be identified most easily. Put the film in, and allow it to be exposed. The exposure time depends on the signal strength and can extend from 30 minutes to several days.

The **sensitivity** of the x-ray films is a different problem. Blackening of the film is linear and related proportionally to the signal strength of the probe in a very limited area. X-ray films are composed of a plastic foil coated with silver bromide crystals. Normal light, β radiation, or γ radiation can activate these crystals, and the activated crystals can be reduced to black silver granules by developing the film. This activated condition is somewhat unstable, and about five photons are required per crystal to achieve a 50% probability that the crystal will be reduced during development. The stability of the activated condition can be increased by exposing the film at −70°C (−94°F), although this does not solve the fundamental problem that weak signals are always underrepresented because they cannot surmount the activation threshold, whereas strong signals result in no further blackening after a certain exposure time, but instead deliver a broader spot, at best. If there are strong and weak signals on the same membrane, two films must be developed, one with a short and the other with a long exposure time.

A trick to increase the sensitivity of the film is the use of **preexposure**. The silver bromide crystals are excited with just enough light that a small amount of radiation from a weak signal is sufficient to activate them. You must empirically determine the correct amount of light. The optimal condition is enough light that the preexposed film has an $OD_{540 \, nm}$ about 0.15 higher than a preexposed film after being developed. Because of the different types of film with their different sensitivities, you must determine the correct conditions for each film by varying the wavelength of the light used, the distance (at least 50 cm), and the time of the flash of light. All the best to those who have a stroboscope at their disposal.

**Literature**
Laskey RA. (1980) The use of intensifying screens or organic scintillators for visualizing radioactive molecules resolved by gel electrophoresis. Methods Enzymol 65:363–371.

Laskey RA, Mills AD. (1975) Quantitative film detection of $^3H$ and $^{14}C$ in polyacrylamide gels by fluorography. Eur J Biochem 56:335–341.

Laskey RA, Mills AD. (1977) Enhanced autoradiographic detection of $^{32}P$ and $^{125}I$ using intensifying screens and hypersensitized film. FEBS Lett 82:314–316.

## Intensifying Screens

*Intensifying screens* are used extensively. Radiation, which passes through the film without exciting crystals, falls on the foil and is absorbed and released in the form of light. It then passes through the film, where it activates it during this second attempt. For the entire procedure to function, the cassette must be cooled to $-70°C$ ($-94°F$); at room temperature, you can forget the intensifying screen (Lasley and Mills, 1977). Intensifying screens can be obtained from DuPont and Fuji.

**Disadvantage:** The intensifying screen functions only with strong β-emitters such as $^{32}P$ or γ-emitters such as $^{125}I$. The signals become more fuzzy, because the light scatters more intensely. The intensifying screens are quite expensive.

## Fluorography

Signals from weak β-emitters such as $^3H$, $^{14}C$, $^{33}P$, and $^{35}S$ can be strengthened with fluorography. The principle is similar to that for intensifying screens. The radioactive test is brought close to an organic scintillation counter, which releases the energy from particles or quanta that have been activated in the form of light. This method is primarily used for the determination of radioactivity with the aid of scintillation counters. You mix an aliquot of the probe with a suitable scintillation fluid before the measurement. For acrylamide gels, there are special scintillation solutions (e.g., Amplify from Amersham), and the gel is incubated for 30 minutes after electrophoresis. Afterward, the gel is dried and then exposed.

## Storage Phosphor

The **PhosphorImager** is a device that was marketed by Molecular Dynamics in 1989; since then, similar instruments have been marketed by Fuji, Bio-Rad, and Canberra Packard. The technology of storage phosphor is interesting, although the underlying physical principles may prove to be too demanding for a simple molecular biologist. A *screen* coated with BaFBr:Eu crystals is required; the electrons are displaced by ionizing energy into an excited, but stable state ($Eu \rightarrow Eu^{2+}$). After successful exposure with the hybridized membrane, the crystals are scanned by a laser beam and excited to a higher energy state, which is a less stable state ($Eu^{2+} \rightarrow Eu^{3+}$). When the electrons return to their basic condition ($Eu^{3+} \rightarrow Eu$), they release a small quantity of light, which can be measured. The method is suitable for radioactive isotopes (β- and γ-emitters such as $^{14}C$, $^{35}S$, $^{32}P$, $^{33}P$, $^{125}I$, $^{131}I$) and for ultraviolet (UV) light.

The method is 10 to 100 times more sensitive than x-ray films and therefore delivers results more quickly and detects weaker signals. The system is saturated less quickly, so very strong and very weak signals can be recorded through a single exposure. The linear range of signals extends over five orders of magnitude (1 to 100,000) and is therefore better suited for a quantification than x-ray films. The data are recorded with a computer and can be evaluated densitometrically without any further efforts.

**Advantage:** The high-tech method is rapid, sensitive, and quantifiable.

**Disadvantage:** The cost for such an instrument is very high (depending on the particular design, approximately $54,000). Because the PhosphorImager is needed only for the scanning signals stored on the screen, it is sufficient if a nearby team offers you access to such an instrument; the evaluation can be performed in your own laboratory with the use of the corresponding equipment (e.g., analysis program, gray-scale printer). The screens can be used repeatedly, but they are sensitive to contamination by radioactive probes, which remain in the background of subsequent exposures. They are very expensive (starting at about $1000), and you will require several if you use the system often. The resolution (50 μm) is lower than with x-ray film.

**Literature**

Johnston RJ, Pickett SC, Barker DL. (1990) Autoradiography using storage phosphor technology. Electrophoresis 11:355–360.

Reichert WL, Stein JE, French B, et al. (1992) Storage phosphor imaging technique for detection and quantitation of DNA adducts measured by the $^{32}P$ postlabeling assay. Carcinogenesis 13:1475–1479.

## 7.3.2 Nonradioactive Methods of Detection

The greatest similarity between nonradioactive methods is that they involve "no radioactivity." You can distinguish between direct and indirect methods. **Direct evidence** is rare and exists only for fluorescein-labeled probes, which you can see with UV light in the microscope or in the FluorImager (discussed later), but the sensitivity is only moderately high, because the number of the labels per molecule is usually small.

The Lightsmith System from Promega is a **semi-direct method of detection**, in which the alkaline phosphatase is coupled directly with the oligonucleotide probe. The phosphatase is seen directly through evidence of an enzymatic reaction. According to the manufacturers, this system should be as sensitive as radioactively labeled oligonucleotides. Labeling of the probe, however, functions better with amino-modified oligonucleotides, which must be specially synthesized. The system does not appear to have been very successful, because there is no longer any trace of it in the Promega catalog. Amersham has two similar systems; AlkPhos Direct functions with alkaline phosphatase, and ECL Direct functions with horseradish peroxidase.

**Indirect methods of detection** are commonly used. Whether the probe is labeled with biotin, digoxigenin (DIG), or fluorescein, the process is always the same. The label is recognized by a specific antibody (DIG or fluorescein) or by streptavidin (which binds to biotin), to which an enzyme is coupled (depending on the particular system, alkaline phosphatase or horseradish peroxidase). This structure can be detected by one of the enzymatic reactions illustrated in Figure 7-2. Consequently, the method used is equivalent to that used by the researchers involved with proteins.

The weakness of the method lies in its many intermittent steps, which present a problem if you wish to quantify the signals. After hybridizing with the probe and washing, the membrane is blocked, incubated with the antibody, and washed again, and a substrate for the detection is added. The more intermittent steps there are, the more inestimable is the situation. The sensitivity of indirect evidence

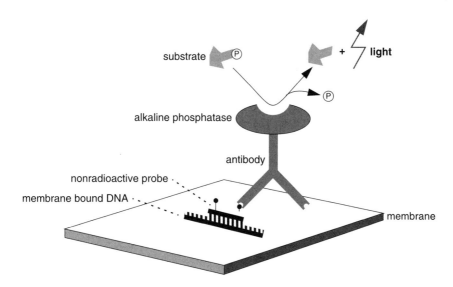

**Figure 7-2. Indirect, nonradioactive methods include the use of color, chemiluminescence, and fluorescence.** After hybridization with a labeled probe, a specific, enzyme-coupled antibody is bound to the probe, and a detection reaction is carried out. According to the substrate used, a dye, light (shown here), or some other substance develops, and a signal is emitted after excitation with laser light.

offers no problems, because with a suitable substrate, thresholds can be achieved that are similar to those seen in radioactively labeled tests.

Roche (formerly Boehringer Mannheim) has developed a comprehensive nonradioactive procedure that includes various applications of DIG-labeled probes. A review of these substances can be found in manuals from Roche. Other manuals include the *Nonradioactive In Situ Hybridization Application Manual* and *The DIG System User's Guide for Filter Hybridization*. Comprehensive information (including all of the package circulars) concerning this and other Roche products can be found on the Internet (http://biochem.roche.com).

## Color Detection

**Alkaline phosphatase** is detected with 5-bromo-4-chloro-3-indolyl-phosphate (**BCIP**) and nitroblue tetrazolium (**NBT**), and **horseradish peroxidase** is detected with **chloronaphthol** by means of a dark color precipitation that can easily be seen. The corresponding protocols can be found in Maniatis and in other sources (see Appendix 2, Recommended Literature). The detection of color is carried out on nitrocellulose and nylon membranes.

Roche offers a *Multicolor Detection Set* for detecting alkaline phosphatase, with which you can color the bands green, blue, or red. This helps to provide evidence of different probes on the same membrane, although these colors are then more difficult to see.

**Disadvantage:** Among all substrates, those that are chromogenic are the least sensitive.

## Chemiluminescence

In contrast to chromogenic substrates, chemiluminescent substrates produce light, which can be demonstrated with the aid of x-ray films. Alkaline phosphatase and horseradish peroxidase require different substrates, which also demonstrate different kinetics. Whereas the horseradish peroxidase reaction is very intense and rapid (in some systems, everything is accomplished within 10 seconds), the alkaline phosphatase reaction is much slower and, after a one-half hour of incubation, may take about 2 days. The exposure time for the horseradish peroxidase system is in the range of seconds to minutes, and that of the alkaline phosphatase system is between 5 minutes (with very strong signals) and several hours. This may give the impression that horseradish peroxidase signals are more favorable because they are faster and stronger, but who can say whether the film is better exposed for 20 seconds or 2 minutes? Even before the film is developed, the reaction has come to an end. Therefore, you add new substrates and try your luck again. This problem was also recognized by the manufacturers, and most horseradish peroxidase systems have strongly suppressed kinetics, which resemble that of an alkaline phosphatase system.

The production of light is comparable to the blackening of film brought about by radioactivity, and it proceeds more rapidly, allowing you to successively perform the procedure with several films within a shorter period. You can achieve the same results after a 2-hour exposure as observed after an overnight exposure with radioactively labeled probes. With weak signals, however, a certain amount of caution is appropriate. The background in alkaline phosphatase systems appears to be more intense in cases involving radioactive probes; signals that can be observed after 2 hours will presumably not be seen after overnight exposure, because the film is blackened more intensely, although the relationship of background to signal unfortunately remains equal. Whether horseradish peroxidase systems prove to function better on this point is doubtful. The ECL system (a horseradish peroxidase system from Amersham that is frequently employed in protein biochemistry), for instance, delivers lovely results in the case of Western blots, which are extremely nonlinear. Strong signals function remarkably, but weak signals do not function at all. In the area between, a fivefold increase in the signal strength represents a doubling in the quantity of material demonstrated. Quantification is impossible in these circumstances, and this system is suitable only for qualitative evidence.

Fitting substrates can be obtained from Lumigen, Tropix, or the company from which you have obtained the nonradioactive labeling system. The substrates are normally the most expensive part for providing such evidence. With a small trick, you can get along with such small quantities that the eyes of the manufacturers will start watering. The membrane with the DNA should be placed on plastic foil; best suited for this purpose are the plastic sleeves, which can be found in every stationary shop. Depending on the size of the membrane, 0.5 to 2 mL of substrate solution is pipetted onto the membrane before a second membrane is laid down. In this way, the substrate is distributed evenly in a thin film over the entire membrane.

**Advantage:** The systems for detecting chemiluminescence are surprisingly stabile. They can be kept for years by anyone who needs to perform hybridizations occasionally.

**Disadvantage:** The membranes cause problems. Nitrocellulose membranes do not function well with chemiluminescent substrates, because the signal is suppressed. The DIG system requires positively charged nylon membranes, and not all manufacturers and all production batches are equally good. For this and for other systems, it is important to personally test an entire series of membranes or swallow the bitter pill and purchase the membranes from the manufacturers of the system that are specially designed for that system.

## Fluorescence

The newest product is the fluorescence substrate (VISTRA, Amersham). The enzyme (a substrate currently is available only for alkaline phosphatase, but one for horseradish peroxidase should be available soon) transforms the fluorogenic substrate into a fluorescing product, which does not shine like the chemiluminescent substrates, but it fluoresces after being excited by a laser. The light that results is measured point by point over the entire membrane and transformed to provide a picture. To enjoy the features of this wonderful method, you require a scanner. Molecular Dynamics offers the FluorImager for this purpose for about $67,000.

**Advantage:** The fluorescing product is only moderately bleached during the scanning process, and the membrane can be scanned repeatedly. The fluorescence product is so stabile that the membrane can be preserved in a dry condition and then scanned again weeks later.

**Disadvantage:** The price for such a scanner is high.

# 7.4 Screening of Recombinant DNA Banks

A **DNA data bank** (i.e., library) is a mixture of clones obtained when you clone a variety of DNA fragments in a vector. Because there are many kinds of DNA fragments and vectors, you can create the banks according to a particular interest:

- According to the origin of the DNA, such as genomic and cDNA banks
- According to the vector used, as in plasmid, phage, cosmid, PAC, BAC, or YAC banks
- According to the characteristics of clones. Expression banks contain cDNA fragments that are cloned in a vector with the aid of a promoter. The expression can be induced if required, and the clones are screened based on the characteristics of the expressed protein, such as by a functional test.

This section briefly describes the individual processes that occur during the screening of a bank to provide an idea of the labor involved in this procedure. The production of such a bank is intentionally left out because it is a quite laborious process, for which a great deal of experience is required to obtain a satisfactory result. A number of individuals have wasted their time with doctoral theses attempting to perform such a venture. In practice, you will primarily make use of DNA banks that can be obtained from other teams or that you buy from a gene technology company. From these companies, you can also obtain a detailed screening record.

## 7.4.1 Plating Out the Bank

Most often, you will be working with phage and cosmid banks. It is astonishing how little initial material is required. For phage banks, less than $100\,\mu L$ of phage suspension is sufficient to find each available clone.

The complexity of the DNA bank is decisive. To have a realistic chance to find the desired sequence, you should plate out three to five times as many clones as the bank has independent clones. In this way, you have the statistics on your side, and each clone may be represented at least once. The number of clones plated out also depends on the frequency of the sequence being looked for. If you search in a cDNA bank for a sequence that is encountered frequently, you will be able to remain below this standard value because the work required is less extensive and fewer filters must be juggled. In the normal case, however, expect to use about 20 14-cm agar plates and a corresponding number of filters.

## 7.4.2 Filter Transfer

After successfully plating out, the clones must be transferred to the membranes for hybridization in a process known as *filter transfer*. The procedure differs somewhat for phages and bacteria.

Phages are plated out together with bacteria. The phages lyse the bacteria and form *plaques* in the bacterial lawns. In these plaques, the phage particles swarm in this way, so that it is adequate to place a fitted nylon or nitrocellulose membrane on the agar plate for a few minutes. This membrane then is carefully removed, and a sufficient number of phages remain attached to the membrane to deliver a good hybridization signal later. The membrane is denatured like a Southern blot and then baked or crosslinked with UV light. From an agar plate, you can easily transfer two to three filters, although this is difficult with more because the membranes become fully saturated with fluid each time so that the agar becomes too dry and begins to stick to the membrane. The disks are preserved because they always contain a sufficient amount of particles to pick out the phages during the next round of screening.

For cosmids, filter transfer is somewhat more difficult, because the DNA is within the bacteria, and they grow in the form of colonies. Colonies can become fairly large, and, if you place a membrane on the plate, the bacteria may smear one half of the plate. Consequently, you use a different approach; you plate out the bacteria onto an agar plate, place the membrane on top, and then incubate the bacteria at 37 °C (99 °F). Because the membrane has large pores, the bacteria grow on the agar, wander through the membrane, and form colonies on the upper surface of the membrane. Some of the bacteria remain stuck to the membrane, which you can strip off and treat as described above earlier. You can then place a new membrane on the plate and incubate it again. One of these membranes is frozen at −70 °C (−94 °F). From this "master" membrane, after hybridization, you can obtain the bacteria for the next round of screening.

It is important that you place **marks** on the plate and the membrane, so that you can later bring them into conformity with one another. This is especially important when using round filters, because it will otherwise be impossible to successfully make the positive signal of the membrane and that of the agar plate or of the master congruent. The classic method consists of using a canula to make a hole in the agar that extends through the membrane. Prick through these layers heartily to make the holes for later identification. The pattern of the holes must be asymmetrical, or this action will prove to be quite useless.

## 7.4.3 Filter Hybridization

Filter hybridization is similar to that for Southern and Northern blots, but the number of filters is usually higher, and the filters are larger. The problems are related to the technical procedures, because a sufficiently large hybridization vessel is often lacking. The number of filters also is a problem. When working with more than 10 filters, they tend to stick to one another as a result of their own weight. Consequently, the probe penetrates into the outer regions of the filter, but the largest part of the membrane remains untouched.

Washing is carried out as for other hybridizations, with $0.1 \times$ to $2 \times$ SSC and 0.1% SDS at 65°C (149°F) if you wish to obtain a high specificity or at lower temperatures if you are looking for sequences that are less homologous. You should always wash with a sufficient volume of fluid because the filters must float individually.

## 7.4.4 Filter Exposure

Providing evidence of positive clones is complicated by the dimensions of and the number of membranes. It is simplest to work with radioactive probes, because the only requirement is a sufficient number of exposed cassettes. The exposure usually is carried out using an intensifying screen (see Section 7.3.1) at −70°C (−94°F) to save time. It is important to ensure that the films do not become wet. If you are impatient and open the cassettes while they are still in a frozen condition, the damp films remain stuck to the support; if you then tear them off with force, the film will have ugly black spots, which may mask all other signals.

With nonradioactive probes, the next step is antibody hybridization and detection. Because of the large quantities of liquids, this can be extremely expensive. You can save on the substrate solution. Pipette a few milliliters of it directly onto the membrane and cover it with plastic wrap. In this way, the small amount of liquid distributes itself evenly over the entire membrane. It is then exposed.

For the **exposure**, use old films for support of the membranes; they are stabile and flexible, and they normally are the right size for the cassettes. Cover the films with plastic wrap to prevent the membranes from sticking irreversibly. Place markings on the support, so that you can bring the exposed film and the membranes into congruence. **Highlighters** are suitable for this, and they can be found in toy stores in the form of a dinosaur or of Mickey Mouse. Amersham offers extremely practical, fluorescing, self-adhesive foil strips (TrackerTape), which can be cut to fit, and on which you can write.

The supports and the membranes must be packaged in another layer of plastic wrap, because x-ray films do not tolerate dampness very well. The fight with the plastic foil is among one of the most difficult things to handle in this procedure, because it usually requires assistance from a colleague. Roth has a foil that is expensive, but it is less adhesive and substantially more suitable for such purposes.

## 7.4.5 Clone Detection

After developing the film, the real work begins. On the film, the position of the membranes must be marked before you look more earnestly for signals. After finding a signal, the film and the agar plate must be brought into congruence for phages out the region of agar, in which the signal is expressed quite generously, and the phages in the phage buffer are then eluted. For bacterial banks, cut out the corresponding region of the master membrane, and make a small culture of it.

In both cases, you end up with a mixture of phages or bacteria, because you cannot perform the stab or the cut so exactly that you obtain only the desired clone. Because the risk of missing a desired clone increases with the attempt to be as exact as possible, you should perform these stabs or cuts somewhat more generously. In time, you will obtain more experience, and the improved precision will reduce the number of mandatory rounds for the screening (Figure 7-3).

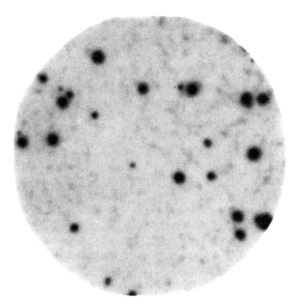

**Figure 7-3. Phage screening.** An autoradiography signal is emitted from a phage cDNA bank after the second round of screening. There is no doubt that it indicates a positive clone.

The clone being looked for in this mixture is clearly enriched, and you can begin with the next round of screening. Eluting the clones, plating them out, transferring the filters, hybridizing, and other methods, depending on your experience and particular talents, requires three to five screening rounds until an individual clone is identified. You can recycle the hybridization probe if you work rapidly, and you may get along with a single round of labeling.

You may experience **screening-specific problems**. The procedure is simple if you receive a single signal after the first round of screening. Usually, however, there are between 10 and several hundred. All of them must be isolated and plated. How do you manage that?

According to experience, some signals are **artifacts**. You can initially extract two filters from each disk and hybridize both with the same probe, or better still, you can proceed with two different probes as far as possible. Only the signals that appear on both filters are then considered. Otherwise, you can find criteria according to which you can select a signal (e.g., with regard to the stronger or weaker rounds), or you can find an alternate, simple screening method to differentiate between good and poor clones. For instance, you can use the eluate from the phages or the bacterial mixture as a template for a PCR if you recognize sequences from the clone being sought. Another tactic consists of pooling potential clones by plating out one with a very promising signal and two to five with less promising signals.

## 7.4.6 Bank Screening in the Future

I do not know how long researchers will continue to search for clones by screening banks, but they may belong to a dying race. The future belongs to mechanically dotted **high-density filter libraries**. The clones are transferred, individually and in organized patterns, onto the membranes by special robots. Screening in this manner becomes a true pleasure. It is sufficient to hybridize the membrane once, to determine the coordinates of the signals, and to transmit these to the institution from which you have received the membrane; the desired clone then follows by means of the postal services.

In Germany, the resource center in Berlin is a good address where you can receive membranes from various banks at a reasonable price. More information can be obtained over the Internet (http://www.rzpd.de).

In the foreseeable future, after most of the genomes have been sequenced, you will need only a computer that can be used to search through the gene banks, and the greatest difficulty will then consist of finding out where you can order the desired clone. Only people who deal with research "animals" that are a not part of the standards usually employed will have to get their hands dirty in the future. Frankly, though, has anyone who is searching for genes in the leopard slug deserved better?

# 7.5 Two-Hybrid System

The search for the function of a protein is not really the responsibility of the molecular biologist, whose right to exist lies more in making the necessary DNA constructs available to others. Nevertheless, I do not consider it to be an acceptable position to refrain from considering other fields of work, nor do I believe that such an attitude will continue to exist, even though it has been quite common. The human genome has been sequenced, many other genomes will follow, and the time is near when there will be nothing new to clone. The job of the molecular biologist will then be to combine different, existing sequences with one another. This work will be carried out by technical assistants and trainees (nonacademics) in the future. If you should be included as one of this unfortunate group of university graduates, you will have to add another skill to defend your right to exist.

Why not use the *two-hybrid system*? Each person has approximately 25,000–30,000 genes, although this information helps about as much as knowing how many components are required to construct the Eiffel Tower. The decisive question in the coming years will be which proteins react with which others in which form and with what results. One method to answer the question about interactions is provided by the two-hybrid system. It allows identification of binding partners for proteins, for which you must know only that they must bind well with another protein to be active. It is the most popular of screening systems for selecting a specific protein from a special expression bank that will react with the targeted protein. The system is based on the use of yeast.

You construct a probe (i.e., bait) by cloning a fusion protein, which is made up of the essential parts of the desired protein and the DNA-binding domains of the LexA bacterial protein. LexA is a repressor that binds specifically to the LexA operator. Other proteins can be used for the construct of a *reporter* as long as a previously still signal system can be activated; examples include transcription factors, kinases, and phosphatases.

A special yeast mutant is required that has a suitable reporter system that can be activated by LexA. There are several possibilities. The EGY 48 stem has two selection systems: a the *LEU2* gene, whose promoter region was replaced by a LexA operator, and a β-galactosidase gene, which is guided by a LexA operator. The LEU2 protein is essential for leucine synthesis, and the yeast cannot grow in leucine-free medium if the enzyme is lacking.

The yeast stem is transformed through the construct that was manufactured previously and that expresses the desired LexA fusion protein. In an ideal case, the new yeast cell line primarily differs functionally from the original stem. The yeast must be transformed with a special cDNA bank, which you have purchased previously, organized, or produced personally. The point is that it deals with a bank of fusion proteins, in which the cDNAs were coupled with the activation domains (i.e., *acidic activators* or *acid blobs*) of an especially active protein, such as VP16 or Gal4-AD. The fusion proteins are under the control of a promoter, such as a galactose-dependent GAL1 promoter, which allows you to control the expression.

If the doubly transformed yeast grows on a special medium that contains galactose but no leucine, the expression of the cDNA constructs is induced, but the yeast cells experience problems, because their supply of leucine is interrupted. Only cells that express a fusion protein from the cDNA bank and whose cDNA binds part of the desired protein will survive this treatment. In this way, the DNA-binding domains of the probe and the activator domains of the fusion proteins are in an adequate spatial proximity to activate the expression of the genes that are under the control of the LexA operators: the *LEU2* gene, which reconstitutes the leucine provision of the yeast cells, and the *LacZ* gene, which codes for β-galactosidase. Application of two reporter systems permits a twofold selection to reduce the number of false-positive results. In 2 to 5 days, the yeast cells that contain the "good" fusion proteins continue to grow and form colonies despite the leucine-free medium. If these are plated on an X-Gal–containing medium, the "especially good" colonies are dyed blue. Figure 7-4 shows details of the two-hybrid system.

From the positive colonies, the plasmid with the cDNA construct is isolated and analyzed in detail. If everything has run well, this construct codes for the binding partner of the probe protein being searched for.

The genes mentioned are only examples, and the available selection of DNA-binding protein domains (e.g., LexA, GAL4), activator domains, and reporter genes has become substantially larger over the years. A system must be found that delivers the strongest possible signal-background relationship; it must demonstrate the smallest possible background expression of the reporter gene, but in the presence of the fitting binding partner, lead to the strongest possible expression of the reporter protein.

In practice, this procedure is somewhat troublesome. In the first step, it must be examined to be sure that the probe does not activate the reporter gene alone. Under favorable conditions, this requires a week. If necessary, different constructs and different systems must be tested, which can be a very

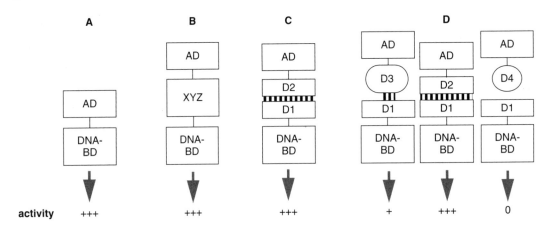

Figure 7-4. The detailed two-hybrid system can be confusing, and it may help to imagine that it deals with only a classic activator with a DNA-binding domain and an activator domain (**A**). Domains are surprisingly stable and independent components of proteins, which can be separated spatially and almost arbitrarily from one another. In a two-hybrid system, instead of placing an extremely long amino acid sequence (XYZ) between the two domains (**B**), two domains that bind together because of their mutual affinity are inserted. The activator decomposes into two proteins; one binds to the DNA, and the other activates the transcription (**C**). The DNA-binding domain (e.g., LexA) is fused with the desired protein (D1) with various cDNAs (D2 to D4), and all that remains is to find the clone that leads to maximum activity, producing the best binding to the desired protein (D1) (**D**). AD, acidic activation domain; DNA-BD, DNA-binding domain; D1–D4, domains 1 to 4.

time-consuming venture. Ensure that all of the components work together; the cDNA bank must be compatible with the other elements.

The second step is the search for binding proteins. Several selections are carried out to maintain workable dimensions. The transformed yeast cells are selected on cells in a suitable, minimal medium that contains a plasmid from the cDNA bank. After that, the surviving yeast is selected in cells using a leucine-free minimal medium that expresses a binding partner for the probe. The remaining colonies are tested for β-galactosidase expression. If everything functions properly, this second step takes approximately a week.

Afterward, the plasmids are isolated from the positive yeast colonies, transformed to bacteria to produce sufficient quantities of DNA and further yeast transformations, and tests are performed with this material to demonstrate the specificity of the interaction. This part of the process can be carried out in the course of a week. The times given depend on everything working properly. In practice, however, everything in molecular biology lasts about twice as long as anticipated, assuming that there are no larger technical problems.

As with all of the screening methods, the chances of success cannot be estimated. One problem is that the cDNA fished out does not always represent the physiologically relevant partner of the probe proteins, although it leads to positive signals. Some proteins are fished out with the most varied of probes, such as ribosomal proteins, ferritin, and ubiquitin. For some, the reason is clear (*heat-shock* proteins play a role in the folding of the probe fusion proteins), but for others, the relationship is uncertain. A helpful list of false positives can be found on the Internet (http://www.fccc.edu/research/labs/golemis/main_false.html). The list demonstrates how frequently false positives occur.

Because a large number is plated out, the number of positive clones can occasionally be very large, although only a handful of different plasmids are behind it. In this case, it is profitable to sort out the plasmid isolated from the yeast cells at the end, such as through restriction digestion. The band pattern allows the clones to be differentiated without much trouble into groups of clearly related plasmids.

It is also unclear how many of the physiologically relevant partners of a protein can be found with this method. Calculations indicate that it should be possible to fish out pairs with a slight affinity for one another, but in some cases, it was not possible to successfully find even known binding partners with the aid of a two-hybrid system. Consequently, you should not start with the goal of finding all interacting proteins, if you would not like to end in utter frustration.

Well-designed systems are available from Clontech, Invitrogen, Display Systems Biotech, Origene Technologies, Stratagene, and many other firms. Depending on what you are looking for, particular searches will result in finding one or another well-polished varieties and, occasionally, a substantial improvement in the system. In addition to the two-hybrid system, there are also *one-hybrid* systems for discovering proteins, which bind to specific DNA sequences, and *three-hybrid* systems to use when a third partner is involved in the binding, such as RNA or another protein (two of which you must know beforehand). The system from Invitrogen expresses a zeomycin-resistant gene, which offers an additional freedom in choice in the selection of yeast stems and cDNA banks, although working with zeomycin is somewhat nerve-wracking.

Some interesting web pages on this topic, which address some known stumbling blocks, can be found on the Internet (http://www.uib.no/aasland/two-hybrid.html; http://www.fccc.edu/research/labs/golemis/InteractionTrapInWork.html).

### Literature

Bartel L, Fields S. (1997) The Yeast Two-Hybrid System. New York, Oxford University Press.

Fields S, Song O. (1989) A novel genetic system to detect protein-protein interactions. Nature 340:245–246.

# 8 DNA Analysis

If you have found a new clone, the next step is analysis. This can be carried out to different degrees, depending on how much time and energy you are willing to invest.

The simplest method is **mapping by means of restriction fragment analysis**. The DNA is digested by different restriction enzymes—individually and in pairs—and separated in the agarose gel. The resulting fragment lengths are then determined (Figure 8-1). Combining the individual fragments with one another until a reliable restriction map of the clone is obtained is like doing a puzzle. With DNAs longer than 10 kb, some intelligence is required to put together the small pieces into a meaningful whole. Nevertheless, the analysis usually is carried out quite smoothly; digestion and letting the gel set can easily be performed in half a day. This provides a good idea of what you have. If you blot the gel and hybridize with a suitable, known probe, you can acquire further information, on the basis of which you can describe longer clones, a process that requires more time.

Do not underestimate the value of a good restriction map. It can help you isolate or subclone individual fragments and use the probe without having to sequence the entire clone. If you find several clones after the screening, the restriction pattern shows whether the clones are related to each other or are identical.

In the end, however, the experimenter must perform sequencing to be happy. Only that which looks similar to GAACTTGCT is sufficient to make him or her content.

Perhaps you do not have a new clone but are instead looking for mutations in a known sequence. Sequencing is the best choice, although not the quickest and certainly not the cheapest. Later sections of this chapter consider various possibilities for detecting mutations.

## 8.1 Sequencing

Sequencing has established a certain history. Perhaps you will find an advisor with experience (e.g., your boss) who will tell you about the wild times involving Maxam-Gilbert sequencing (Maxam and Gilbert, 1977), as if the reading of sequences was a kind of intellectual puzzle. The details are found in biochemistry books. Other predecessors can describe the exciting times they had been, sequencing according to Sanger (Sanger et al., 1977), when the Klenow polymerase and experience in interpreting the band patterns was considered to be priceless.

Those times are gone. Sanger sequencing with T7 or Taq polymerase has become the standard method, and experimenters now distinguish only between procedures requiring much work and those requiring little, with the choice depending on the available budget. Researchers with little money must perform more work manually. More or less clever devices for the classic sequencing of gels can be obtained from BioRad, AGS, Hoefer, and other manufacturers, but you can also have something built in a workshop you trust.

Anyone with adequate finances can purchase an automatic sequencer, devices that now are found almost everywhere, because of the genome projects that have been shooting up like mushrooms. Look around to see whether you can make contact with a team that has such an instrument. It can save you

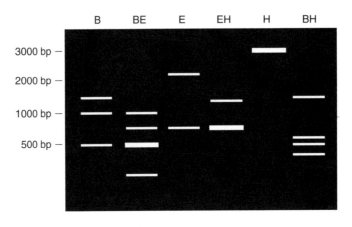

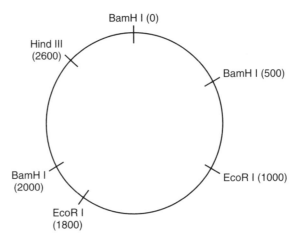

**Figure 8-1. Restriction fragment analysis of a fictional 3-kb plasmid.** The example shows digestion with *Bam*HI, *Eco*RI, *Hind* III, and their combinations, as well as the resulting restriction map. The gel is easiest to analyze if the following sequence is observed. 1, Determine the length of all fragments. 2, Determine the number of cleavage sites per restriction enzyme. 3, Determine which fragments contain further cleavage sites (i.e., which are cut up in a double digestion). 4. Determine how the fragments can be combined to form a logical whole. This step requires some combining abilities and offers a nice contrast to daily pipetting. The example shown can by used for practice.

work, because the sequencing reaction is generally accomplished more easily, it is not radioactive, and you do not have to worry about the service of the instrument.

If you sequence only rarely and are not restricted by finances, you can let someone else do the sequencing. These companies supply sequences at a fixed price per base and with complete documentation.

**Literature**

Maxam AM, Gilbert W. (1977) A new method for sequencing DNA. Proc Natl Acad Sci U S A 74:560–564.

Sanger F, Nicklen S, Coulson AR. (1977) DNA sequencing with chain-terminating inhibitors. Proc Natl Acad Sci U S A 74:5463–5467.

## 8.1.1 Radioactive Sequencing

Radioactive sequencing is the classic system of manual sequencing. The DNA to be sequenced is denatured, hybridized with a primer, and lengthened with the help of a polymerase, similar to the procedure used in polymerase chain reaction (PCR) (see Chapter 4, Section 4.1). In addition to the four usual $2'$-deoxynucleotides, the preparation contains a certain amount of a fifth nucleotide, which differs from the other in so far that the necessary hydroxyl group of the third C atom of the ribose, to which the next nucleotide is usually bonded, is lacking. If a $2',3'$-dideoxynucleotide is inserted instead of a dNTP, the chain can no longer be lengthened any further. Installation occurs according to the laws of statistics, depending on whether the polymerase of the required base meets the deoxy or the dideoxy variant precisely. The new DNA chains synthesized in this preparation have very different lengths. Whether the fragments are more likely to be shorter or longer depends on the concentration of the dideoxynucleotide. This is of interest when you crystallize a sequence mix yourself; previously you would have used different sequencing kits to obtain short or long sequences.

The dideoxynucleotide that is inserted (i.e., ddATP, ddCTP, ddGTP, or ddTTP) determines with which base the newly synthesized DNA fragments end. Four preparations are made, each with a respective dideoxynucleotide, and the fragments synthesized are differentiated according to their length on a denatured acrylamide gel. Because the fragments are labeled radioactively (discussed later), it can be determined how far they have run by laying x-ray film on top and developing it after a sufficient exposure time (usually overnight). The running stretch corresponds with the length of the fragment, and the trace of the preparation reveals with which nucleotide the fragment ends. In this way, you "climb" from below to above, with a film on the left and a notepad on the right, until the difference in the runs is so small between the fragments that you can no longer decide which is the next base. The distance of the run is determined for the most part by means of the gel (e.g., concentration, tension, electrophoresis time), a topic that is discussed later.

Radioactive labeling of the synthesized fragments can be performed in three different ways:

- Use a **5′-labeled primer** (see Chapter 7, Section 7.1.1).
- Add **radioactively labeled nucleotides**, which are inserted during the synthesis.
- Label the **dideoxynucleotide**.

Each of these processes has advantages and disadvantages. The strongest signals are received through insertion of labeled nucleotides, because each new fragment is labeled multiply, thereby shortening the exposure times considerably, and because time is of the essence, sequencing usually is done by this method. Although the first bases after the primer are missed in this way, none or very few labels are built into these very short fragments. You can counteract this with labeled primers. Because the labeling is located in the primer, the signal strength for all fragments is identical, regardless of their length. If you use labeled dideoxynucleotide instead, only fragments labeled correctly end with a dideoxynucleotide. This process also has its advantages, because structures that are difficult to dissolve prematurely give up the polymerases, and the synthesis is interrupted. That also occurs with the application of labeled dideoxynucleotides, although these fragments are not seen later because they are not labeled. Nevertheless, these fragments carry only a single label, and four radioactively labeled dideoxynucleotides are required; this process is applied rarely in radioactive sequencing.

Labeling usually is performed using $^{32}$P or $^{35}$S. The choice is a matter of taste. $^{32}$P has a higher radiation energy, and the **exposure times** are consequently shorter. You can use intensifying screens, which shorten the exposure even further. Unfortunately, this also leads to blurred bands, because the film surface lying directly adjacent is blackened. The **resolution** is also poorer, and the sequence can be read to only a reduced level. With $^{35}$S, labeling functions in a converse manner. The radiation energy is so weak that even a layer of plastic between the gel and the film eliminates a large part of the radiation; the exposure time is increased somewhat, and the bands become sharper. If you do not carry out sequencing procedures daily, the longer half-life ($^{35}$S: 87 days; $^{32}$P: 14 days) may be an important aspect, because you will not have to constantly purchase new nucleotides. [33]P has become somewhat more fashionable over the past few years; its characteristics are somewhere between those of [35]S and [32]P, although its radiation energy and half-life are more similar to the characteristics of [35]S.

You can use every arbitrary, double-stranded DNA as a template, as long as it is clean enough. Single-stranded DNA functions somewhat better and provides longer sequences. The work required to produce them is quite extensive (see Chapter 6, Section 6.2.2 and Chapter 2, Section 2.5.5).

Sequencing was previously accomplished using Klenow polymerase, until T7 DNA polymerase was discovered and cloned, and it proved to have a better proportioned band pattern so that it was easier to interpret the results. People without any experience were able to read such sequences. A modern alternative is **cycle sequencing** with the omnipresent Taq DNA polymerase, which is practically another application using PCR. Because only one primer is used, only linear (not exponential) amplification is obtained, although this is frequently sufficient to substantially reduce the amount of template required. Another advantage is the higher temperatures used in *cycle sequencing*, with which even stubborn, secondary structures and GC-rich sequences can be overcome. Such structures are feared, because they can lead to compressions (i.e., regions of 2 to 20 bases that cannot be read) or to terminations in the synthesis. To solve this problem, different techniques were developed, such as substitution with 7-deaza-dGTP, which causes such structures to become destabilized, although success in this matter is by no means guaranteed.

The four sequencing preparations are applied side by side to a denatured polyacrylamide gel (4% to 6% [v/v] acrylamide/8 M urea in TBS [see Appendix 1, Standard Solutions], mostly in the GATC or ACGT sequences), and the synthesized fragments are separated according to their size by means of electrophoresis at approximately 3000 V. The gel must be denatured, because the single-stranded DNA, just like RNA, tends to form secondary structures so that the migrational behavior in the gel is delayed (an effect taken advantage of with the SSCP [see Section 8.2.2]), making it impossible to carry out a clean separation according to size. The gel is fixed in 10% (v/v) acetic acid, transferred from the glass plate to filter paper, dried, and ultimately autoradiographed (i.e., an x-ray film is exposed). Overnight exposure is normally sufficient, as long as the sequencing has functioned to some extent.

Reading the sequence is the next step. You begin at the lower edge of the film and must decide in which of the four tracks the next largest fragment can be found (i.e., which of the four bases follows next) (Figure 8-2). Poor sequences can be read for about 200 nucleotides, but good sequences can be read for 400 nucleotides, and very good sequences can be read for 600 nucleotides. However, you usually must run two gels with different electrophoresis times. Kits for the sequencing are available from Promega, Pharmacia, and Amersham.

## 8.1.2 Nonradioactive Sequencing and Automatic Sequencing Units

There is a nonradioactive alternative for sequencing available from Promega (SILVER SEQUENCE) that demonstrates the DNA fragments by the classic means of **silver staining**. The gel is first watered down in acetic acid to precipitate the DNA and to wash out the urea and electrophoresis buffer. It is then incubated in a silver nitrate–formaldehyde solution and developed in an alkaline sodium carbonate solution with formaldehyde and sodium thiosulfate. The reaction is terminated using acetic acid. The gel can be blotted on filter paper and then dried, producing good documentation of your work.

The silver staining is faster than autoradiography, which usually requires 12 to 24 hours, although it also demands more manipulations and therefore more work. Another problem is the purely practical aspect. Gels of these dimensions cannot be flooded in a Tupperware box.

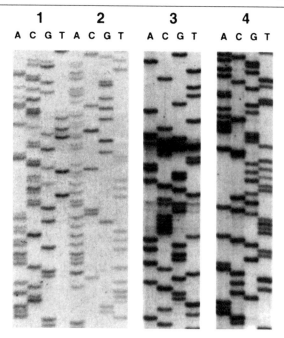

**Figure 8-2. Radioactive sequencing.** In addition to many lovely bands, sequence 3 demonstrates a small compression.

Roche offers a nonradioactive sequencing kit (*DIG Taq DNA Sequencing Kit*) which, like all of the Roche systems, is based on a **digoxigenin labeling** of the synthesized fragments. To carry out detection, the DNA must be blotted from the gel onto a nylon membrane. This is a simple process, although membranes of this size are quite expensive, and there is a problem of oversize.

These are somewhat exotic experiments. Nonradioactive sequencing usually begins with **automatic sequencing** (Figure 8-3). The DNA is labeled with **fluorescent dyes**, which are excited by a laser and measured during electrophoresis. The reading lengths are substantially longer than with manual sequencing. Reading lengths under optimal conditions are realistically 500 to 1000 nucleotides, whereas they are about 300 to 500 nucleotides in normal procedures.

The labeling is performed in the primer (which must be ordered with a corresponding modification such that the freedom in sequencing is somewhat limited) or in the dideoxynucleotides. Labeled deoxynucleotides are not used, because fluorescent dyes are large molecules that have a considerable effect on the migrational behavior of the fragments. The differences in migration can be corrected using a computer, but only if the fragment contains a defined number of labeled sites.

Labeled dideoxynucleotides offer a substantial advantage. The four dideoxynucleotides can be labeled using different dyes so that the sequencing reaction can be carried out in a single vessel and a single track on the gel is required per sequence, which saves work and increases the throughput. Because of patent restrictions, four-color sequencing has been available only from Applied Biosystems, but Pharmacia has recently caught up with them in this field.

As with radioactive sequencing, you can use T7 or Taq polymerase. After the sequencing, the preparation is briefly purified and then given to the person who is responsible for the sequencing device. They are too expensive to be able to use them personally, a situation that may frustrate individual researchers, but it also spares them a lot of work.

You can pick up the results in the form of a four-color print, one color for each base. The evaluation is substantially more simple than with autoradiograms, but it is still a matter of experience. You

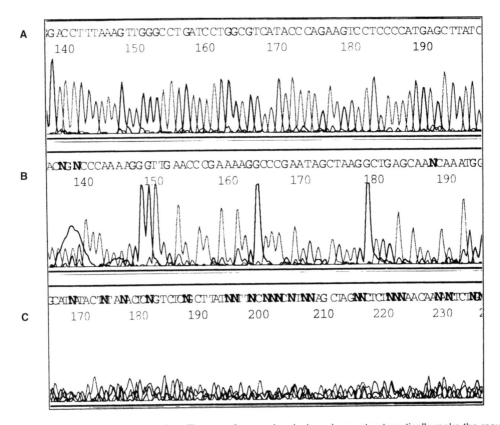

**Figure 8-3. Nonradioactive sequencing.** The use of expensive devices does not automatically make the results better. **A,** A lovely sequence. **B,** A sequence with only a moderate quality, but one that can still be evaluated. **C,** An unattractive result.

cannot avoid checking the printed sequence and will soon ascertain that peculiarities appear. In the *Dye Terminator Cycle Sequencing Kit* from Applied Biosystems, for instance, it is occasionally extremely difficult to work with the Gs. Labeled nucleotides or primers that have not been inserted may lead to an intensive background, which complicates reading of the first 100 bases. That is why the manufacturers of the kits recommend that you purify the preparations before application, using techniques such as ethanol precipitation. On other points, you should not blindly trust the manufacturers. It may be, for instance, that much more template DNA is required than indicated for a good result; in my case, it has been as much as 2 to 5 µg. You must determine what works.

Automatic sequencing units can be obtained from ABI (Applied Biosystems) and from Pharmacia.

## 8.1.3 Minisequencing

With **minisequencing** (i.e., *primer extension method* or *solid-phase minisequencing*), you must free yourself from the classic interpretation of the term *sequencing*. It is not a method for determining unknown DNA sequences, but rather a means for providing evidence of very specific, exactly defined mutations with relatively little effort. For this reason, the term *primer extension* represents the heart of the matter substantially better. In the simplest version of the method, the DNA to be examined is amplified by PCR, and one of the two primers is labeled on the 5′ end with biotin. Thanks to the biotin markers, you can bind the PCR product to streptavidin, which itself is bound in an arbitrary, stable phase. Through denaturation with an alkaline solution and subsequent washing, the unlabeled DNA chains are removed, and a single-stranded DNA template is obtained onto which the sequence primer is hybridized. These primers are formed in such a manner that the base to be examined directly follows the last 3′ base of the primer.

By means of DNA polymerase and a labeled dideoxynucleotide, a DNA synthesis is performed. It consists of only one step. If the dideoxynucleotide added (e.g., ddATP) is complementary to the next base of the DNA template (e.g., T), a single nucleotide is added, and no insertion occurs otherwise. After that, you must wash it and examine the product of the synthesis. If it contains a label, the first base behind the primer was then evidently complementary to the inserted nucleotide. If not, it was one of the other three bases.

Initially, this approach sounds absurd, but the method is very interesting for investigating a large number of probes for a specific mutation. In the case of *single-nucleotide polymorphisms* (SNPs [see Chapter 9, Section 9.8]), there is question about how frequently they are represented in a certain population. If you use streptavidin-coated 96- or 384-well plates, this question can be answered in a substantially shorter time and reduced cost than if you sequence 500 DNA probes in a classic manner.

There are several varieties to this method. If the labels can be differentiated (for instance, using $^{3}$H and $^{32}$P, or digoxigenin and fluorescein, or fluorescent dyes), you can easily insert more than one labeled nucleotide per preparation and consequently reduce the number of preparations or clearly differentiate heterozygous from homozygous genotypes. If one does not bind the template in a stable phase, you can then develop multiplex preparations in which more than a single mutation can be explored by uniting several templates with their respective sequence primers in a preparation; the primers must differ in length and the "sequencing reaction," and depending on the labeling of a sequencing gel or a sequencer, analysis can determine whether a nucleotide has been added.

Especially interesting for minisequencing is the application of **DNA chips** (see Chapter 9, Section 9.1.6). By reversing the order, the sequencing primer is fixed on the DNA chip, and the template is hybridized on the primer. Hundreds of bases can be sequenced rapidly and with little effort. This is probably the most promising transformation brought about by minisequencing.

**Literature**

Pastinen T, Kurg A, Metspalu A, et al. (1997) Minisequencing: A specific tool for DNA analysis and diagnosis on oligonucleotide arrays. Genome Res 7:606–614.

Sokolov BP. (1990) Primer extension technique for the detection of single nucleotide in genomic DNA. Nucleic Acids Res 18:3671.

Syvänen A-C. (1999) From gels to chips: "Minisequencing" primer extension for analysis of point mutations and single nucleotide polymorphisms. Hum Mutat 13:1–10.

# 8.1.4 Pyrosequencing

A Swedish company known as Pyrosequencing is marketing a new system with which short DNA fragments can be analyzed relatively rapidly. Pyrosequencing is a neologism, made up of *pyros* (Greek for "fire," because light is produced) and *sequencing*. The method has little to do with sequencing as it is known in classic molecular biology, because the limits are approximately 20 to 30 bases, which is a completely different order of magnitude from that of approximately 800 bases, which can be achieved with optimal Sanger sequencing.

The method differs considerably from that employed in Sanger sequencing, because there are no labeled nucleotides and no dideoxynucleotides are installed. In a certain way, this is a "back to the roots approach", because the qualities of a classic, coupled reaction have been remembered in the experimental design.

By means of PCR, a test fragment from the probe material is amplified, and one of the two primers used is labeled with biotin. The DNA fragment is coupled to streptavidin-bound beads (from sepharose or paramagnetic Dynabeads) and denatured with NaOH. The biotin-labeled chain remains bound to the pellets, and the complementary chain is washed; ultimately, a single-stranded template remains. Up to this point, the experimental principles are identical to those used in minisequencing (Figure 8-4).

The template is pipetted onto a microtiter plate, and the sequencing primer, the necessary enzyme mixture (i.e., DNA polymerase, ATP sulfurylase, luciferase, and apyrase), and the substrate mixture (i.e., adenosine-5′phosphosulfate [APS] and luciferin) are added. The preparation is placed in the machine. The primer is first hybridized on the template, and this is followed by a gradual DNA synthesis. The machine adds one of the four necessary nucleotide triphosphates and lets the reaction take its course. What happens then? If the dNTP added is complementary to the next base of the template, it is inserted by the DNA polymerase, and a pyrophosphate (PPi) molecule is released. The ATP sulfurylase synthesizes a molecule of ATP from the PPi and APS, which is transformed in another reaction into light and oxyluciferin by the luciferase together with luciferin. For almost every installed dNTP, a light quantum emerges. The light production is recorded during the entire procedure by means of a CCD camera; it is quantified, evaluated, and presented in the form of signal peaks. If two or three of the same nucleotides follow one another in the template, several nucleotides will be installed in a single round, and the light production will be twofold or threefold as high as normal; if no nucleotide is inserted, it remains pitch black in the *well*.

To remove surplus dNTPs and to complete the reaction rapidly, the dNTPs, parallel to the detection reaction described previously, can be broken down into dNDP and phosphate by adding apyrase. The entire dNTP is exhausted within a minute, and the cycle is completed. Five seconds later, the next nucleotide can be added, and a new cycle begun. At first, the apyrase reaction appears to be absurd because, aside from the dNTPs, the freshly formed ATP is also disassembled, which is required for light production. This juxtaposition is possible, however, because the degradation resulting from apyrase occurs more slowly than the detection reaction.

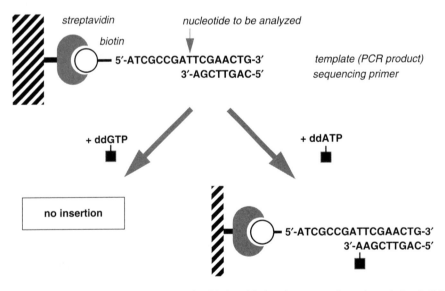

**Figure 8-4. Minisequencing.** A single dideoxynucleotide is added to the sequencing primer, but only if the inserted nucleotide is complementary to the next base of the template (here, a T). The insertion can be shown by labeling of the dideoxynucleotide.

With known sequences, it is not always beneficial to stubbornly work with the four bases in the same sequence, as is generally performed with unknown sequences, because other methods can help to save time and material. Anyone who has had experience with advertising material for pyrosequencing (which is rather unsatisfactory from a scientific aspect) will have some difficulty in understanding the sequence of the cycles or the addition of the dNTP, as indicated in the examples provided. For instance, a nucleotide is employed at the beginning of the sequencing that is not inserted at all, and in this way, it is possible to make a statement concerning the signal background. If the sequence contains several of the same bases next to one another, the fitting nucleotide is added successively, twice (to control whether the insertion had been carried out fully during the first addition). This is confusing. You can imagine how complex the affair becomes for performing multiplex analyses (it should be possible to perform up to three SNPs in a single preparation). Luckily, there is a software product available that takes such problems into account.

**Advantages:** According to the manufacturers, their equipment allows a rapid method for the analysis of individual base polymorphisms with a high throughput—up to 4500 tests can be analyzed—and with a second, related system, you can sequence up to 48,000 probes per day. Because the light signal can be quantified very well, you can easily determine the frequency of alleles by pooling the probe to be tested and, at the end, determining the frequency of the two expected alleles from the size of the signal peaks. Because the polymorphism does not have to lie directly adjacent to the sequencing primer, however, you have a better chance with primer designs. Polymorphisms that are close to one another can be analyzed in a single process.

**Disadvantages:** This is not a procedure for simple research laboratories. The acquisition of such a system is profitable only for routine laboratories in which a large quantity of standardized analyses must be carried out, such as for the characterization of microorganisms, diagnosis of diseases, or analysis of allele frequencies. It is only moderately useful for the determination of unknown sequences. The work required is considerable, and special software is required for the instrument and for the analysis. A complete system costs $130 to $200,000. An attempt to analyze an SNP, for instance, costs about $1.25.

**Literature**

Ronaghi M, Karamohamed S, Petersson B, et al. (1996) Real-time DNA sequencing using detection of pyrophosphate release. Anal Biochem 242:84–89.

Ronaghi M, Uhlen M, Nyren P. (1998) A sequencing method based on real-time pyrophosphate. Science 281: 363–365.

## 8.1.5 Idiosyncrasies in Sequencing: Octamers

Everyone has experienced identifying a clone and, after short time, being plagued by serious doubts concerning the identity of the plasmid because attempts at cloning or amplification simply will not work. The best approach is to sequence the *corpus delicti* quickly to clarify the matter. It is stupid only if you have no suitable primer at hand. Perhaps the following method can be useful.

Instead of the usual primers of 16 to 25 nucleotides, the trick is to use **octamers**. Octamers are the shortest oligonucleotides that are still reliable for sequencing because the annealing temperature is reduced sufficiently (i.e., to approximately 30°C [86°F]). However, a given octamer occurs purely randomly in about every 65,500 nucleotides; if both chains are considered, the chance of finding such a structure is 1 in 33,000. If you have purchased a few dozen octamers, you should find one among them with the sequence of the questionable clone, and if you try all of the octamer primers, you may discover unknown plasmid sequencing information.

Some rules with regard to the formation of the octamer primer must be considered. More details can be found in the sources listed in the Literature section that follows.

**Literature**

Ball S, Reeve MA, Robinson PS, et al. (1998) The use of tailed octamer primers for cycle sequencing. Nucleic Acids Res 26:5225–5227.

Hardin SH, Jones LB, Homayouni R, McCollum IC. (1996) Octamer-primed cycle sequencing: Design of an optimized primer library. Genome Res 6:545–550.

Jones LB, Hardin SH. (1998) Octamer-primed cycle sequencing using dye-terminator chemistry. Nucleic Acids Res 26:2824–2826.

# 8.2 Methods of Analyzing DNA for Mutations

Even if sequencing provides the clearest results, it is not always the method of choice when searching for mutations Luckily there are alternatives. Some are simpler and others more complicated; some are cheaper and others more expensive.

## 8.2.1 Restriction Fragment Length Polymorphism

The search for restriction fragment length polymorphisms (RFLPs) is the simplest of research methods and the grandmother of mutation analysis. The principle is rather banal. The DNA to be examined is digested with different restriction enzymes, separated by means of gel electrophoresis, and the band pattern compared with normal and examined DNA (Figure 8-5). Changes in the band patterns are indications of mutations.

**Advantage:** You can also analyze very complex mixtures of DNA fragments by blotting the gel and hybridizing with a suitable probe. In this way, sections of genomic DNA that are several kilobases long can be analyzed without previously isolating, amplifying, or purifying the corresponding sections in some manner.

**Disadvantage:** Only relatively large mutations (e.g., deletions, insertions) can be verified. The change must be large enough that normal and mutated fragments differ visibly in their migrational behavior. You can demonstrate point mutations only if they destroy or create a new cleavage sites, and this is purely a matter of luck.

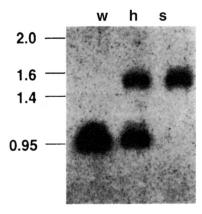

**Figure 8-5. Southern blot with restriction fragment length polymorphism.** The photograph reveals the hybridization of the genomic DNA of three mice with different genotypes. h, heterozygous mouse; s, homozygous mutant; w, wild-type mouse.

## 8.2.2 Single-Strand Conformation Polymorphism

*Single-strand conformation polymorphism* (SSCP) is a conformational polymorphism on a single strand. Nucleic acid chains have a strong tendency for base pairing. Because of the lack of a complementary strand, RNA and single-stranded DNA pair with themselves and thereby assume strange conformations, which are highly complex and cannot be determined beforehand. The conformation depends on many different factors, including the sequence and the temperature. With a bit of luck, you can find a temperature at which the mutation of only a single base leads to a conformational change, which alters the migrational behavior of the fragment in (polyacrylamide) gel (Figure 8-6). After you have discovered these particular conditions, you can examine many DNA probes with this method in a short time. The most frequent use of this method is in screening patients for mutations that may prove to be responsible for specific diseases.

For SSCP, an arbitrary fragment of the DNA of approximately 250 bp is amplified by means of PCR. Between 50 and 100 ng of the product in a 10-μL volume is mixed with 1 μL of 0.5 M NaOH/10 mM EDTA and 1 μL of blue marker (50% [w/v] sucrose with bromophenol and xylene cyanol blue) and denatured for 5 minutes at 55°C (131°F). Immediately afterward, the probe is applied to a nondenatured 5% to 10% polyacrylamide gel, and the DNA is separated electrophoretically. At the end of the migration, the DNA is dyed, and the gel is evaluated.

The temperature of the electrophoresis device can be set so that a **definable temperature** can be maintained constantly throughout the entire course of the gel migration. Usually, several gels are run at different temperatures for each sample, such as at 10°C, 15°C, and 25°C (41°F, 59°F, 68°F, and 77°F, respectively) with the hope of observing a sign of a polymorphism at one of these temperatures. When looking for a specific mutation, you have to determine the optimal conditions once and then proceed with a single gel.

The DNA can be demonstrated in different ways. In the original protocol from Orita and colleagues (1989), the DNA was labeled **radioactively** and identified by autoradiography. An alternative that is

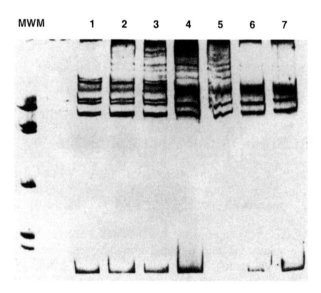

**Figure 8-6. Single-strand conformation polymorphism (SSCP).** In a silver-stained, real-life SSCP gel, tracks 1 to 3 show DNA from heterozygous individuals with a normal and a point-mutated allele. Tracks 4 to 7 represent DNA from people with two normal alleles. MWM, molecular-weight marker. (Courtesy of Nicoletta Milani and Cord-Michael Becker.)

somewhat more troublesome but faster and not radioactive is **silver staining**. Silver-stained gels can be dried and preserved.

Another possibility is staining with ethidium bromide, as described by Yap and McGee (1992), although its sensitivity leaves something to be desired, and some of the weaker bands may be overlooked. Other dyes, such as SYBR Green I from Molecular Probes, may be more suitable because of their higher sensitivity.

A useful "abuse" of the SSCP was described by Suzuki and coworkers (1991). They separated a mixture of alleles of a DNA fragment by means of SSCP, isolated the individual bands from the gel, amplified them again, and separately sequenced each allele.

**Advantages:** For the analysis, arbitrary primers can be used. The method can be relatively easily adjusted to suit large sample quantities.

**Disadvantages:** The length of the fragments that can be analyzed by means of SSCP is limited to about 250 bases. With longer fragments, the probability of detecting a mutation decreases substantially. Because each mutation behaves differently with regard to denaturing and the migration pattern, several gels must be run at different temperatures for each fragment analysis, a situation that can increase the work required appreciably (if you are not working with a specific mutation whose exact conditions are known). Generally, the probability of discovering an arbitrary mutation by means of SSCP is between 60% and 85%. Reproducibility, however, is somewhat problematic often only approximately 80%.

**Literature**
Hayashi K, Yandell DW. (1993) How sensitive is PCR-SSCP? Hum Mutat 2:338–346.
Orita M, Iwahana H, Kanazawa H, et al. (1989) Detection of polymorphisms of human DNA by gel electrophoresis as single-strand conformation polymorphisms. Proc Natl Acad Sci U S A 86:2766–2770.
Orita M, Suzuki Y, Sekiya T, KaayashiK. (1989) Rapid and sensitive detection of point mutations and DNA polymorphisms using the polymerase chain reaction. Genomics 5:874–879.
Suzuki Y, Sekiya T, Hayashi K. (1991) Allele-specific polymerase chain reaction: A method for amplification and sequence determination of a single component among a mixture of sequence variants. Anal Biochem 192:82–84.
Yap EP, McGee JO. (1992) Nonisotopic SSCP detection in PCR products by ethidium bromide staining. Trends Genet 8:49.

## 8.2.3 Denaturing Gradient Gel Electrophoresis

The idea of denaturing gradient gel electrophoresis (DGGE) is similar to that used with SSCP, because you would like to change the migrational behavior of the fragments through the application of suitable conditions. An important advantage over SSCP is that you can also analyze fragments more than 500 bp.

With DGGE, you can take advantage of the fact that DNA fragments are not denatured all at once, but can instead be subdivided into **melting domains** with different melting temperatures. Partially melted DNA fragments demonstrate an altered migrational behavior during electrophoresis. The melting behavior of the domains depend on their particular sequences. This means that mutations change the melting behavior of the affected domains. Even a point mutation can have an effect.

In principle, the phenomenon is effective for every DNA fragment, although the effect can best be shown if the fragment contains an area with a substantially higher melting temperature, which holds the two chains together while other areas melt away. You can also insert such an area artificially by supplying one of the two PCR primers with a **GC clamp** on the 5′ end, a GC-rich sequence of 25 to 30 bases.

The next difficulty is to maintain the fragment in a partially melted form to take advantage of the altered migrational behavior of the DNA fragment. In practice, temperatures of 50°C (122°F) to 60°C (140°F) are used for DGGE in combination with a denatured substance, such as urea or formamide.

How much urea should be used? The people from BioRad, who have developed a gel apparatus for the DGGE, recommend the following procedure. The wild-type probe is run on a gel with isovertical gradients (i.e., from the left to the right) with 0% to 80% urea to determine the conditions at which partial denaturing of the DNA fragments results (Figure 8-7B). Under these conditions, the probability

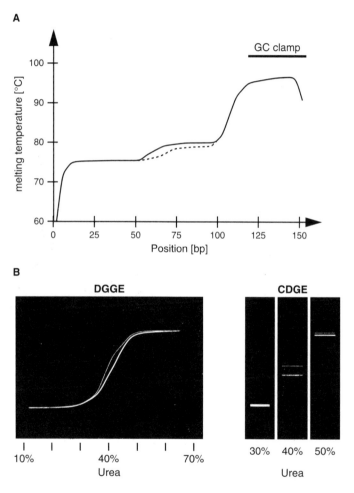

**Figure 8-7. Denatured gradient gel electrophoresis (DGGE) and constant denaturing gel electrophoresis (CDGE). A,** Theoretical melting curve of a 150-bp fragment with a GC clamp (a GC-rich segment 25 to 30 bases long). The *dotted line* shows the change in the melting curve caused by a point mutation. **B,** DGGE of the two fragments in one 10% to 70% urea gradient. The mutated DNA is represented with a *thinner line*. CDGE of the same fragments at 30%, 40%, and 50% urea. With a 40% urea gel, an experimenter would be most likely to discover the mutation; this would not be possible with 30%, and at 50%, this result would be a matter of luck.

of observing a change in the migrational behavior brought about by mutation is greatest. The melting behavior is constant for each fragment, and a single gel is sufficient for determining the melting curve for a given PCR product. For the analysis, there are two possibilities:

- The probe is applied to a gel with isovertical gradients, although the experimenter concentrates on the region that demonstrates a better resolution because of the development of partial denaturation (e.g., 30% to 60% urea). In this way, there will be a good chance of discovering a mutation probability (allegedly as high as 95% to 99%). However, one gel is required per sample.
- The alternative is to use **constant denaturing gel electrophoresis** (CDGE), which is a DGGE gel with a constant urea or formamide concentration in the area previously determined with a DGGE gel. The result looks similar to that seen with an SSCP gel, because the DNA fragments run as bands, and the mutations lead to altered migrational behavior (see Figure 8-7C). The partially denatured conditions are constant over the entire gel, improving differentiation of normal from mutated fragments.

**Advantage:** The number of the probes that can be analyzed with a gel is substantially larger with CDGE.

**Disadvantage:** Because the area of the melting curve in which the mutation demonstrates its effects on the migrational behavior is unknown, a mutation may be missed if the concentration of urea chosen is incorrect. Several gels can be run with different concentrations of urea, but this increases the amount of work.

### Literature

Borresen AL, Hovig E, Smith-Sorensen B, et al. (1991) Constant denaturant gel electrophoresis as a rapidly screening technique for p53 mutations. Proc Natl Acad Sci U S A 88:8405–8409.

Fischer SG, Lerman LS. (1983) DNA fragments differing by single base-pair substitutions are separated in denaturing gradient gels: Correspondence with melting theory. Proc Natl Acad Sci U S A 80:1579–1583.

## 8.2.4 Temporal Temperature Gradient Electrophoresis

Another variant of DGGE is temporal temperature gradient electrophoresis (TTGE), in which the concentration of the denatured substance is maintained constant and a **temperature gradient** is used instead. BioRad describes the method as being a "hot" alternative to DGGE and to SSCP.

For TTGE, as for DGGE, you need a PCR fragment with a GC-rich segment that holds the double chain together at increasing temperatures. You also need a device with which to run a temperature gradient; it should increase the temperature continuously by about 1°C to 3°C (34°F to 37°F) per hour. At a certain temperature, which depends on the sequence, the DNA begins to melt and to change its migrational behavior. Mutations lead to changes in the melting behavior and consequently to a change in the migrational behavior, which is seen as a difference in the band pattern.

To be successful, you should have an idea of the conditions required for denaturing so that you can keep the urea concentration and the temperature gradient within a sensible range. BioRad offers a program that can calculate the optimal conditions based on the specific sequence. Without this, you can try various combinations or use an isovertical DGGE.

**Advantage:** High throughput is possible because many samples can be applied to a gel. The rate of success is about 95% to 99%. Because it is not a gradient gel, it is easy to pour.

**Disadvantage:** As with DGGE, requirements for TTGE include a special primer and a gel device for controlling the temperature. BioRad offers such an instrument (DCode Universal Mutation Detection System) with a gel-pouring stand for approximately $6000. The software for the analysis of the melting behavior can be obtained for about $200. The gel run time, however, may be quite long.

**Literature**

Riesner D, Steger G, Zimmat R, et al. (1989) Temperature-gradient gel electrophoresis of nucleic acids: Analysis of conformational transitions, sequence variations, and protein-nucleic acid interactions. Electrophoresis 10:377–389.

Wiese U, Wulfert M, Prusiner SB, Riesner D. (1995) Scanning for mutations in the human prion protein open reading frame by temporal temperature gradient gel electrophoresis. Electrophoresis 16:1851–1860.

## 8.2.5 Heteroduplex Analysis

Two different methods are applied under the concept of heteroduplex analysis. With the molecular biologic method, a standard DNA and a DNA to be analyzed are amplified separately, mixed, denatured, and slowly cooled (approximately 1°C to 2°C [34°F to 37°F] per minute) to permit the formation of heteroduplices from standard and sample DNA. The preparation is then applied to a polyacrylamide gel and subjected to electrophoresis for several hours. Slight differences in the migrational behavior can produce mutation-specific band patterns.

The method is quite popular in genetics for the analysis of extremely polymorphic genes. It is technically simple, although it is far more suitable for providing evidence of larger changes than for point mutations. Success depends on many factors, such as the fragment length or the specific acrylamide used. More information is available in the articles by Zimmermann and colleagues (1993) and D'Amato and Sorrentino (1995).

**Literature**

D'Amato M, Sorrentino R. (1995) Short insertions in the partner strands greatly enhance the discriminating power of DNA heteroduplex analysis: Resolution of HLA-DQB1 polymorphisms. Nucleic Acids Res 23:2078–2079.

Zimmermann PA, Carrington MN, Nutman TB. (1993) Exploiting structural differences among heteroduplex molecules to simplify genotyping the DQAl and DQB1 alleles in human lymphocyte typing. Nucleic Acids Res 21:4541–4547.

## 8.2.6 Amplification Refractory Mutation System

The *amplification refractory mutation system* (ARMS) is a PCR procedure that is especially suitable for providing evidence of known mutations. With the aid of a specific primer (each mutation requires its own), you can examine a DNA probe for the presence of a mutation. The method is described in Chapter 4, Section 4.3.9.

## 8.2.7 Enzyme Mismatch Cleavage

Enzyme mismatch cleavage (EMC) is a practical method for the detection of point mutations and more extensive changes. Mutated (or the sample to be analyzed) and normal DNA are amplified separately. The normal DNA is labeled radioactively through phosphorylation of the primer or the product. The mutated and normal DNA are mixed, heated, and cooled very slowly. DNA heteroduplices emerge, which are made up of one mutated and one normal chain. If a mutation is present, mismatches develop that change the structure of the double strand. **Endonuclease VII** (of the T4 phage) recognizes these mismatches relatively specifically and cleaves the DNA at these sites. The cleavage products are separated electrophoretically and then autoradiographed. If a mutation exists, the radioactive fragment is smaller than before this process (Figure 8-8).

A problem is the relatively intense background. Endonuclease VII cuts in regions that demonstrate faulty pairs and cuts normal double-stranded DNA, although at a substantially slower rate. Because

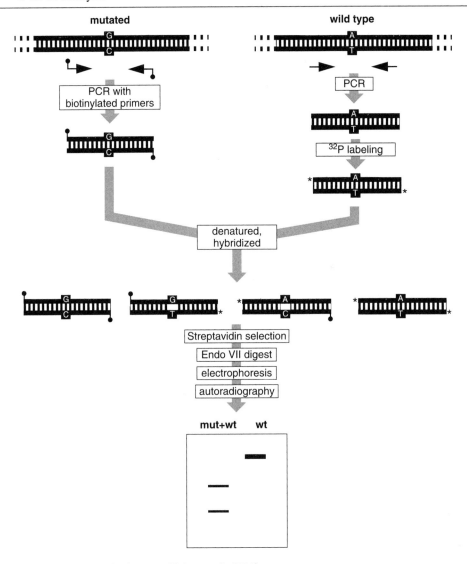

**Figure 8-8. Enzyme mismatch cleavage** (Babon et al., 1995).

more homoduplices than heteroduplices develop after rehybridization, a rather intense background pattern is obtained from the digested homoduplices, which can cause the interesting bands to disappear.

Babon and colleagues (1995) propose a modified protocol that leads to a substantially less intense background. For amplification of the DNA being analyzed, a **biotin-labeled primer** is used; the normal DNA is amplified with unlabeled primers. After the heteroduplex formation, you can then purify the biotin-labeled double strand from the radioactively labeled normal homoduplices, which are responsible in large part for the background, by binding them to streptavidin-coupled, paramagnetic pellets (Dynabeads, Dynal). After the endonuclease VII digestion, a substantially better signal-to-background relationship is obtained.

**Advantage:** In addition to identifying the presence of a mutation, information concerning the site may be obtained based on the band pattern.

**Disadvantage:** The signals are not always intense, because some mismatches are cleaved poorly. Occasionally, the signals are so weak that you can overlook mutations.

**Literature**
Babon J, Youil R, Cotton RGH. (1995) Improved strategy for mutation detection—a modification to the enzyme mismatch cleavage method. Nucleic Acids Res 23:5082–5084.

## 8.2.8 Protein Truncation Test

The *protein truncation test* (PTT) (essentially evidence for protein shortening) is somewhat unusual because it deals with proteins, a substance the case-hardened molecular biologist prefers not to touch.

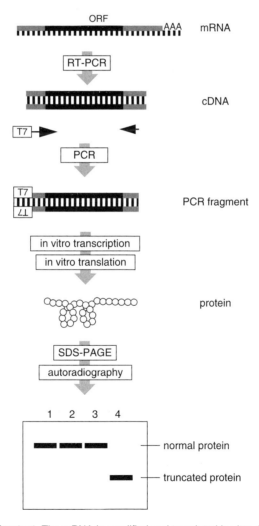

**Figure 8-9. Protein truncation test.** The mRNA is amplified and translated in vitro. In this way, all mutations of the gene can be recorded that lead to a change in the size of the coded protein.

RNA is isolated, cDNA is produced from this, and the fragment of your dreams is amplified from this. One of the two primers contains a T7 promoter sequence at the 5′ end (Figure 8-9). The PCR product can be transcribed in vitro and then translated. The length of the synthesized protein is determined by sodium dodecyl sulfate–polyacrylamide gel electrophoresis (SDS-PAGE).

The method is popular for the analysis of some genes, such as *DMD* (Duchenne's disease, spinal muscular atrophy) or *BRCA* (breast cancer), for which it is known that a mutilation (*truncation*) of the protein frequently occurs. Promega offers a kit for the coupled amplification and translation (TNT T7 Quick Coupled Transcription/Translation System).

**Advantage:** PTT is suitable for the analysis of long DNA fragments, and information is obtained about the site of the mutation.

**Disadvantage:** PTT is suitable for providing evidence of mutations when the length of synthesized protein is altered substantially; point mutations, which lead to amino acid exchange, remain undetected, as well as smaller insertions and deletions, so long as they do not change the reading pattern. Many steps in the work process, substantial background intensity, and radioactivity must be considered.

### Literature

Roest PA, Roberts RG, Sugino S, et al. (1993) Protein truncation test (PTT) for rapid detection of translation-terminating mutations. Hum Mol Genet 2:1719–1721.

Hogervorst FBL. (1997) Promega Notes 62:7.

# 9 Investigating the Function of DNA Sequences

*Herr Doktor, nicht gewichen! Frisch!*

*Now, Doctor, courage! Don't give way!*

Cloning of DNA usually is not an end in itself, but only a means. This is unfortunate, because cloning is relatively fast and clean. Bacteria are patient, and DNA is even more so. The problems usually begin when you want to do something with your constructs.

When working with cDNA, you must first discover in which tissues the pertinent gene is expressed. It is least complicated if you take a specific antibody for the protein involved and perform a bit of histologic investigation with this. The problem in practice is that you typically have no specific antibody, it is not suitable for the histology, or the quantity of protein is too small to recognize it properly.

Molecular biologists are generally unwilling to concern themselves with such imponderable objects as proteins. How much more beautiful is it to demonstrate expression indirectly by means of the existence of mRNA by **ribonuclease protection**, **in situ hybridization**, or **in situ polymerase chain reaction** (PCR).

The next step consists of cloning the DNA in an **expression vector**, a plasmid that, in addition to the usual plasmid components such as the bacterial replication origin and antibiotic resistance, contains a gene with a promoter or a coding area. Depending on whether you want to examine a promoter or a cDNA, you replace the region of the DNA that interests you. These have become two kinds of problems for which there are different vectors. Anyone examining **promoters** wants to know when and how intensely active they are. Consequently, a vector is chosen with an easily demonstrable reporter gene, before which you can insert your promoter. Enhancer sequences can also be explored in this manner by using a vector that contains a basic promoter and whose activity in the cloning of regulatory elements can be upregulated or downregulated.

When determining the section in which the DNA activity can be found, you will return to mutagenesis to remove sequences, here and there, until you see which way the wind blows. The choice of a proper reporter gene is decisive in this kind of investigation, and some possibilities are presented in Section 9.4.8.

When working with **cDNAs**, you typically want to find out something about the function of the proteins. For this purpose, a multitude of expression vectors are available that deliver a promoter and consequently wait to be associated with an open reading pattern. With promoters, you have a free selection. Bacterial promoters permit the expression of proteins in bacteria, a satisfying procedure if you want to obtain large quantities of protein, because bacteria can be bred in almost arbitrary quantities. Viral promoters such as that of the cytomegalovirus (CMV) or from Simian virus 40 (SV40) permit a quite intense expression in mammalian cells, and you can frequently accomplish functional investigations immediately. Cell-typical promoters permit expression based on certain tissue limitations, such as in the manufacture of transgenic animals, and controllable promoters already exist (see Section 9.6).

The topic of mutagenesis frequently arises, and with the help of smaller mutations, it is possible to better define which areas of the protein are functionally significant and which are not or to "rapidly" insert an antibody epitope and to carry out a bit of proteinology.

# 9.1 Investigating Transcription in Tissues

Depending on whether you have more interest in how intensely a gene is expressed in a certain tissue or would like to know where expression occurs, different methods are used. Quantification can be carried out better with a *ribonuclease protection assay* or by means of *real-time quantitative PCR*, whereas in situ hybridization and in situ PCR provide a much better overview about where this occurs. Anyone who is interested in differences in gene expression between the different tissues or cell lines will probably find the *arrays* to be more interesting.

## 9.1.1 Ribonuclease Protection Assay

With the ribonuclease protection assay (RPA), you can provide evidence of and quantify a particular mRNA in the total RNA that you have isolated from your favorite tissue. The trick consists of hybridizing the RNA with a radioactively labeled RNA probe. All of the RNA that has not been hybridized is degraded through the digestion of a specific single-stranded RNase, the fragments are differentiated electrophoretically, and the quantity of remaining samples is determined, which is proportional to the originally available mRNA quantity.

Production of the **probe** is somewhat laborious because it requires a suitable DNA *template* from which to synthesize antisense RNA by means of in vitro transcription (see Chapter 5, Section 5.5). The probe is labeled by replacing one of the four nucleotides with a corresponding radioactively labeled nucleotide during synthesis. Subsequently, the DNA *template* is removed through DNase digestion, the sample is purified, and the activity is determined. Because it is concerned with RNA, the usual regulations concerning RNA are applicable.

The yield of the labeled probe is normally quite high, but even if it should turn out to be low, everything is not lost, because the specific activity is high because every fourth nucleotide that has been inserted is labeled radioactively.

The probe is then incubated overnight with the total RNA at about 45°C (113°F). The entire preparation is incubated with an RNase A/RNase Tl mixture for 1 hour at 30°C (86°F), purified with phenol-chloroform, precipitated with ethanol, and separated electrophoretically in a denatured polyacrylamide gel. The remaining quantity of undigested probe is verified by autoradiography. The signal can be quantified by scanning the film and then evaluating it or by cutting out the signal and then measuring the activity with a scintillation counter. The signal quantity provides information concerning the quantity of mRNA examined in the initial material.

You can avoid the trouble of producing the assay. Ambion offers a series of different RPA kits.

**Literature**
Williams DL, Newman TC, Shelness GS, Gordon DA. (1986) Measurement of apolipoprotein mRNA by DNA-excess solution hybridization with single-stranded probes. Methods Enzymol 128:671–689.

## 9.1.2 Real-Time Quantitative Polymerase Chain Reaction

Real-time quantitative polymerase chain reaction (RTQ-PCR) may be the most popular method for the quantification of nucleic acids, at least in places where a fitting PCR instrument is available. Experience has shown that these instruments usually are overloaded.

This popularity is based on the fact that much less effort is required to prepare an RTQ-PCR than is required for the preparation of a *ribonuclease protection assay*. The somewhat more lengthy evaluation required in front of the computer, especially for those who are less experienced, is no longer frightening for those active in this field, because everyone is already used to spending hours in front of the monitor every day. The RTQ-PCR is faster and allows you to examine substantially more preparations at once. It also is much more flexible; whether you would like to perform 96 tests on only a single mRNA or 48 tests on one and 48 on another, the work required is not substantially more extensive. A disadvantage is that the evidence is obtained indirectly, because the quantity of cDNA that you have won from the RNA through reverse transcription is measured, not the quantity of mRNA in the preparation. This is a small, but fine difference, because this additional step plays a considerable role in the margin of error. Exact results must be reproducible, and the deviations from one attempt to the next must be small.

In this case, the *ribonuclease protection assay* continues to be superior, because the RNA is quantified with this procedure. The technical background of the RTQ-PCR is described in Chapter 4, Section 4.3.5.

## 9.1.3 In Situ Hybridization

With in situ hybridization, you can provide evidence for mRNA within the cell (in situ). Another kind of in situ hybridization is fluorescence in situ hybridization (FISH), whereby localization of a specific DNA sequence is determined on a chromosome. The goal is to make the mRNA of a specific gene visible, to follow the tissue or cell type-specific expression of a protein and occasionally follow the temporal course of development.

A preparation is needed for demonstrating the mRNA. It can be cells from a cell culture, and it becomes especially interesting when first using tissue microsections (Figure 9-1). As with Southern

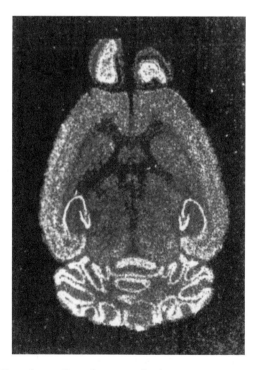

**Figure 9-1. In situ hybridization of a section of a mouse brain.**

blot hybridization, a probe is required that has been labeled in some way. Oligonucleotide or in vitro synthesized RNA fragments are used that are labeled radioactively or nonradioactively. The remainder is simple. You hybridize, wash, and identify your probe according to the specific labeling.

A cryostat is required for the production of **tissue microsections**. Cryostats are normally available for diagnostic purposes in university clinics, such as in the pathology departments, but the equipment is rarely available, especially to novices. The instruments are generally in use by diagnosticians during the normal working hours, and their work primarily takes precedence. Occasionally, you can find someone who is willing to collaborate and will make a slice. Otherwise, it is helpful to search for a group of researchers with a cryostat who have hearts of gold.

If this problem has been solved, the 10- to 20-μm-thick microsections are mounted on a slide coated with poly-L-lysine, which is then fixed with glutaraldehyde or formaldehyde and dehydrated. Preserved in this manner, the sections can be stored for a longer period in 95% ethanol.

The success of hybridization depends on the choice of a **probe**. Oligonucleotide probes are the easiest to manufacture. You simply place an order for them. To obtain a sufficient hybridization, the oligonucleotide should be about 50 nucleotides long, should have a GC content of about 50%, and as a precaution, should be purified using high-performance liquid chromatography (HPLC). The sequence should be examined with a corresponding computer program for possible secondary structures that could make your life difficult. Because of their short length, oligonucleotide probes do not permit any deviations in the sequence, and you should be very sure of your sequence. The oligonucleotides can be labeled using a kinase to transfer a radioactive phosphate remnant to the 5′ end of the oligo, although this allows insertion of a maximum of one label per molecule, the evidence for which is insensitive. Consequently, a terminal deoxynucleotide transferase (TdT) is used. It hangs a nonspecific nucleotide onto the 3′ end of single- or double-stranded nucleic acids (see Chapter 7, Section 7.1.1). The number of nucleotides added depends on the labeling behavior and the time the enzyme is permitted to work. The length of the nonspecific tail is a critical factor, because the number of labels increases with increasing length; the proportion of nonspecific sequences in the oligo also increases, as does the danger of a faulty hybridization. The labeling functions can be completed within an hour. Radioactively and nonradioactively (e.g., biotin, digoxigenin [DIG]) labeled nucleotides can be used. The choice of a suitable radioisotope is somewhat problematic in radioactive labeling. The larger the emission energy, the more quickly a result is obtained, although the picture seen with autoradiography is somewhat blurry. For this reason, $^{32}$P is suitable only for inaccurate evidence, whereas $^{33}$P and $^{35}$S provide sharper pictures, although the exposure must be carried out for a period four times as long.

Roche has dedicated an entire *Application Manual* to nonradioactive, in situ hybridization using DIG-labeled probes. These and the other DIG-based labeling methods, as well as the hybridization and washing procedures, are described in detail. Give them a call to receive the manual.

Double-stranded DNA probes can be used. The fragments used are normally substantially longer, and double-stranded DNA can easily be obtained. The disadvantage is that the probe must be denatured before hybridization and consequently hybridizes with itself. The risk of nonspecific background signals is also higher. The labeling is carried out as with samples for membrane hybridization by means of nick translation or *random-primed labeling* (i.e., oligo labeling) with radioactively or nonradioactively labeled nucleotides. The simplest way to manufacture double-stranded probes is through the amplification by means of PCR. The labeling occurs through the insertion of labeled nucleotides during the course of the amplification.

The best results are obtained with singled-stranded RNA probes, because RNA-RNA hybrids are more stable than DNA-RNA hybrids. You can decompose single-stranded RNA probes that have bonded nonspecifically by means of RNase digestion and wash them to reduce the background signal. The disadvantage lies in the larger amount of work required to produce samples. RNA probes are produced by in vitro transcription, although the DNA draft must be cloned in a suitable vector that contains a transcription start for RNA polymerases. RNA probes are more sensitive to RNase contamination.

After labeling, the probe should be purified with a suitable gel-chromatography column or by ethanol precipitation, and the preparation should be examined for the installation of labeled nucleotides. For radioactive probes, this is done with a scintillation counter, and for nonradioactive preparations, it is done according to the suggestions of the manufacturer. Considering the amount of work required and the duration of the exposure, which takes at least a few days and frequently requires weeks for radioactive probes, the procedure is worth the trouble.

The **hybridization** is carried out on the slide with the use of a formamide-containing buffer and in a damp (steam-saturated) chamber overnight. The hybridization temperature depends on the type of sample used, but it generally lies between 37°C (99°F) and 50°C (122°F). Subsequently, the sections are washed and dried. A detailed description concerning the choice of the probes and the conditions for hybridization has been provided by Wisden and colleagues (1991).

The **evidence** in the probe depends on the type of labeling. Radioactive probes are easiest to use. Put the film on top, and throw it in a corner; depending on the sample and the quantity of mRNA being sought, for 1 week or longer. A higher resolution is obtained if you dip the slide, coating it with a film emulsion, and then expose it. The procedure, however, is not very easy, and because you can only develop the preparation once, you need some experience to determine when the exposure should be stopped. You can test by first developing a film and then estimating the signal strength from this, but this is also a matter of experience. The advantage of this method is that the slide can be observed using a microscope, and the resolution is almost as good as that seen at a cellular level.

For nonradioactively labeled samples, the selection is larger. You can see the probes labeled with fluorescent dye directly with the ultraviolet (UV) microscope, and all of the others are seen after incubation with specific antibodies or binding proteins (e.g., avidin, streptavidin). They are coupled with a fluorescent dye, which can be observed with a UV microscope or with an enzyme, such as horseradish peroxidase or alkaline phosphatase, through the transformation of a chromogenic substrate, such as diaminobenzidine, 5-bromo-4-chloro-3-indolyl-phosphate (BCIP), or nitroblue tetrazolium (NBT).

A great advantage of nonradioactive probes is that you can demonstrate evidence from several samples using the same preparation. The brown precipitates from the horseradish peroxidase reaction (as long as they are strong enough) can be easily distinguished from the blue of the alkaline phosphatase, and neither of these interferes with fluorescent-labeled probes.

**Literature**

Polak JM, McGee JOD (eds). (1991) In situ Hybridization: Principles and Practice. New York, Oxford University Press.
Wisden W, Morris BJ (eds). (1994) In Situ Hybridization Protocols for the Brain. New York, Academic Press.

# 9.1.4 Fluorescence In Situ Hybridization of Chromosomes

The **fluorescence in situ hybridization** (FISH) of chromosomes does not belong in a section dealing with transcription, although it is fairly closely related to in situ hybridization. The **interphase nuclei** or **metaphase chromosomes** are hybridized, not the mRNA in the tissues. Blood or tumor cells from short-term cultures are left to swell in a hypotonic solution, fixed, and then dripped onto a slide. The cells burst and distribute their contents onto the slide. This is then fixed, and the DNA on the slide is denatured and hybridized with a nonradioactively labeled sample. After washing, a fluorochrome-coupled antibody binds to the remaining probe, and you can evaluate the entire thing using a fluorescence microscope.

This description sounds banal, although its implementation is anything but that, and managing the whole process is a matter of experience. The manufacture of the chromosome preparation is critical, because it determines whether the procedure will be successful. The chromosomes should not lie too densely in a prepared metaphase so that they cannot overlap and extend somewhat into the marginal areas. They may not lie too far removed from each other either, because the microscopic evaluation will otherwise be prolonged, and individual chromosomes can be "lost." If chromosome spreading is not adequate, the hybridization functions poorly, and the chromosomes can be identified only with difficulty. Remnants of the cytoplasm and of cellular structures may inhibit the combination of the probe with the chromosomes and cause a strong background signal.

Because you have a substantially smaller amount of target material than with the detection of mRNA, the **probe** must be appreciably longer. For a clearly identifiable signal, you should use a DNA several kilobases long that is nonradioactively labeled using *nick translation* or *random priming*. The fluorochrome-labeled antibody used to provide evidence appears later in the fluorescence microscope as a light signal, and the chromosomes are made visible with differential staining (4′,6-diamidino-2-phenylindole dihydrochloride [DAPI], which also makes it possible to identify the individual chromosomes. The nonradioactive labeling is a great advantage in this case, because it allows the formation of double staining, such as using digoxigenin and biotin-labeled probes, which can be demonstrated with various fluorochromes, such as green FITC and red Cy3. By layering the DAPI, Cy3, and FITC dyes, an appealing blue-red-green figure is obtained, which enables identification of the location of a gene on a chromosome band and is guaranteed to convince everyone (Figure 9-2).

**Disadvantage:** You should not attempt FISH on your own. If you are convinced that FISH is needed, turn to an expert for help.

**Literature**
Lichter P, Tang CJ, Call K, et al. (1990) High-resolution mapping of human chromosome 11 by in situ hybridization with cosmid clones. Science 247:64–69.

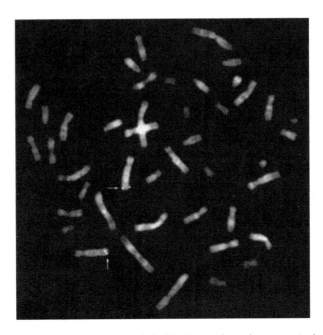

**Figure 9-2. Fluorescence in situ hybridization (FISH).** This image shows human metaphase chromosomes after hybridization with an FITC-labeled sample (→) and a DAPI dye. The inexperienced eye can see only lighter and darker spots, but a more detailed analysis of this experiment reveals that the sample has hybridized chromosomal bands 4q33-34. (Courtesy of Ruthild G. Weber, Zeljko Nikolic, Cord-Michael Becker, and Peter Lichter, University of Heidelberg, Heidelberg, Germany.)

Lichter P, Bentz M, Joos S. (1995) Detection of chromosomal aberrations by means of molecular cytogenetics: Painting of chromosomes and chromosomal subregions and comparative genomic hybridization. Methods Enzymol 254:334–359.

## 9.1.5 In Situ Polymerase Chain Reaction

In the field of in situ RNA evidence, the PCR is beginning to make the use of classic hybridization debatable. In situ PCR combines the cellular resolution of in situ hybridization with the higher sensitivity of PCR evidence.

Amplification is carried out directly on fixed cells or tissue microsections. Evidence results by means of the insertion of labeled nucleotides or through a subsequent in situ hybridization (proving that the newer methods do not always make the older ones superfluous). More information about in situ PCR can be found in Chapter 4, Section 4.3.10.

## 9.1.6 Microarrays

I have been undecided about where **high-density DNA microarrays**, also known as **DNA chips** or **gene chips**, should be classified—with the methods of hybridization or with the methods for DNA analysis? Ultimately, they have landed among the methods for transcription analysis, because this is their most frequent application.

The basic idea is a dot blot performed by a different means. You can define the beginning even earlier. An *array* is an organized quantity, an arrangement to some extent, and the Southern blot therefore remains the forerunner of all *DNA arrays*. From this, the screening of DNA banks on filter membranes developed quickly, and it has been systematized in recent years to save time and to gradually increase the rate of success. The results were the *gridded libraries*, in which every clone has its defined position on the filter membrane, so that fishing for new clones is reduced to the operating sequence of "hybridize, develop film, order your clone." From there, it was only a small step to change the target direction. Instead of using an unknown clone that is hybridized with a known sample, you can dot with known cDNAs that have been hybridized using unknown samples (i.e., mRNAs from certain cells or organs), and using these signals, you can derive which, where, and how strongly the genes have been expressed (i.e., **reverse Northern**). Such membranes are also designated as **macroarrays** and are of great interest for normal molecular biologic laboratories (see Chapter 3, Section 3.3.3), because the most expensive procedure is the manufacture or purchasing of the membrane. Usually, the equipment is available for hybridization and quantification of the signals, with the exception of the software necessary for quantification. Nevertheless, *macroarrays* rapidly reach their limits, because the DNA spots on the nylon membranes cannot be made smaller than $300\,\mu m$, and only a few thousand cDNAs can be put on a single membrane, a quantity that is still manageable ($22 \times 22\,cm$). With 30,000 to 40,000 human genes, about 20 membranes would be required.

The first great breakthrough was the use of solid carrying material such as coated glass or aminated polypropylene instead of flexible nylon membranes (Schena et al., 1995). The smooth surface of these materials offers several advantages. In this way, diffusion processes during hybridization present a far smaller problem than with porous membranes (Livshits and Mirzabekov, 1996), because the hybridization and the washing process proceed more rapidly. Because all molecules are ordered in a single plane, you can quantify the signal strength of the individual points in the *array* more accurately. The solid carrier material facilitates the construction of the smallest of hybridization chambers and the automation of the entire process.

With the aid of robots (i.e., *arrayers*), it has become possible to manufacture **high-density DNA microarrays** with up to 30,000 cDNAs on a single slide. Not all DNA microarrays are so small, however, and it is common to find up to 10,000 cDNAs per slide. Between 0.25 and 30 nL of DNA solution is applied so that the resulting dots have a size between 50 and 250 μm and contain up to 15 ng of DNA. In principle, every type of nucleic acid can be inserted, although cloned cDNAs or PCR products 0.6 to 2.4 kb long typically are used, and synthetic oligonucleotides of approximately 80 bp are being used more frequently. In this way, the manufacturers can eliminate the time-consuming search for cDNA clones. Splice variants of a gene or greatly homologous proteins can be distinguished by this method. These arrays are primarily hybridized classically with radioactively labeled samples, and the evaluation is then best performed with a PhosphorImager (see Chapter 7, Section 7.3.1).

The method first became truly limitless by means of a second development, the synthesis of different oligonucleotides directly on the carrier material in the smallest available space. With the aid of photolithographic techniques, as they are used in the production of semiconductors (Fodor et al., 1991) (Figure 9-3), it has become possible to manufacture high-density synthetic oligonucleotide arrays with 300,000 different oligonucleotides, each of which occupies less than 10 μm². Some make use of a kind of ink jet printer or develop other clever ideas. Arrays with a million oligonucleotides or more appear to be conceivable in the near future.

The oligos are only 20 to 25 nucleotides long, and the danger of faulty hybridizations increases considerably compared with that seen in cDNA arrays. To receive reliable results, 16 to 20 specific oligonucleotides are synthesized for each gene being examined. To be able to distinguish better between a genuine signal and a faulty hybridization, a *mismatch oligo* is placed beside each of these *perfect match oligos*; these oligos differ by only a single nucleotide. This approach reduces the number of genes that can be explored with such an array to about 8000.

The true strength of oligonucleotide arrays lies less in the number of the measurable genes (although increases are expected soon) and more in the fact that they can easily be drafted on a drawing board without any problems. The need to lay out extensive banks with cDNA clones is obviated.

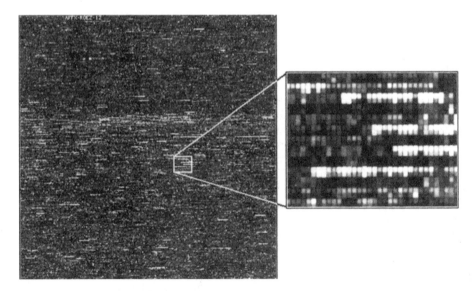

**Figure 9-3. A high-density synthetic oligonucleotide array after successful hybridization.** Between 16 and 20 oligonucleotides are aligned next to each other per gene, and the corresponding *mismatch oligos* are located directly below them. The signal can be calculated from the difference in signal intensity between the *perfect match oligos* and the *mismatch oligos*. (Courtesy of Christophe Grundschober and Patrick Nef. Hoffmann-La Rache, Basel).

The production of targets (in contrast to the classic Southern blot hybridization, the sample is located on the chip, while the *target* DNA floats in the fluid) generally results by obtaining RNA from the object being examined (usually cultivated cells or tissues) and synthesizing cDNA from this. Labeling occurs through the addition of fluorescently labeled nucleotides during reverse transcription. The *microarray* is then hybridized with the *target*, and the signal strength is determined for each point of the array. If you use a different fluorochrome to manufacture a second target, you can then hybridize with both targets simultaneously and need only to determine the difference between the signals of both fluorochromes to obtain information concerning the differences in the gene expression pattern for both of the examined test conditions.

The method also has disadvantages. Because the excitation and emission spectra of fluorochromes overlap extensively, strong signals of a dye may lead to a certain background signal for the measurement of the other signals that limits measurement of the differences in intensity. With the increasing application of microarrays, it will soon no longer be sufficient to compare only two conditions with one another.

There is a broad palette of methods available for the manufacture of targets or for providing evidence of hybridized fragments (Figure 9-4). As an example, **minisequencing** is considered here (see Chapter 8, Section 8.1.3). An *oligonucleotide array* is hybridized with an unlabeled *target*, and a type of sequencing reaction is performed in which a single, labeled dideoxynucleotide is made available. Only a single nucleotide is added as long as the 3′ region of the oligonucleotide on the array and the target DNA are exactly complementary to one another (Nikiforov et al., 1994; Pastinen et al., 1997). If the minisequencing is repeated with another, differently labeled dideoxynucleotide, several rounds of detection can be performed with only a single hybridization.

The applications of microarrays concentrate on two areas, the analysis of expression and the evidence of DNA variants, such as in the search for mutations and polymorphisms or the analysis of genotypes. **Expression analysis** accounts for most uses of microarrays. The sensitivity of the method is sufficiently high to determine the expression of genes in mammalian cells, in which only a few transcript copies exist per cell; less than one transcript per cell is available in yeast cells. In smaller, thoroughly sequenced organisms such as *Saccharomyces cerevisiae*, you can follow the expression of all 6000 genes with only a single microarray, which represents a quantum leap in the analysis of this unusual organism.

Many advancements in knowledge will be based on this type of analysis. Work involving the simple comparison of two conditions is already progressing to comparison of many different experiments. *Clustering* analyses (Eisen et al., 1998) can show which groups of genes can be upregulated and downregulated, knowledge that permits understanding of the function of genes whose functions were previously obscure.

**Genotyping** is still in its infancy. The genomes of higher organisms are very large and are too complex to provide clear and reproducible results. Producing subfractions is difficult and associated with many errors. To improve the techniques, researchers would gradually have to progress from organisms with smaller genomes to the larger mammalian genomes. However, because the large financial resources flow to work carried out on humans, the dedicated researchers and money are still lacking in work with lower organisms, and the needed technical developments are proceeding slowly. Investigations of single-celled organisms, such as bacteria, yeasts, or mitochondria, are possible today. With *oligonucleotide arrays*, you can determine the distribution of alleles and the presence of polymorphisms of the individual nucleotides (*single-nucleotide polymorphisms [SNPs]*). You can even carry out sequencing, instead of inserting only a single *mismatch oligo*, by inserting three of these and covering a specific position for each of the four possible nucleotides. Ideally, only one of the four oligos will deliver a strong signal, which allows a conclusion about the sequence of the target. In this way, point mutations can be determined, although the discovery of other mutations, such as insertions, deletions, or rearrangements, may be difficult with this method.

However, the spectrum of use may expand considerably in the future. We are still at the beginning of these developments. Microarrays are becoming less expensive, and their distribution will soon

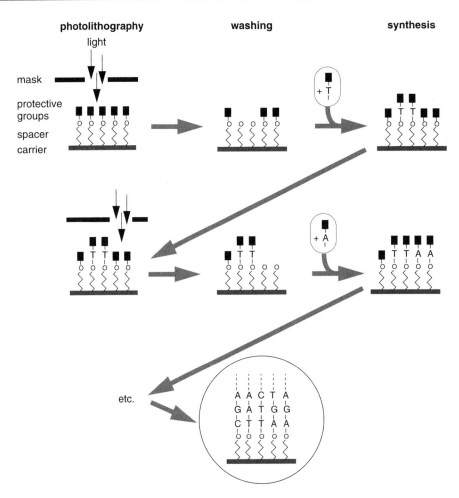

**Figure 9-4. Photolithography.** In a first step, the carrier material is irradiated with a laser. Through the application of a photolithographic mask, a certain pattern emerges; at sites where the light hits, photolabile protective groups are split off, and reactive groups are set free. A nucleotide is then coupled to them in a second step. Because the new nucleotide is furnished with a photolabile protective group, the first step of the next synthesis can begin. With only 4 × N synthesis cycles (N is the length of the synthesized oligonucleotide), you can manufacture hundreds of thousands of different oligonucleotides in this manner, because the number of different oligos on a chip is limited only by the resolution of the exposure device. Depending on the carrier material, the yield is between 0.1 pmol (glass) and 10 pmol of (polypropylene) primer per 1 mm$^2$.

increase. Clever minds will develop improvements and new applications, as has already happened with the PCR. For example, because all questions can be answered through the use of suitable methods of fractionation that employ nucleic acids, it is possible to isolate RNA containing ribosomes and thereby explore which of the mRNAs have been translated in a cell and which changes have occurred in response to certain external influences. With the classic methods, you could at best have obtained information concerning a handful of genes, but you could cover the entire genome with a couple of small arrays.

**Disadvantages:** The RNA quantities that are required for evidence can be considerable, usually between 100 ng and 20 μg. This is a problem, especially when quantities of the initial material are limited, as in biopsy material or in the investigation of smaller regions of the brain.

Genotyping, or the search for mutations in genomes, has not been possible in higher eukaryotic organisms because of the complexity of the genome. Suitable methods for fractionation or for the segmental amplification of the material must be developed.

Costs must be considered. The costs have been as enormous as the potential uses. Autonomous work requires an *arrayer*, a *reader*, a computer with software for evaluation, the starting materials, and ultimately people who can work on these experiments. At a total cost of more than one third of a million dollars, only larger institutes are likely to participate. Even if the necessary equipment is available, a single microarray costs a few hundred dollars, and one is not enough. The prices should decrease in the coming years.

This technology is still in its infancy, and experimenters working in this realm must be willing to carry out developmental work. Currently, researchers are still fighting with reproducibility of the results obtained. Because the method is sensitive, even small variations in the performance of the experiments are significant; the same experiment performed a few weeks later can deliver very different results. Analysis of the data obtained is also in its initial stages. You can determine the differences between two hybridization experiments quite easily, but the really interesting knowledge is derived from a comparison of many attempts, such as when different cell lines have been treated for different periods with different growth factors. This produces a vast amount of data that is ordered in multidimensional matrices and must then be evaluated to somehow filter out particular routine characteristics. Biologists traditionally have a fear of mathematics, and for them, the work is beginning to get worse.

Another problem is the limit of this technology. Can you imagine measuring citizens at certain points and determining what they are doing? Can you understand the conception of an individual, during which you are aware of which of its neurons fires and when? Can you understand the operational function of a cell simply by following the expression intensity of its genes? This is a problem that we will have to chew on for some time to come.

Ethical and moral problems will also develop. With a *microarray*, everybody can be examined for the most frequent 1000 or 10,000 hereditary diseases—just what the insurance companies or eugenicists are waiting for. The same technology will allow us to describe diseases in great detail, such as the viral subtypes of HIV infections or the type and developmental stage of cancerous diseases, which will make it possible to provide custom treatments. Because the genetic variability of an individual is quite small (Gagneux et al., 1999), we may soon be able to manufacture microarrays demonstrating clear relationships between the genotype and behavioral conduct, enabling the preparation of a genetically determined psychogram that could determine the choice of a partner, the hunt for a job, and the right to keep pets. How should a responsible researcher behave when considering these perspectives?

**Suggestion:** More information on the topic of microarrays is available in the special issue of *Nature Genetics* (January 1999).

**Literature**

Eisen MB, Spellman PT, Brown PO, Botstein D. (1998) Cluster analysis and display of genome-wide expression patterns. Proc Natl Acad Sci U S A 96:14863–14868.

Fodor SPA, Read JL, Pirrung MC, et al. (1991) Light-directed, spatially addressable parallel chemical synthesis. Science 251:467–470.

Gagneux P, Wills C, Gerloff U, et al. (1999) Mitochondrial sequences show diverse evolutionary histories of African hominoids. Proc Natl Acad Sci U S A 96:5077–5082.

Livshits MA, Mirzabekov AD. (1996) Theoretical analysis of the kinetics of DNA hybridization with gel immobilised oligonucleotides. Biophys J 71:2795–2801.

Nikiforov TT, Rendle RB, Goelet P, et al. (1994) Genetic bit analysis: A solid phase method for typing single nucleotide polymorphisms. Nucleic Acids Res 22:4167–4175.

Pastinen T, Kurg A, Metspalu A, et al. (1997) Minisequencing: A specific tool for DNA analysis and diagnostics on oligonucleotide arrays. Genome Res 7:606–614.

Schena M, Shalon D, David RW, Brown PO. (1995) Quantitative monitoring of gene expression patterns with a complementary DNA microarray. Science 270:467–470.

The Chipping Forecast (special issue). (1999). Nature Genet 21:1–60.

# 9.2 Mutagenesis

Mutagenesis is an important idea in the work with nucleic acids. Because every kind of mutation is allowed, ultimately resulting in an end with double-stranded DNA, there is no universally valid protocol. The conception of sensible mutagenesis is one of the most intellectually demanding tasks in this profession.

There are several reasons why you should make at least one attempt with mutagenesis. A variety of scenarios follows:

- You consider the cDNA that you have to be a protein and wish to examine the function of individual domains of this protein. The individual amino acids that are considered to be important are mutated to see what happens.
- You examine patients for mutations and have discovered that the mutation is the cause of the illness or that this mutation has something to do with the function of the protein.
- You would like to exchange a part of a protein for a part of another to verify a theory.
- You have the cDNA for a protein but no specific antibody; a rough estimate indicates that you can more quickly insert an epitope, for which an antibody exists, than you can generate a specific antibody against the protein.
- You want to insert a marker that allows you to later purify the specific protein.
- You have a piece of DNA that evidently regulates the expression of a gene, although this piece is more than 5 kb long, and somebody has raised a question about how much of this substance is necessary.
- You have limited the truly important area of this regulatory sequence to 30 bases. You mutate each base from the beginning to the end and look to see what happens.

Everyone can probably add more examples stemming from their own experiences. Whatever the motive might be, introducing a target-oriented mutation in an existing DNA is commonly required.

The second step consists of ordering an **oligonucleotide** with a suitable mutation. Oligonucleotides can be synthesized and delivered within 3 to 4 days, and it does not even cost much because prices for oligonucleotides have gradually decreased over the years. The order is so quick that you may forget the first step—conception of the oligonucleotides.

The oligonucleotide with the mutation is used as a primer for a **DNA synthesis**. T4 or Klenow DNA polymerase is commonly used, although more experimenters turn to thermostabile polymerases because of the substantially greater yield. Taq polymerase is not recommended in this case because of the high rate of error; you are better off using polymerases with a corrective activity such as Pfu. After the synthesis reaction, the template DNA is frequently destroyed to increase the proportion of mutated clones, and the mixture is then transformed in bacteria. The emerging colonies are **screened** with the clones that demonstrate the proper mutation.

To insert a restriction cleavage site, you need the amino acid sequence of the area involved. In Table 9-1, you first search for the second amino acid in the second column and then search for the first amino acid in the first column. If necessary, try to find the third amino acid. If all cases have been successful, the fourth column (Sequence) indicates which restriction cleavage site can be inserted.

Although many methods and manufacturers promise miracles, the share of clones with mutations usually is substantially lower than you would hope to find. Yields of 90% mutated clones or more

**Table 9-1.** Insertion of a Restriction Cleavage Site in a Protein-Coded DNA Sequence

| First AA | Second AA | Third AA | RE | Sequence | First AA | Second AA | Third AA | Sequence | RE |
|---|---|---|---|---|---|---|---|---|---|
| G | A | | NarI | (GGC′GCC) | ATGKLMPQRSTVW* | L | T | (G′TTA′AC) | HpaI |
| GR* | A | HLPQR | SacI | (GA′GCT′C) | H | M | | (CAT′ATG) | NdeI |
| GRW | A | HLPQR | ApaI | (GG′GCC′C) | CGRS | M | HLPQR | (GC′ATG′C) | SphI |
| EKQ* | A | CFLSWY* | HindIII | (AA′GCT′T) | APST | M | ADEGV | (CC′ATG′G) | NcoI |
| APST | A | ADEGV | PstI | (CT′CA′G) | V | N | | (GTT′AAC) | HpaI |
| | | | SacII | (CC′GCG′G) | ATGKLMPQRSTVW* | N | S | (G′AAT′TC) | EcoRI |
| | | | PvuII | (CA′GCT′G) | G | P | | (GGG′CCC) | ApaI |
| EKQ* | A | CFLSWY* | StuI | (AG′GCC′T) | R | P | | (AGG′CCT) | StuI |
| APST | A | ADEGV | EagI | (CG′GCC′G) | | | | (CGG′CCG) | EagI |
| FILV | A | IKMNRST | NruI | (TC′GCG′A) | | | | (TGG′CCA) | BalI |
| DHNY | A | CFLSWY* | MluI | (AC′GCG′T) | W | P | | (C′CCG′GC) | NaeI |
| CGRS | A | HLPQR | ApaLI | (GT′GCA′C) | ATGKLMPQRSTVW* | P | A | (C′CCG′GG) | SmaI |
| LMV | A | IKMNRST | BalI | (TG′GCC′A) | ACDFGHILNPRSTVY | P | G | (T′CCG′GA) | BspEI |
| ATGKLMPQRSTVW* | A | P | NarI | (G′GCG′CC) | ACDFGHILNPRSTVY | P | DE | (CTG′CAG) | PstI |
| A | C | | SphI | (GCA′TGC) | L | Q | | (CCG′CGG) | SacII |
| ACDFGHILNPRSTVY | C | RS | PstI | (C′TGC′AG) | P | R | | (TCG′CGA) | NruI |
| ATGKLMPQRSTVW* | C | T | ApaLI | (G′TGC′AC) | S | R | | (TCT′AGA) | XbaI |
| I | D | | ClaI | (ATC′GAT) | | | | (ACG′CGT) | MluI |
| V | D | | SalI | (GTC′GAC) | T | R | | (GC′CGG′C) | NaeI |
| ACDKLMPQRSTVW* | D | P | BamHI | (G′GAT′CC) | CGRS | R | HLPQR | (GG′CGC′C) | NarI |
| AEGIKLPQRSTV* | D | L | BglII | (A′GAT′CT) | GRW | R | HLPQR | (AT′CGA′T) | ClaI |
| ACDFGHILNPRSTVY | D | R | PvuI | (C′GAT′CG) | DHNY | R | CFLSWY* | (CT′CGA′G) | XhoI |
| ACDFGHILNPRSTVY | D | HQ | BclI | (T′GAT′CA) | APST | R | ADEGV | (CC′CGG′G) | SmaI |
| L | E | | XhoI | (CTC′GAG) | FILV | R | IKMNRST | (TC′CGG′A) | BspEI |
| E | F | | EcoRI | (GAA′TTC) | CGRS | R | HLPQR | (GT′CGA′C) | SalI |
| A | G | | NaeI | (GCC′GGC) | ACDFGHILNPRSTVY | R | G | (C′CGC′GA) | SacII |

*(continued)*

**Table 9-1.** Continued

| First AA | Second AA | Third AA | Sequence | RE |
|---|---|---|---|---|
| P | G | | (CCC'GGG) | SmaI |
| S | G | | (TCC'GGA) | BspEI |
| ATGKLMPQRSTVW* | G | P | (G'GGC'CC) | ApaI |
| AEGIKLPQRSTV* | G | L | (A'GGC'CT) | StuI |
| ACDFGHILNPRSTVY | G | R | (C'GGC'CG) | EagI |
| ACDFGHILNPRSTVY | G | HQ | (T'GGC'CA) | BalI |
| V | H | | (CTG'CAC) | ApaLI |
| ATGKLMPQRSTVW* | H | A | (G'CAT'GC) | SphI |
| ACDFGHILNPRSTVY | H | G | (C'CAT'GG) | NcoI |
| D | I | | (GAT'ATC) | EcoRV |
| GR* | I | HLPQR | (GA'ATT'C) | EcoRI |
| GRW | I | HLPQR | (GG'ATC'C) | BamHI |
| EKQ* | I | CFLSWY* | (AG'ATC'T) | BglII |
| APST | I | ADEGV | (CG'ATC'G) | PvuI |
| LMV | I | IKMNRST | (TG'ATC'A) | BclI |
| ATGKLMPQRSTVW* | I | S | (G'ATA'TC) | EcoRV |
| ACDFGHILNPRSTVY | I | CW* | (C'ATA'TG) | NdeI |
| E | L | | (GAG'CTC) | SacI |
| K | L | | (AAG'CTT) | HindIII |
| Q | L | | (CAG'CTG) | PvuII |
| ATGKLMPQRSTVW* | L | A | (G'CTA'GC) | NheI |
| ACDFGHILNPRSTVY | L | DE | (T'CTA'GA) | XbaI |
| AEGIKLPQRSTV* | L | V | (A'CTA'GT) | SpeI |

| First AA | Second AA | Third AA | Sequence | RE |
|---|---|---|---|---|
| ACDFGHILNPRSTVY | R | DE | (T'CGC'GA) | NruI |
| AEGIKLPQRSTV* | R | V | (A'CGC'GT) | MluI |
| A | S | | (GGA'TCC) | NheI |
| G | S | | (GCT'AGC) | BamHI |
| R | S | | (GGA'TCC) | BglII |
| | | | (AGA'TCG) | PvuI |
| T | S | | (ACT'AGT) | SpeI |
| * | S | | (TGA'TCA) | BclI |
| ATGKLMPQRSTVW* | S | S | (G'AGC'TC) | SacI |
| AEGIKLPQRSTV* | S | IM | (A'TCG'AT) | ClaI |
| AEGIKLPQRSTV* | S | FL | (A'AGC'TT) | HindIII |
| ACDFGHILNPRSTVY | S | RS | (C'TCG'AG) | XhoI |
| ACDFGHILNPRSTVY | S | CW* | (C'AGC'TG) | PvuII |
| ATGKLMPQRSTVW* | S | T | (G'TCG'AC) | SalI |
| G | T | | (GGT'ACC) | KpnI |
| ATGKLMPQRSTVW* | V | P | (G'GTA'CC) | KpnI |
| P | W | | (CCA'TGG) | NcoI |
| GR* | Y | HLPQR | (GA'TAT'C) | EcoRV |
| GRW | Y | HLPQR | (GG'TAC'C) | KpnI |
| APST | Y | ADEGV | (CA'TAT'G) | NdeI |
| CGRS | * | HLPQR | (GC'TAG'C) | NheI |
| FILV | * | IKMNRST | (TC'TAG'A) | XbaI |
| DHNY | * | CFLSWY* | (AC'TAG'T) | SpeI |
| CGRS | * | HLPQR | (GT'TAA'C) | HpaI |

To insert a restriction site, you first need the amino acid sequence of the region of interest. Then you look first in the second column for the second amino acid of your sequence, second in the first column for the first amino acid and finally, if necessary, you try to find the third amino acid in the third column. If you were successful on all behalves, the fourth column will tell you which restriction site you can insert in your DNA without modifying the amino acid sequence.

Asteriks(*) represent stop codons.

AA, amino acids; RE, restriction enzyme.

are extolled, although in practice, 50% is considered to be quite good, and occasionally, a rate of only 10% and sometimes even less is obtained. You are well advised to plan for possibilities by which you can differentiate mutated from nonmutated clones. This can also be useful later, because it happens repeatedly in practice that an experimenter erroneously exchanges a couple of clones and then unsuspectingly works for weeks with a completely different variant.

If you insert larger DNA segments, you can use them as probes for screening bacterial colonies; the procedure is comparable with the screening of a bank (see Chapter 7, Section 7.4). For smaller mutations, it is best to insert a new restriction cleavage site, based on which you can later examine the clone within 1 to 2 hours by simple restriction digestion. This is possible because the genetic code is degenerated, and the introduction of passive mutations, which alters the nucleotides but not the amino acid sequences, is allowed. One or several bases of a triplet can be mutated in such a manner that there is no change in the amino acid.

The cleavage site must not directly affect the mutation, because it is also sufficient if it is in the vicinity. It is best to destroy the cleavage site if one is available. However, you should consider the possibility that one of the cleavage sites may be missing later, and nothing is more troublesome in the course of cloning than clones without any favorable cleavage sites. The insertion of a restriction cleavage site is therefore more significant, although it is also more difficult to figure out the fitting mutation. Shankarappa and colleagues considered this problem in 1992, resulting in an article and a rather graceful computer program known as **SILMUT**. The program is somewhat primitive, but it is extremely helpful in the search for silent mutations to introduce the cleavage sites for restriction enzymes. This little program exists in a DOS and a Mac version, and it can also be found in the Internet (http://iubio.bio.indiana.edu/soft/molbio/ibmpc/). Anyone without access to computers or to the Internet can find some assistance in Table 9-1.

## Techniques of Mutagenesis

Considering the many different requirements, it is difficult to bring some order to the procedure for causing mutation. The following discussion addresses many of the issues.

**1. I would like to perform** *site-directed mutagenesis.* In variant 1A, a primer is ordered that contains the desired mutation in the middle, and a second primer is ordered that is exactly complementary to the first and that has a recognizable site to the left or to the right of the first (i.e., upstream or downstream) (Figure 9-5). With one mutant and one outer primer, you separately amplify the left and the right half, purify both PCR products, and then use them as a primer in a second PCR, with few cycles and without any further template. Because the products in the area of the mutated primer overlap, they can hybridize with one another and are then elongated. The completed PCR product, which now carries the desired mutation, can be cloned in the classic manner. The method is simple and functions reliably, although it demands that suitable cleavage sites for the cloning are available to the left and the right of the mutated site. You can increase the yield of the second PCR by adding the two flanking primers and carrying out the normal number of cycles.

You should preferably use polymerases with corrective activity for mutagenesis by means of PCR (see Chapter 4, Section 4.1). The **rate of errors** is lower by a factor of about 10 compared with the use of Taq polymerase. However, they produce smooth ends, whereas Taq usually causes detachment at a nonspecific base and thereby produces a 3′ overlap. This is disturbing in the direct cloning of PCR products if you would like to employ a PCR product such as a primer. *Proofreading* polymerases also have their disadvantages. The yield of a product is frequently lower, and the mutated primer is occasionally recognized as being faulty and then "corrected." The consequence is that it frequently results in the complete degradation of the primer, and you should therefore use primer concentrations that are higher than usual.

**In variant 1B,** a primer is ordered that contains the desired mutation in the middle. A single-stranded template DNA is generated by means of an M13 vector (see Section 9.6) in an *Escherichia*

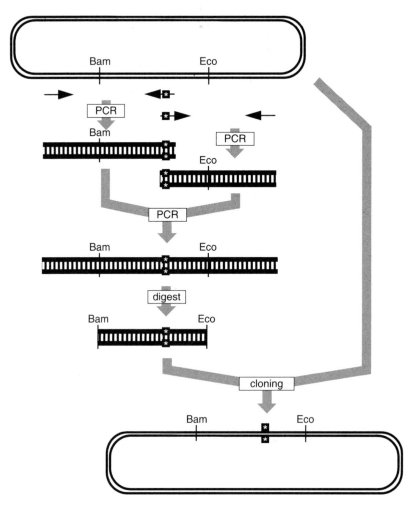

**Figure 9-5. Site-directed mutagenesis with the help of available restriction cleavage sites.**

*coli dut⁻ung⁻* F′ strain (e.g., CJ236). It is important to use the right strain in this method, because a certain amount of uracil rather than thymidine is found in the DNA of *ung⁻* bacteria. The template DNA is hybridized together with the mutated primer, filled to form a normal double-strand using T4 DNA polymerase and T4 DNA ligase, and the product is transformed in an arbitrary *E. coli* strain, which should be *ung⁺*. These bacteria contain uracil-*N*-glycosylase, which removes the uracil from the template strand, which is recognized as defective and then disassembled. A more detailed description of the method can be found in the article by Kunkel (1985). The method can increase the share of mutated clones, although it presupposes that the bacteria are suitable and that the template DNA can be found in an M13 vector, which is not usually the case.

In variant 1C, you can get along without any special bacterial strains (Figure 9-6) (Weiner et al., 1994). The degradation of the template strand is transferred from the cell to the test tube. It also functions with normal double-stranded plasmid DNA. You need two primers—a strand primer and a complementary strand primer—that both contain the desired mutation. With a *proofreading* polymerase, you perform a few PCR cycles and digest the PCR product with *Dpn*I, a *4-cutter* with an

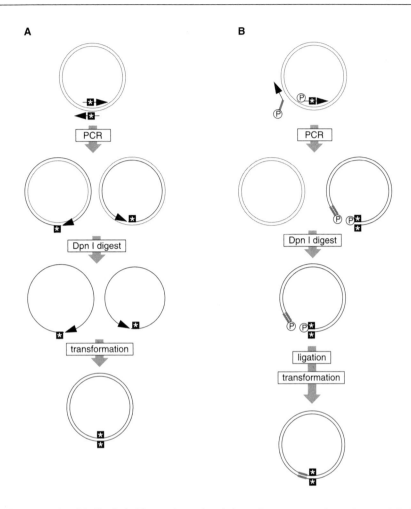

**Figure 9-6. Mutagenesis with *Dpn*I. A,** The easier variant is based on two complementary, mutated primers.
**B,** The second variant demonstrates greater flexibility and enables production of point mutations, small insertions, and even larger deletions. However, it is more practical to employ previously phosphorylated primers for the polymerase chain reaction.

unusual characteristic. *Dpn*I specifically cuts methylated and hemimethylated DNA, and it does not cut unmethylated DNA. Most bacterial strains demonstrate a *Dam* methylation system, so that the plasmid DNA is normally methylated and is therefore degraded. You can also methylate in vitro. If the template DNA is chopped, you transform the mixture into bacteria, which ligates the PCR fragments together to form a neat plasmid for their multiplication. The yield of mutated clones in variant C is encouraging. Stratagene offers the necessary components together in a kit that they market as the QuickChange Site-Directed Mutagenesis Kit.

**2. I want to insert some mutations that are located close to one another.** This approach principally functions as described for site-directed mutagenesis in point 1. However, the fully homologous region of the 3′ end of the oligonucleotide must be sufficiently long (>8 bases), because the polymerase will not otherwise recognize the oligonucleotide as a primer.

**3. I would like to introduce one or several random point mutations.** Sometimes you want to perform a mutation without knowing exactly which nucleotide must be turned around. It is better to give the mutation a chance to work on its own and rely on a good screening system from which you can fish out the desired mutant. Subsequent sequencing then shows where you should have started or in what region you can proceed with a more targeted mutation.

**In variant 3A,** if the region for the mutation is small, you are better off to work with degenerated oligonucleotides. When ordering the primer, you give an N instead of a defined base for the position of the mutation. For the synthesis, all four bases are added in the step of synthesis, rather than only one. According to random distribution, one is then added per oligo. Because four different oligo variants develop per N, the test tube, which you are sent, contains a mixture of $4^n$ oligos (n is the number of Ns that have been ordered). Instead of all four bases, you can make use of partial amounts, such as any one base other than C or some kind of purine; the proper code for your order can be obtained from the oligo manufacturer.

If there is a restriction cleavage site to the left and to the right of the mutated region, you can order a complementary oligo to the degenerate oligo, hybridize both together, and insert this fragment in the desired clone. If the clone has no such cleavage sites, you can add them by carrying out the procedure described earlier under point 1. If no cleavage sites can be added in this vicinity, you must also insert the mutations as described earlier for site-directed mutagenesis in point 1.

**In variant 3B,** if the region in which the mutations are to be found are somewhat larger, you are better off using something that is avoided at all cost: the high rate of error of Taq polymerase. By changing the PCR conditions, the rate of error can be elevated somewhat. A change in the dNTP concentration and the addition of manganese (up to $640 \mu M$ $MnSO_4$) demonstrate a clear effect. An increase in the quantity of Taq polymerase, the magnesium concentration, and the pH have a negative effect on precision. Unfortunately, these changes work out differently on the kind mutations that occur. High concentrations of dGTP, for instance, lead preferably to mutations of A or T to G or C and consequently result in an increase of the GC proportion in the sequence. A more comprehensive investigation of this topic can be found in the article by Cadewell and Joyce (1992), who have come to the conclusion that nucleotide concentrations of 0.2 mM dGTP, 0.2 mM dATP, 1 mM dCTP, and 1 mM dTTP (with 7 mM $MgCl_2$, 0.5 mM $MnCl_2$, and 5 units of Taq polymerase; add $MnCl_2$ before the enzyme) lead to a fourfold higher rate, without demonstrating any preference for certain mutations. The mutation rate was about 0.66% per base. If you prefer working with kits, you can take advantage of the Diversify PCR Random Mutagenesis Kit from Clontech. According to their claims, the rate of mutation, depending on the specific conditions, is 2 to 8 mutations per 1000 bases.

The rate of errors can be increased further by increasing the number of cycles. If two normal PCR reactions are performed successively, the product of the first cycle is purified using an agarose gel, and is then employed as a template in the second cycle. In this way, you can avoid finding a nonspecific smear rather than a band after the second amplification.

**4. I would like to insert an additional triplet.** The procedure is the same as in variant 1A. A homologous piece that is sufficiently long should lie to the left and to the right of the inserted triplet so that the primer hybridizes properly.

**5. I would like to insert even more bases.** The principle for variant 5A is similar to that seen for variant 1C. You order two primers whose 5′ halves are made up of the inserted sequence, and the 3′ halves are homologous to the template sequence. The two fragments are amplified with two primers lying externally. From these and based on their homologous ends, a large fragment is forged in a second PCR (Figure 9-7A). Care must be taken that the homologous 3′ portion of the primer is sufficiently long so that it will hybridize with the template (i.e., 15 to 25 bases, depending on the specific sequence). The 5′ end can be as long as the oligonucleotide synthesis allows. Fundamentally, primer lengths of 100 nucleotides and more are possible, although the yield drops considerably for primers longer than 50 nucleotides. Because the oligonucleotide synthesis proceeds in a 3′ to 5′ direction, it may occur that precisely the portion that you would like to insert is missing. Very long primers should therefore be purified using HPLC. At least the second PCR must be performed using

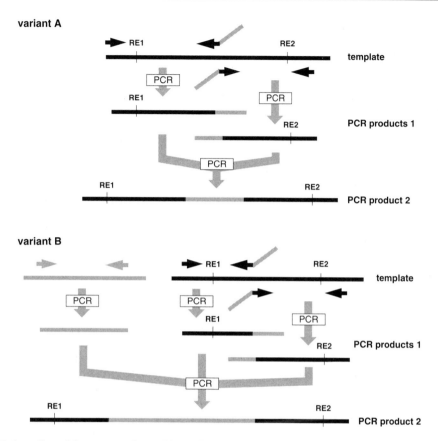

**Figure 9-7. Insertion of fragments of an arbitrary length.** Variant A is technically quite simple, although the length of the insertion is limited by the length of the primer; up to 100 bp are considered realistic. Variant B allows the insertion of arbitrary fragment lengths, but it is technically quite demanding; the yields are frequently modest when three fragments are joined by means of the polymerase chain reaction. RE, restriction cleavage site.

a polymerase with correction activity, because the Taq polymerase produces nonspecific overhangs that cannot be elongated.

**In variant 5B,** you can insert fragments of any length, even if it is technically somewhat tricky (see Figure 9-7B). In the first round, you separately amplify the three required fragments (5′ half, insertion, 3′ half), and you must take care in the selection of the primer to ensure that the sequences of the 5′ half and the insertion of the 3′ half demonstrate an overlap of at least 20 bases. The fragments are then purified (gel purification is usually recommended to increase the chances for success) and used jointly in a second PCR as a template. In principle, a fragment then serves as a primer for the other one, and the new fragment that is desired should appear, including the insertion. In practice, the matter is usually somewhat more difficult, and the experimenter must frequently play around a bit until it functions. The three fragments, for instance, can be inserted together and without additional primer in a second PCR within a few cycles, so that the fragments function simultaneously as primer and template, and the product is then used as a template for a third PCR, in which the two flanking primers are inserted as usual. You can immediately add small amounts of primer to the second PCR, purify the PCR product using the gel, and then insert the fragment with the right length as a template in a third PCR, or you can mix everything together—fragments and flanking primers—and amplify the entire mixture with the hope that it will function correctly.

The method offers many possibilities, because you can insert fragments and perform deletions at the same time. You can even unify three arbitrary fragments or do whatever you want, as long as the sequences at the ends of the fragments fit to one another.

**6. I would like to perform a targeted deletion.** You can achieve this with variant 1A. The simplest approach is to use a mutated primer that is approximately 40 nucleotides long and that is homologous to the 20 bases before and behind the deletion site. Sufficient as a second "mutated" primer is the use of a 20-mer, which is complementary to the 5′ half of the 40-mer polymerase. In this way, you can produce deletions of an arbitrary size.

Variant 1C is well suited for this purpose and has the advantage of being faster. It is the most versatile method because, depending on how you select the primer, it can be used to perform deletions, insertions, point mutations, or a combination of all three. A strand and a complementary strand primer are used, although they do not overlap this time. Amplification is performed, and digestion is carried out with *Dpn*I. The remaining PCR product is phosphorylated and ligated in a third step, before it is ultimately transformed. This process is manufactured by Stratagene as a kit known as the ExSite PCR-Based Site-Directed Mutagenesis Kit.

**7. I would like to carry out a nontargeted deletion.** That sounds a bit strange, because in certain promoter studies, for instance, you usually want to obtain the largest number of clones with different and more extensive deletions requiring the smallest amount of work. The method of choice is the **nested deletion**, which was first described by Henikoff in 1984. The template clone is cut with two restriction enzymes. One produces a 3′ overlap (Table 9-2), and the other produces a 5′ overhang (if needed, a smooth end can be used). The cut template is digested with **exonuclease III** (*Exo*III), which has the wonderful characteristic of munching away at only the 3′ ends of double-stranded DNA, while single-stranded 3′ ends, as found in 3′ overlaps, remain unaltered, because the reaction is very rapid and difficult to control. It is best to remove aliquots at different times (between 1 and 30 minutes) and then stop the reaction. A single strand remains at the digested end, and it can be degraded with a single-strand–specific nuclease, such as *mung bean nuclease*. A double strand with smooth ends remains, which can be ligated and transformed. The size of the deletion depends on the incubation period; for an exact characterization, the resulting clones must be sequenced, because the effectivity of the reaction can vary. It is important that the template DNA used has not been washed with phenol, because nothing will run.

Even if the template has no cleavage sites for enzymes with a 3′ overhang, all is not lost, because ends which are filled in with α-thio dNTPs are also protected from *Exo*III degradation. However, the amount of work required is substantially greater, because you must first cut with a restriction enzyme, fill in, and cut again with the restriction enzyme at the site where the deletion should begin. It is simpler to purchase a corresponding kit for the *nested deletion*, such as from Promega or Stratagene.

I hope that this list of suggestions can provide you with a couple of useful ideas. Otherwise, you must take advantage of your own creativity or pump it out of one of your colleagues. After all, there are so many beautiful varieties of mutagenesis.

**Table 9-2.** Restriction Enzymes with a 3′ Overhang

| Cutters | Four-Base Overlap | Two-Base Overlap |
|---|---|---|
| 4-cutter | *Cha*I, *Nla*III, *Tai*I | *Hha*I |
| 6-cutter | *Aat*II, *Apa*I, *Bbe*I, *Kpn*I, *Nsi*I. *Pst*I, *Sac* I, *Sph*I | *Pvu*I, *Sac*II |
| 8-cutter | *Fse*I, *Sse*8387I | *Pac*I, *Sgf*I |

**Literature**

Cadwell RC, Joyce GF. (1992) Randomization of genes by PCR mutagenesis. PCR Methods Appl 2, 28–33.

Henikoff S. (1984) Unidirectional digestion with exonuclease III creates targeted breakpoints for DNA sequencing. Genes 28:351–359.

Kunkel TA. (1985) Rapid and efficient site-specific mutagenesis without phenotypic selection. Proc Natl Acad Sci U S A 82:488–492.

Shankarappa B, Sirko DA, Ehrlich GD. (1992) A general method for the identification of regions suitable for site-directed silent mutagenesis. Biotechniques 12:382–384.

Shankarappa B, Vijayananda K, Ehrlich GD. (1992) SILMUT: A computer program for the identification of regions compatible for the introduction of restriction enzyme sites by site-directed silent mutagenesis. Biotechniques 12:882–884.

Weiner MP, Costa GL, Schoettlin W, et al. (1994) Site-directed mutagenesis of double-stranded DNA by the polymerase chain reaction. Genes 151:119–123.

# 9.3 In Vitro Translation

In vitro translation is not a part of the work of the molecular biologist, who does not even know what can be done with these synthesized proteins. In vitro translation can be differentiated into three parts.

In the first step, a template DNA must be generated that codes for a protein. Usually, a cDNA is cloned in a vector with Sp6, T3, or T7 RNA polymerase promoter (Table 9-3), although a PCR fragment that carries a corresponding promoter sequence at the 5′ end is also suitable. Such a fragment can be manufactured easily by attaching the promoter sequence to the 5′ end of one of the primers. The sequences indicated can be found in many vectors with RNA polymerase promoters. In Table 9-3, the underlined segments behind the carets ($^\wedge$) represent the starting site for transcription; afterward, they are a part of the transcript (*leader sequence*).

In a second step, RNA is made from this template by means of in vitro transcription (see Chapter 5, Section 5.5). The DNA must be sufficiently clean, or the RNA yield will be small. It is best to use cesium chloride gradients, anion exchange columns, or glass milk–purified plasmid DNA. PCR products must be purified accordingly. Cut plasmids at the 3′ ends (i.e., behind the coding sites) before the transcription to ensure an efficient termination of the transcription and a uniform RNA population. If the GTP is replaced by 90% diguanosine GTP [$m^7G(5′)ppp(5′)G$] and 10% GTP for the RNA synthesis, RNA with a 5′ cap structure will be produced. In this way, you can increase the efficiency of the translation (Song et al., 1995). In the event of longer RNA transcripts ($>1\,kb$), you should increase the proportion of GTP, or the rate of RNA synthesis will suffer. Promega, for instance, suggests a 5-to-1 ratio for transcripts between 0.3 and 3 kb long.

The success of the RNA synthesis should be controlled by means of gel electrophoresis, because the results may turn out to be quite varied. You can remove an aliquot and carry out a denatured

**Table 9-3.** Frequently Used RNA Polymerase Promoter Sequences

| Promoter | Sequence |
| --- | --- |
| SP6 promoter | 5′-ATT TAGGT GACAC TATA$^\wedge$GAATAC-3′ |
| T3 promoter | 5′-TTA TTAAC CCTCA CTAAA$^\wedge$GGGAAG-3′ |
|  | 5′-AAA TTAAC CCTCA CTAAA$^\wedge$GGGAAT-3′ |
| T7 promoter | 5′-TAA TACGA CTCAC TATA$^\wedge$G GGCGA-3′ |
|  | 5′-TAA TACGA CTCAC TATA$^\wedge$G GGAGA-3′ |

gel or even a Northern blot (see Chapter 3, Section 3.3.2). However, it is generally sufficient if you apply the aliquot to a normal agarose gel. Although you cannot control the size of the RNA with this, you can obtain an idea of the quantity of synthesized RNA and can see whether a uniform band is produced.

In the third step, translation of RNA and subsequent protein synthesis are performed using a *wheat germ extract* or a reticulocyte lysate (i.e., with one of the two eukaryotic translation systems). You can manufacture such extracts, but the work required and the chances of success are not comparable to the money saved (although such extracts are extremely expensive), and most experimenters purchase them instead (e.g., from Promega). Both extracts function in principle, although one or the other may prove to be better in individual cases. For implementation, follow the instructions provided by the manufacturer. The protocol is generally carried out as follows. All components are pipetted together, and the preparation is incubated for 60 minutes at 30°C (86°F). Success is controlled by sodium dodecyl sulfate–polyacrylamide gel electrophoresis (SDS-PAGE). The translation usually is performed in the presence of $^{35}S$-methionine, and the protein can then be demonstrated using autoradiography. If a protein-specific antibody is available, you can demonstrate this immunologically by performing a Western blot hybridization. If you have synthesized an enzyme, you may carry out a procedure to provide functional evidence.

A more detailed description of in vitro translation and its pitfalls can be found in the article by Krieg and Melton (1987) and in the *Promega Application Guide*.

The cell-free protein synthesis has the advantage of using a specific RNA as a template. In this way, the background signal is substantially reduced compared with the heterologous expression seen in cells. In vitro translation makes it possible to label a certain protein, whose subsequent development you may wish to pursue in another experiment. You can also synthesize proteins in this way to examine their activity, or you can see whether a cDNA codes for a protein at all. There is also a molecular biologic application in the search for mutations with the help of the *protein truncation test* (see Chapter 8, Section 8.2.8). Only a single protein has been synthesized and labeled, although the mixture is full of proteins that originate from the cell extract, and these make up most of the proteins in the preparation.

**Literature**

Krieg PA, Melton DA. (1987) In vitro RNA synthesis with SP6 RNA polymerase. Methods Enzymol 155:397–415.

Song HJ, Gallie DR, Duncan RF. (1995) m7GpppG cap dependence for efficient translation of *Drosophila* 70-kDa heat-shock-protein (Hsp70) mRNA. Eur J Biochem 232:778–788.

Promega. (1996) Protocols and Applications Guide, pp 260–276. (http://www.promega.com/paguide/)

# 9.4 Expression Systems

For the expression of proteins, selection of the proper system is the largest problem. If you want to perform functional investigations directly with the transfixed cells, the degree of freedom is limited, because you will sensibly use a system from which the protein you wish to explore originates. Consequently, the electrophysiologic derivations of ionic canals of the mouse to bacteria make little sense. Such measurements are also carried out on *Xenopus* oocytes, although the African clawed frog is not the nearest relative of the laboratory mouse.

Otherwise, you have a free selection. There is no ideal expression system, all of them have their advantages and disadvantages. You can always experience unexpected problems, because none of the problems is predictable. There is the torment of making a choice of the correct transfection system, which must be optimized at the beginning to obtain a maximum yield, and the entire affair is not a trivial matter.

## 9.4.1 Bacterial Expression Systems

*E. coli* is one of the most proficiently explored organisms. The organism was used quite early for evaluating the heterologous expression of proteins and to achieve large quantities of recombinant proteins.

The cDNA becomes cloned to a bacterial expression vector, which looks more or less like a normal plasmid, except for the fact that a bacterial promoter sits before a *multiple cloning site* that controls the expression of the cloned gene. The construct is then transformed into a bacterium.

Usually, an additional coding sequence is inserted at the beginning or at the end of the cDNA to generate a fusion protein with a specific, helpful characteristic. The popular histidine tag (a polyhistidine sequence of 6 to 8 amino acids long) permits a simple purification of the expression of protein by means of *immobilized metal affinity chromatography* (IMAC) columns. If a protease cleavage site has been inserted, such as for thrombin, you can remove the *tag* (a word used to signify labeling) after the purification. Instead of polyhistidine, you can use other tags, such as *c-myc*. Such markers, however, are not typical for bacterial expression systems, but more an expression of the general problem of fishing the desired protein from the vast number of proteins in the expression system after its successful expression. The fusion proteins with thioredoxin, which are used to solve the problem of insoluble proteins (discussed later), are more bacterially specific. A nice selection of prokaryotic expression systems is offered by Invitrogen.

**Advantage:** Bacteria are simple to manipulate, can be maintained simply, and are inexpensive to use.

**Disadvantage:** The expression frequently causes difficulties, and each protein creates its own problems. The overexpressed proteins are frequently insoluble in the bacterium. They then accumulate in the **inclusion bodies**, which are purified (a simple procedure) to attempt to restore the activity of the proteins. This approach is rather difficult and usually does not function successfully in the case of larger proteins. Marston and Hartley (1990) have dedicated a comprehensive article to this problem. Sometimes, the rate of expression is not particularly high, or it decreases rapidly over time, or the protein cannot be purified.

**Literature**
Marston FAO, Hartley DL. (1990) Solubilization of protein aggregates. Methods Enzymol 182:264–276.
Schein CH. (1989) Production of soluble recombinant proteins in bacteria. Biotechnology 7:1141–1148.

## 9.4.2 Baculoviral Expression Systems

The baculoviral system is somewhat unusual, although it has been successfully used for about 15 years. It is popular because it has the advantages of a viral transfection system (with a high transfection rate and high yields) and is simple to operate.

Baculoviruses are double-stranded DNA viruses with a genome approximately 130 kb long. They specifically transfix **insect cells**. *Autographa californica* nuclear polyhedrosis virus (AcMNPV) or modified variants typically are used. For humans, baculoviruses are not dangerous because the virus promoters are largely inactive in mammals (Carbonell et al., 1985), which substantially reduces the problem of safety, and this work can be done in an S1 laboratory. Because the viral genome is large, it clones larger DNAs. Baculoviruses require no wheat viruses and produce a higher titer.

Because the expression occurs in insect cells and insects belong to the eukaryotes, the proteins demonstrate many eukaryotic protein modifications such as glycosylation, phosphorylation, or acylation, which must be dispensed with in bacterial expression systems. The processing and transport of the proteins in insect cells proceeds similar to that observed in mammalian cells, and the expression of cytoplasmic, secretory, and membrane-related proteins is possible, which may cause difficulties in bacteria. Insect cells grow more rapidly than mammalian cells and require no $CO_2$ incubator.

As is typical for most viral transfection systems, the **expression rate** in the baculovirus system is very high, permitting production of 1 to 500 mg of protein per liter of culture. Most of the overexpressed proteins remain dissolved in the cell, an inestimable advantage compared with bacterial expression systems, in which this causes the greatest difficulties. Because you can cultivate the cells as a suspension, you can to a large degree produce preparations without any further difficulties (Figure 9-8).

The production of recombinant viruses is simpler than expected, considering the gigantic genome. The cDNA is cloned in a special transfer vector that is approximately 5 kb long and that contains longer virus-specific sequence segments to the left and the right of the cloning site. This construct is transfixed with the wild-type virus DNA in the insect cells. Through recombination, there is an exchange of virus and construct DNA along the homologous sequences. You find the desired viruses, isolate them, and then examine whether they correctly express the desired protein. After that, there is nothing to stand in the way of boundless protein expression.

Initially, the polyhedrin gene in the virus genome was exchanged with the cDNA; the recombinant viruses could then be recognized in that the plaques that they formed no longer demonstrated any visible **occlusion bodies** and consequently no longer appeared to be so white under the microscope. A recombination rate of 0.2% to 5% results in quite a bit of work. Understanding that linearized virus DNA recombines far more frequently increased the recombination rate to 30%, although the greatest relief came from the discovery that **protein 1629** was essential for the replication of the virus. If parts of the 1629 gene are deleted and inserted instead into the transfer vector, protein 1629 is again complete in the recombined viruses, but the nonrecombined viruses resent this and refuse to grow (Kitts and Possee, 1993). In this way, the probability that the surviving viruses will contain the desired cDNA can be increased to more than 90%. An accordingly modified baculovirus DNA and the fitting transfer vectors can be obtained from Clontec. Invitrogen constructed a baculovirus that also contains the 3' end of **β-galactosidase**. The 5' end is stuck in the transfer vector so that the correctly recombined clone expresses an intact β-galactosidase, and the plaques can then be easily detected by the blue coloration.

**Disadvantage:** Baculovirus is not the perfect expression system. Many of the typical mammalian modifications are seen in insect cells, but unfortunately, not all and not always exactly as

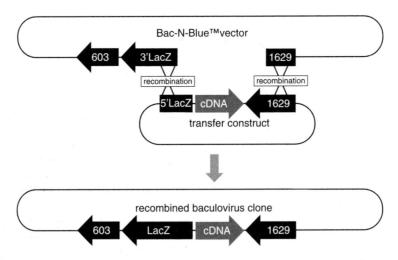

**Figure 9-8. Manufacture of recombinant baculoviruses based on the example of the Bac-N-Blue system from Invitrogen.**

in mammalian cells. Glycosylation is found, for instance, but the composition of the sugar differs considerably. In individual cases, such differences can lead to considerable difficulties.

**Literature**

Carbonell LF, Klowden MJ, Miller LK. (1985) Baculovirus-mediated expression of bacterial genes in dipteran and mammalian cells. J Virol 56:153–160.

Kitts PA, Possee RD. (1993) A method for producing recombinant baculovirus expression vectors at high frequency. Biotechniques 14: 810–817.

O'Reilly DR, Miller LK, Luckow VA. (1994) The Baculovirus Expression Vectors: A Laboratory Manual. New York, WH Oxford University Press.

Richardson CD (ed). (1995) Baculovirus Expression Protocols. Totowa, NJ, Humana Press.

## 9.4.3 Additional Expression Systems

### Yeasts

You can pursue the process of yeast genetics over the course of decades without coming to the idea of using them as expression systems. Invitrogen uses *Pichia pastoris*, a methylotrophic yeast, because the expression rate is very high (depending on the toxicity of the protein, between 1 and 12,000 μg/L of culture), and it is controlled through the addition of methanol. The breeding of yeasts is quite cheap and can easily be carried out at a large scale. A more detailed description has been provided by Brandes and colleagues (1996). Clontech offers a few different yeast expression vectors that make use of a $Cu^{2+}$-inducible promoter for the control of the expression (Macreadie et al., 989; Ward et al., 1994).

### *Drosophila*

Invitrogen has a newer expression system (*Drosophila* Expression System) that makes use of *Drosophila* cell lines. Unlike in the baculovirus system, normal expression vectors are used; these transfection methods are also used in mammalian cells. In contrast to mammalian cells, however, insect cell lines require no $CO_2$ incubator.

The system is suitable for transient and for stable transfections (see Section 9.4.7). The latter should proceed substantially more quickly than in mammalian cells, because the cell line used integrates several hundred copies of the constructs into the genome. The expression rate in the cells is continuously high, and you can forsake clonal selection of the cells. According to the manufacturer, it should be possible to obtain stabile, transfixed cells within a month.

A disadvantage to this system is the patent protection, which limits its use. Every user must be registered and is obligated not to transfer the system to others, not to use it commercially, and to make notice of every invention or discovery to the ladies and gentlemen from SmithKline Beecham (the patent holders) so that they may decide whether they want to make some profit from it. Most researchers will not be very happy with this situation, because it substantially limits the possibilities for cooperation, and it can hardly be considered pleasant in these times of increasing commercialization to dispense with making use of some new, pioneering discovery.

### *Xenopus* Oocytes

An exotic member among the expression systems is the oocyte from the African clawed frog, *Xenopus laevis*. It employs a classic method that is experiencing a small revival. Frog oocytes are as large as the head of a pin and are therefore quite accessible experimentally. A portion of transcribed RNA can be inserted in vitro by microinjection, which they then loyally convert into protein; ordinary plasmid DNA functions in this way as well, as long as this material is injected into the cell nucleus. Because frogs are eukaryotes, the proteins demonstrate many of the necessary modifications, unlike the results from bacteria. These little animals are especially popular in electrophysiology, because electric leads in animals of this size are quite simple. The system is also suitable for obtaining protein, because the oocyte is a single, gigantic cell that is programmed completely for protein synthesis. Ten oocytes are sufficient to deliver yields that are adequate for many experiments.

### Literature

Brandes HK, Hartman FC, Lu TY, Larimer FW. (1996) Efficient expression of the gene for spinach phospho-ribulokinase in *Pichia pastoris* and utilization of the recombinant enzyme to explore the role of regulatory cysteinyl residues by site-directed mutagenesis. J Biol Chem 271:6490–6496.

Macreadie IG, Jagadish MN, Azad AA, Vaughan PR. (1989) Versatile cassettes designed for the copper inducible expression of proteins in yeast. Plasmid 21:147–150.

Ward AC, Castelli LA, Macreadie IG, Azad AA. (1994) Vectors for Cu(2+)-inducible production of glutathione *S*-transferase-fusion proteins for single-step purification from yeast. Yeast 10:441–449.

## 9.4.4 Heterologous Expression in Mammalian Cells

A few terms need to be defined for those, like myself, who are at loggerheads with such ideas. An expression is *heterologous* when a protein is produced that does not occur in a cell in this form; the counterpart is *a homologous* expression.

There are two main possibilities for expression in mammalian cells. The cells can be treated with naked DNA or with viruses. Another option is to transfix with RNA, although such techniques are exceptions.

Naked DNA usually means that special expression plasmids are used that can easily be increased in bacteria to receive sufficient DNA and that contain everything that a DNA needs to become expressed in a mammalian cell: a eukaryotic promoter, an open reading frame that codes for the desired protein, an intron that can later cleave the cell, and a polyadenylation signal so that the RNA, which transcribes the cell of this plasmid, later looks like a true *messenger* RNA (mRNA). The plasmid also may contain other useful tomfoolery. The plasmid is transcribed into the desired cells using one of the methods described in Section 9.4.5, and it is then expressed readily.

Expression plasmids are not a safety problem. All work can be performed in S1 laboratories, as long as the cDNA to be expressed does not code for something that is poisonous or infectious. This is an essential difference from the **viral expression systems** (i.e., retroviruses, adenoviruses, adeno-associated viruses, herpes simplex virus, and alphaviruses such as Sindbis and Semliki Forest viruses), which, with only a few exceptions, require S2 laboratories. Every virus is a small, highly efficient transfection system. This is advantageous for the work in a laboratory. If you have selected the correct cell line, you can infect up to 100% of the cells, whereas the rate of success is 10% to 20% with other transfection methods (see Section 9.4.5), representing a large safety problem. Because mutation, recombination, and contamination can be observed in many cases when working with wild-type viruses, which may multiply and spread in an uncontrolled or reckless manner, the legal aspects related to mammalian-specific viral systems are restrictive. That is also positive, because nobody wants to be infected with herpes simplex virus, adenovirus, or retroviruses. Nevertheless, viruses are interesting expression systems, because in addition to the high transfection rate, they frequently have high expression rates. Some viruses, such as retroviruses, integrate into the host genome independently, a characteristic that greatly simplifies the manufacture of stably transfixed cells.

To reduce the risk associated with laboratory viruses, work is being carried out on systems that contain one, two, or three security systems to minimize the risk of "retrograde mutations" and the unintentional production of pathogenic viruses. The efforts are substantial because viral systems are being examined as possible carriers in the area of gene therapy. For therapeutic application, the systems must be so reliable that the risk of releasing dangerous viruses is near zero.

Another disadvantage of viruses is the size of their genomes, which are considerably larger than those of plasmid expression vectors, often making the manufacture of constructs quite laborious. Usually, the quantity of DNA that can be inserted is limited.

**Advantage:** For heterologous expression of mammalian proteins, mammalian cells are the best choice, because modifications and pleating usually appear as they should. If you choose the correct cell line, all of the proteins in the cell will interact with your own protein, greatly simplifying functional investigations. There is no limit to the number of cell lines, which also represents a problem, because each one of them demonstrates its own peculiarities that must be identified in detail. *American Type Culture Collection* (http://www.atcc.org) offers a very good selection of many different cell lines. A comparable European source is the *European Collection of Cell Cultures*, which is based in England (http://www.ecacc.org.uk/).

**Disadvantage:** The expression in mammalian cells presupposes that you have access to a functioning **cell-culture laboratory**. The goal of cell culturists is to protect their sensitive cells from every injustice of this world; in practice, that also means from bacteria. Anyone who wants to tackle cell culturing, if only for the purpose of performing an occasional transfection, is well advised to recognize the problems of the cell culturists and to take them seriously. It is not a cheerful job to spend weeks or months testing different batches of media and plastic material to find out the optimal conditions for the growth for your cells. Nor is it pleasant to have to negotiate with a colleague about holidays because mammalian cells may not prefer to dispense with encouragement for longer than 4 days. Frustration can become hatred if the fruits of your week-long labor lands in the autoclave merely because some inattentive but inquisitive rapscallion has contaminated the entire incubator with vile bacteria and fungi. Such events occasionally lead to individuals being kicked out of the cell-culture laboratory and can deteriorate the working climate in a long-lasting manner. The experimenter should therefore be well trained for work in a cell-culture laboratory and should assume a fastidious neatness and demonstrate a sterile mode of operation. Despite these efforts, you will be considered the principal suspect during the next series of contaminations, but it always helps to have a good reputation to endure such a crisis. Working with cell culturing is even more troublesome than handling of RNA!

**Literature**

Boris-Lawrie KA, Temin HM. (1993) Recent advances in retrovirus vector technology. Curr Opin Genet Dev 3:102–109.

Martin BM. (1994): Tissue Culture Techniques: An Introduction. Boston: Birkhäuser.

Miller AD. (1996): Retroviral vectors. In: Dracopoli N et al. (eds): Current Protocols in Human Genetics, pp 12.5.1–12.5.19. New York, John Wiley & Sons.

Morgan SJ, Darling DC. (1994): Kultur tierischer Zellen. Heidelberg, Spektrum Akademischer Verlag.

Pollard JW, Walker JM (eds). (1997): Basic Cell Culture Protocols. Methods in Molecular Biology, vol 75. Totowa, NJ, Humana Press.

Riviere I, Sadelain M. (1996) Methods for the construction of retroviral vectors and the generation of high-titer producers. In: Robbins P (ed): Methods in Molecular Medicine: Gene Therapy Protocols. Totowa, NJ, Humana Press.

# 9.4.5 Transfection Methods

The number of methods available to transport DNA into a small, defenseless cell is surprisingly large. Even if the methodical preparations differ greatly, they all suffer from the same problem.

The effectivity (i.e., proportion of transfixed cells) depends on the type of cell and the conditions. Anyone wishing to transfix cells should first clarify which method has been effective for comparable experiments and cell lines.

In principle, every **DNA** can be used to perform a transfection, as long as it is pure enough that the cells are able to survive the procedure. Typically, the DNA has been purified through the use of a cesium chloride gradient (see Chapter 2, Section 2.3.6) or with the aid of commercial anion exchange columns (see Chapter 2, Section 2.3.4); some experts recommend that this DNA be subjected to an additional phenol-chloroform purification. RNA can also be used for the transfection.

For the expression of proteins in eukaryotic cells, a flood of **expression vectors** has been developed that leaves almost no wish unfulfilled. You can find these products in the catalogs from Invitrogen, Clontech, or Promega, and it helps to perform an intense search of the literature to learn about other interesting vectors.

Eukaryotic expression vectors usually are shuttle vectors; the DNA is produced in bacteria, and it requires the usual elements such as a replication origin and resistant genes (Figure 9-9). In the eukaryotic cell, expression of only the desired gene occurs. It requires a eukaryotic promoter and the coding area of the gene to be expressed. Depending on whether you are carrying out the expression of a protein or performing a promoter study, you must search for a vector that controls the other components and has a suitable *multiple cloning site*.

The transfection methods can be differentiated according to the particular mechanism by which the DNA is taken up by the cell:

- Through **endocytosis or phagocytosis** (e.g., DEAE-dextran, CaPO$_4$, transferrin)
- Through **fusion with the cell** (e.g., lipofection, liposome fusion, protoplast fusion)
- Through **diffusion** (i.e., electroporation)
- Through **force** (e.g., microinjection, bombardment)
- Through **viral mechanisms** of transfection (i.e., viral systems)

With most of these methods, plasmid DNA, RNA, and oligonucleotides can be transferred into the cells. The method preferred is based on the particular necessities. In this case, a micromanipulator for the microinjection and someone who knows how to work with it are needed, whereas the operation with most viral systems demands an S2 authorization if the institute or the group does not possess

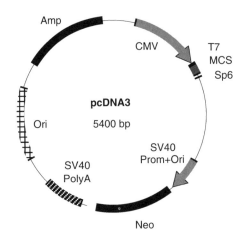

**Figure 9-9. Construction of a typical eukaryotic expression vector (pcDNA3 from Invitrogen).**
Amp, β-lactamase; CMV, CMV promoter; MCS, multiple cloning site; Neo, aminoglycoside phosphotransferase; Ori, bacterial replication start; Sp6, Sp6 RNA polymerase promoter; SV40 PolyA, SV40 polyadenylation signal; SV40 Prom + Ori, SV40 promoter and eukaryotic replication start; T7, T7 RNA polymerase promoter.

such authorization. However, alternatives are available, and you should choose another method if there are no convincing reasons to undergo the administrative struggle and the numerous months of waiting.

## Calcium Phosphate Precipitation

Calcium phosphate precipitation is one of the most common methods, because it is simple, cheap, and requires no special apparatus. DNA, calcium, and phosphate buffer are mixed, and a fine-grained precipitation of calcium phosphate and DNA develops, which is pipetted onto the cells, which resorb these crystals by endocytosis. The details of this process are not known and are unimportant for the implementation of a transfection. The DNA is transported through the cells to the nucleus, where it is transcribed and translated. The proportion of expressed cells is between 5% and 30%, depending on the cell type used.

The $CaPO_4$ precipitation can be accomplished without any extensive technical efforts. The classic protocol makes use of an $N$-2-hydroxyethyl piperazine-$N'$-2-ethane-sulfonic acid (**HEPES**) buffer, although $N, N$-bis(2-hydroxyethyl)-2-aminoethane-sulfonic acid (**BES**) buffer has a higher transfection efficiency.

A total of $5 \times 10^5$ cells are cultivated in a 10-cm plate with 10 mL of medium for 24 to 48 hours (which corresponds to one or two cell divisions). Then, 500 µL of $CaCl_2$ DNA solution (20 µg of DNA in 500 µL of 0.25 M $CaCl_2$ DNA solution) and 500 µL of $2 \times$ BBS (50 mM of BES/280 mM NaCl/1.5 mM $Na_2HPO_4$, pH 6.95, held at pH 6.95 with NaOH and filtrated in a sterile fashion, stored at $-20°C$ [$-4°F$]) are mixed well and incubated for 5 to 20 minutes at room temperature until a fine precipitate develops. The calcium phosphate DNA mixture is then pipetted drop by drop onto the plates with the cells and carefully distributed. The cells are then cultivated for 12 to 24 hours in an incubator with 3% $CO_2$. On the next day, the cells are washed twice with 5 mL of PBS, 10 mL of medium is added, and the cells are cultivated in an incubator with 5% $CO_2$. With transient transfection, the cells attain their expression maximum 48 to 72 hours after the addition of DNA.

Several points are critical:

- The **quality of the DNA**. If the DNA is too dirty, no precipitation will develop, or the cells will die. It is therefore advisable to purify the plasmid DNA by means of a cesium chloride gradient, or one use commercial columns for DNA purification and perform a phenol-chloroform purification. Despite this special treatment, it is not unusual that two DNA preparations precipitate differently with calcium phosphate.
- The **DNA quantity**. The quantity indicated is a standard value, and the optimal amount must be determined individually.
- The **pH of the BBS solution**. The pH must be adjusted to exactly 6.95. Even small deviations can transform the precipitation experiment into a tragedy. You should always adjust the pH meter, but because that is generally not sufficient, you must have ready BBS solution that has functioned well in the past or several solutions with different pH values that are tested. Good BBS solutions can be stored for several months at $-20°C$ ($-4°F$), repeatedly thawed, and frozen again. Poor preparations are useless on the following day.
- The **precipitate**. If the DNA and BBS solution are correct, a fine precipitate develops after 1 to 15 minutes. You can check this by observing a drop of the preparation under the microscope. Alternatively, you can control the entire preparation. If the precipitate becomes visible, the solution appears to be light blue to the skilled eye. To see this difference, you must use crystal-clear vessels (as a reference, use a vessel with an equal volume of water). The transfection efficiency is largest if the calcium phosphate crystals are very small. The precipitate can be frozen in the event of an emergency and used directly for transfection after thawing.
- The **condition of the cells**. The cells should be found in an exponential state of growth. If the cells have increased too densely and no longer divide, it is most likely that the transfection will

not function. The cells should be healthy; cells that are sickly cannot be transfected. The cells can deteriorate if the cell line has been kept in the culture for too long.

- The **CO$_2$ content** of the incubator. A 3% $\pm$ 0.5% level of CO$_2$ is best. Transfection can also be accomplished at 5% CO$_2$, although the effectivity is reduced substantially, because the pH of the medium does not remain in the optimal range.

**Advantage:** Transfection can be performed inexpensively and without extensive technical effort.

**Disadvantage:** The process is very sensitive to small deviations in the pH, to the DNA quality, and to other conditions.

**Literature**

Chen C, Okayama H. (1988) Calcium phosphate-mediated gene transfer: A highly efficient system for stably transforming cells with plasmid DNA. Biotechniques 6:632–638.

## Transfection with Polycations

The oldest of the transfection methods (initiated as early as the 1960s) uses DEAE-dextran, to direct the DNA into the cells. The DNA is bound to polycations, and the dextran-DNA complexes are pipetted onto the cells, where they are taken up by endocytosis and expressed.

In this process, $5 \times 10^5$ cells are plated in a 10-cm plate with 10 mL of medium and cultivated for 24 to 48 hours. You then mix 4 μg of plasmid DNA in 40 μL of TBS (see Appendix 1, Standard Solutions) and 80 μL of DEAE-dextran (10 mg/mL in TBS) and then distribute the mixture dropwise onto the cell-culture plate. The plate is placed in the incubator for 2 to 8 hours, the upper layer is drawn off, and the cells are incubated in 5 mL of DMSO solution (10% [v/v] dimethylsulfoxide in PBS), washed with 5 mL of PBS, and cultivated with 10 mL of medium until they are eventually harvested.

As with calcium phosphate precipitation, the optimal transfection conditions must be determined, particularly concerning the DEAE-dextran concentration, the DNA concentration, and the duration of the transfection. DEAE-dextran is toxic for the cells, and if the incubation is carried out for too long, the cells will die.

**Advantage:** This method is cheap and has no serious technical requirements. It is far more tolerant than the calcium phosphate precipitation. It is easier to apply and to reproduce.

**Disadvantage:** Because of its cytotoxicity, DEAE-dextran is not very suitable for the production of stably transfixed cells.

**Literature**

Lopata MA, Cleveland DW, Sollner-Webb B. (1984) High-level expression of a chloramphenicol acetyltransferase gene by DEAE-dextran-mediated DNA transfection coupled with a dimethylsulfoxide or glycerol shock treatment. Nucleic Acids Res 12:5707.

Sussman DJ, Milman G. (1984) Short-term, high-efficiency expression of transfected DNA. Mol Cell Biol 6:3173–3179.

## Transfection with Transferrin Polycation Conjugates

The principle of transfection with transferrin polycation conjugates is similar to that seen with the previous method, although it is used for the receptor-mediated endocytosis system. **Transferrin**, an iron-transport protein, is coupled with polylysine or alkaline, positively loaded proteins, which can easily bind covalently with the acidic, negatively loaded DNA. This transporter construct is loaded with DNA, and the cells are incubated for 24 hours with this mixture. The procedure has been called *transferrinfection*.

**Literature**

Wagner E, Zenke M, Cotton M, et al. (1990) Transferrin-polycation conjugates as carriers for DNA uptake into cells. Proc Natl Acad Sci U S A 87, 3410–3414.

## Transfection with Cationic Liposomes: Lipofection

Lipofection has become more common than the $CaPO_4$ precipitation, because it is just as simple to use but frequently provides a much better transfection efficiency. It is a relatively modern method (first described by Felgner et al., 1987), which has in this short time experienced substantial changes. The first-generation lipofection was a single-component system, in which only the lipid $N$-[1-(2,3-dioleyl-oxy)propyl]-$N$, $N$, $N$-trimethylammonium chloride (**DOTMA**) was used. The effectivity was improved by mixing a cationic lipid with an unloaded helper lipid, such as L-dioleoyl phosphatidyl-ethanolamine (DOPE), to facilitate fusion with the cell (Rose et al., 1991) (Figure 9-10). The third generation of lipid mixtures contain additional DNA-compacting substances.

The principle of the procedure has for the most part remained the same. The lipid or lipid mixture in an aqueous solution is subjected to an ultrasonic treatment, during which the liposomes develop. They are provided in the form of a lyophilized lipid film, which must be resuspended in water and the cell-culture medium and then mixed with the DNA. The mixture is pipetted to the cells and incubated for an hour or longer. During this time, the liposome-DNA complexes fuse with the cell membrane, and the DNA somehow arrives in the cell nucleus. After that, the medium is changed (or not), and the cells are incubated for another 8 to 48 hours until the expression has reached its maximum.

All experts and the manufacturers are convinced that you cannot avoid **optimizing** conditions in a self-sacrificing series of tests, because the differences from one cell line to the next are tremendous.

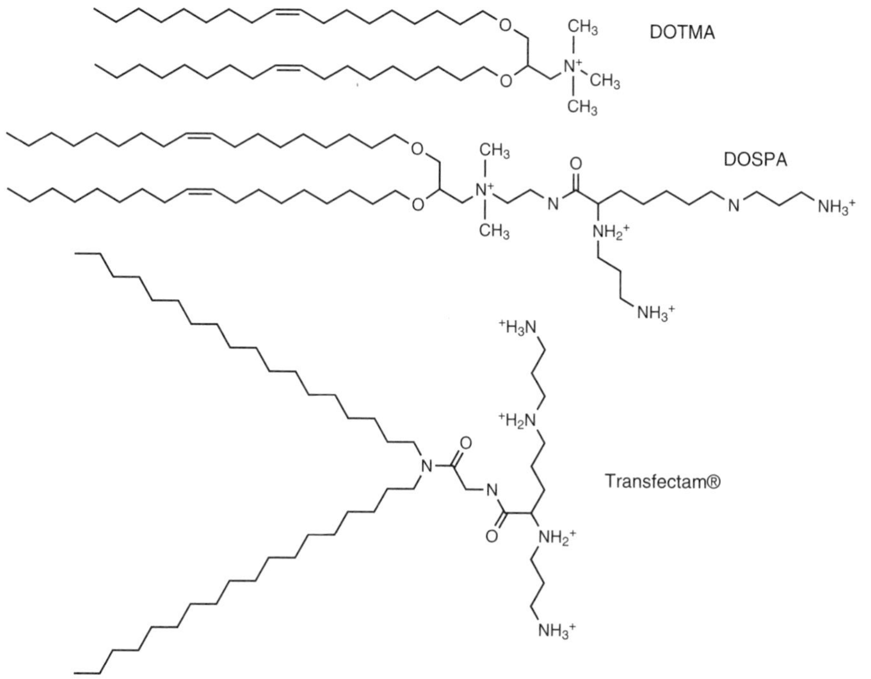

**Figure 9-10. Examples of cationic lipids.**

The parameters to be defined have become quite extensive. It is important to find the optimal lipid, the correct quantity of lipid, the best ratio of loaded to unloaded lipids, and the right DNA quantity. You must find out whether you are better off using the medium with or without serum, how densely the cells should grow, the duration of transfixion, when to harvest, and so on. This is a problem for those who like to work with more than one cell line, to say nothing of working with primary cell cultures! Ultimately, the cells appear to be differentiated into two classes. In one, **transfection rates** of 90% to 95% appear to be thoroughly realistic, whereas you may open a bottle of champagne if the second group attains a level of 20%.

It is not always as dramatic as is described here, especially if you are satisfied with less than the optimal results. Otherwise, the method would not be as popular as it is.

In a typical protocol, $10^6$ cells are plated on a 10-cm plate together with 10 mL of medium and cultivated for 24 to 48 h. Then, 5 μg of plasmid DNA is diluted in 5 mL of serum-free medium; 25 to 50 μL of liposome suspension is added and vortexed well; and the mixture is incubated for 5 to 10 minutes. The cells are washed once with serum-free medium, the medium is removed, and the liposome-DNA mixture is added. After 3 to 5 hours of incubation in the cell incubator, 5 mL of medium containing serum is added and incubated for another 16 to 24 hours. The transfection medium is exchanged for a normal medium, and the cells are cultivated until a final application.

**Advantage:** There is a slight sensitivity to pH changes and a higher efficiency for transfection compared with calcium phosphate precipitation, although these features depend on the specific type of cell. Nonmitotic cells can become transfixed.

**Disadvantage:** The liposome solutions are relatively expensive, requiring about $3.50 per transfection when using commercial lipofection systems.

**Literature**

Felgner PL, Gadek TR, Holm M, et al. (1987) Lipofection: A highly efficient, lipid-mediated DNA/transfection procedure. Proc Natl Acad Sci U S A 84:7413–7417.

Rose JK, Buonocore L, Whitt M. (1991) A new cationic liposome reagent mediating nearly quantitative transfection of animal cells. Biotechniques 10:520–525.

Zhou X, Huang L. (1994) DNA transfection mediated by cationic liposomes containing lipopolylysine: Characterization and mechanism of action. Biochim Biophys Acta 1189:195–203.

## Liposome Fusion

Neutral phospholipids can be used for transfection. A mixture with water, DNA, and phospholipids is processed with ultrasound until liposomes develop. They contain a DNA solution in their interior. The liposomes can be fused with the plasma membrane of the cells using polyethylene glycol (PEG). However, not only cells and liposomes fuse, but also cells with cells. The method is less suitable for the manufacture of stabile, transfixed cells.

**Literature**

ltani T, Ariga H, Yamaguchi N, et al. (1987) A simple and efficient liposome method for transfection of DNA into mammalian cells grown in suspension. Genes 56:267–276.

## Protoplast Fusion

Protoplast fusion is the fruit of an original idea. If you can load and fuse liposomes with DNA, why should you go to the trouble of manufacturing these, because bacteria are really nothing more than large liposomes filled with plasmid DNA.

The bacteria are treated with chloramphenicol to increase the proportion of plasmid DNA, and the lysozyme digests the cell wall. In this way, protoplasts can be fused to the cells like liposomes.

The well-versed cell culturologist develops gray hair merely from the idea that bacteria should voluntarily be added in large numbers to the cells without there being any emergency. The problem is solved through the application of large quantities of antibiotics. The simultaneous transfer of plasmid DNA and bacterial chromosomes is not entirely problem free, and the method has not prevailed.

**Literature**

Sandri-Goldin RM, Goldin AL, Levine M, Glorioso J. (1983) High-efficiency transfer of DNA into eukaryotic cells by protoplast fusion. Methods Enzymol 101:402–411.

## Electroporation

Electroporation is becoming increasingly popular, because it is simple and reliable, and it provides the experimenter with a feeling of employing a high-tech operation. The effectiveness is similar to that seen with calcium-phosphate transfection.

In principle, the electroporation of eukaryotic cells functions like that of bacteria (see Chapter 6, Section 6.4.4). The cell membrane is permeabilized for a short time through an electrical impulse, and the DNA has the chance to pass from the solution into the cell by diffusion. Eukaryotic cells are much more sensitive than bacteria, and far less harsh conditions are used.

Approximately $10^7$ cells are suspended in 0.5 mL of electroporation buffer, such as cell-culture medium without FCS or phosphate-buffered sucrose solution (270 mM sucrose/7 mM $K_2HPO_4$, pH 7.4/1 mM $MgCl_2$ or PBS), electroporated in a 4-mm cuvette at 0°C (32°F) and a field strength of 2 to 4 kV/cm with 20 to 80 μg of DNA, and transferred as quickly as possible into 9.5 mL of fresh medium and cultivated as usual. After 24 to 72 hours, the cells can be harvested and undergo further processing.

The conditions (e.g., intensity of the field strength, duration of the impulse) must be optimized for each cell type. You can consult the extensive documentation material available from every manufacturer; it is usually delivered with the instrument. The conditions are optimal if 50% of the cells are lysed; if fewer cells are lysed, too little DNA is taken in, and the yield decreases when more are lysed.

Anyone who enjoys working at a large scale can take advantage of the 96-well "cuvettes" that are available. Soon, we will be able to use 384-well plates.

**Advantage:** The procedure is simple and functions with many different cell types, even when the calcium phosphate method fails. Through the concentration of the DNA used, you can influence the quantity of DNA in the transfixed cells better than with any other method.

**Disadvantage:** A special electroporation instrument is required, because the field strengths and resistances used lie in another area from that used for the electroporation of bacteria. Many units cover both areas, in part through the application of additional modules. You can find these devices at BTX, BioRad, or Gibco. More cells and DNA are needed than for calcium phosphate or DEAE-dextran transfection.

**Literature**

Baum C, Förster P, Hegewisch-Becker S, et al. (1994) An optimized electroporation protocol applicable to a wide range of cell lines. Biotechniques 17:1058–1062.

Potter H, Weir L, Leder P. (1984) Enhancer-dependent expression of human K immunoglobulin genes introduced into mouse pre-B lymphocytes by electroporation. Proc Natl Acad Sci U S A 81:7161–7165.

## Transfection through Bombardment: Particle Delivery

Transfection through bombardment was originally developed for the transfection of plant cells, but it is now also used for mammalian cells. Small wolfram or gold particles are coated with DNA and then shot onto the cells at a high speed. This can be performed on cells in culture and in the living animal, such as cells of the liver, skin, and spleen. The depth of penetration is slight so that the transfection results only on the surface. The instrument required for this method can be obtained from BioRad under the martialistic name of Helios Gene Gun System.

**Advantage:** This method functions quite well where other methods fail, because the DNA is "forced" into the cells.

**Disadvantage:** Extensive manipulation, large preparations, and lots of time are needed. The efficiency of the transfection is frequently not particularly high.

**Literature**
Johnston SA, Tang DC. (1994) Gene gun transfection of animal cells and genetic immunization. Methods Cell Biol 43:353–365.

## Microinjection into Cell Nuclei

Microinjection is one of the few methods that allows experimenters to see what they are doing. With the aid of an extremely fine glass capillary tube, cells can be punctured and the DNA solution injected directly into the cell nucleus. In practice, the procedure is quite troublesome and is associated with much work. At the end of the day, you have only a comparatively small yield. The application is therefore limited to special cases, such as the manufacture of transgenic animals.

**Advantage:** You can easily control the DNA quantity per cell. The effectivity is high, because the DNA is injected directly into the cell nucleus.

**Disadvantage:** The expense of the apparatus is high and the method demands considerable experimental experience and skill.

## 9.4.6 Cotransfection of Several Genes

For all of the methods presented, the type of the DNA to be transfixed (at least for transfection) is insignificant as long as it is sufficiently clean. This has an advantage in that you can mix different DNAs with one another without any problems and can transfix them jointly. Because the cell takes up many DNA molecules in every case (if they do so at all), the probability that more DNA molecules of each type are taken up from the mixture of DNA is very high. Conversely, a cell that has expressed one type of DNA will also express the others.

A system being examined rarely contains only a single protein (e.g., neurotransmitter receptors). There are often two or more different subunits, so that transfection of only one plasmid using a single gene makes little sense. Several cDNAs can be placed in a single plasmid, but that is troublesome because an individual expression construct is then required for each combination.

More frequently, experimenters want to control the **effectivity** of the transfection before investing further work in the attempt, even though the emerging protein cannot be demonstrated at all or only with a great deal of difficulty. The solution is to cotransfix it with another cDNA construct, such as with a β-galactosidase–expressing plasmid. Successfully transfixed cells can be made visible by dyeing them blue. Such a control is evident and can be accomplished easily, although the cells are killed. For that reason, *green fluorescent protein* (GFP) is becoming increasingly more popular, because the process can be carried out without great technical effort and without any loss in material by using an (inverse) fluorescence microscope. GFP is available in different varieties (see Section 9.4.8).

Another application of cotransfections is to provide an internal standard. Evidence of the activity of a promoter is frequently provided by the use of a tandemly arranged reporter protein such as chloramphenicol transferase or luciferase (see Section 9.4.8). To eliminate the effects of different transfection efficiencies, the experimenter standardizes against a second reporter protein construct with a constitutive expressing promoter that has been cotransfixed.

# 9.4.7 Transient and Stabile Transfections

Regardless of the transfection method used, the serviceable life of the DNA taken up is always limited to a few days. After that, the foreign DNA is fully degraded in the cell, and the method is therefore called a **transient transfection**. Rarely the DNA is integrated into the genome of the cell and then protected there from further degradation. The probability for this depends on the particular transfection method, the cells, and the other factors, although it demonstrates a maximum on the order of $10^{-4}$; only a single cell from 10,000 becomes **stably transfected**.

The conformation of the DNA influences the process. For the transient transfection, you are better off using *supercoiled* plasmid DNA, because the DNA in this conformation is transcribed with a higher effectivity. The stabile transfection functions better with linearized DNA, because it is better integrated into the genome, with a higher effectivity. The fragments during transfection in the cell are ligated to form larger units, which then integrate into the genome. In later analyses, you therefore find several copies of the original DNA in the genome.

If the DNA construct contains a selection marker, you can perform a selection on solidly transfixed cells. The selection of stably transfixed cells principally occurs in two ways: through complementation or through dominant markers. With **complementation**, you use cells with a defined defect that grow only in a special manner, whereby the defect is compensated. Together with the gene of your dreams, a gene is transfixed whose product removes (complements) the defect. The gene can lie with it on an equivalent plasmid or on a second, independent plasmid. The cells used have defects in the nucleotide metabolism, primarily in thymidine kinase (TK), hypoxanthine guanine phosphoribosyl transferase, adenine phosphoribosyl transferase, or dihydrofolate reductase (DHFR). The selection of **dominant markers** is similar, except for using normal cells and cotransfixing with dominant reporter genes, typically a resistance gene. The selective conditions are obtained through the addition of a cytotoxic substance in the medium, which is made milder through the action of the resistant gene. The most common dominant markers are aminoglycoside phosphotransferase (APH, *neo*; neomycin/G418 resistance), xanthine-guanine phosphoribosyltransferase (XGPRT, *gpt*), hygromycin B phosphotransferase (HPH; hygromycin B resistance), and chloramphenicol acetyltransferase (CAT).

The great advantage of dominant markers is that you are not dependent on special cell lines with specific defects, but can instead transfix whatever you desire. The neomycin-resistant aminoglycoside phosphotransferase (APH), generally known as *neo*, is very popular.

After the transfection, the cells are cultivated for two divisions in nonselective medium and then converted to a selective medium in which only those cells survive that have taken in DNA. From them, only those survive for the next few weeks that have transfixed DNA integrated into the genome and expressed genes. After 10 divisions, the clonal selection begins. Individual cells or colonies are isolated and separately cultivated further to obtain genetically uniform cell lines, which are examined for their fitness in later attempts. Because the DNA constructs are integrated by chance into the genome and the particular environment, the characteristics of the cell lines can be quite different. Sometimes, it is difficult to find a clone that expresses the desired protein to a sufficient degree.

Access to a **fluorescence-activated cell sorting (FACS) device** allows another approach to selection. The transfixed cells are labeled with fluorescent dyes (through incubation with a desired protein demonstrating fluorescently labeled antibodies or through cotransfection with GFP) and separated from nonexpressed cells with the cell sorter.

Establishment of stabile cell lines takes approximately 2 months. However, the difficulties have not been eliminated, because a characteristic of stabile, transfixed cell lines is that they are not stable. This problem usually arises in all cell lines, because changes in the genome develop over the course of time, and the characteristics of the cells are altered. In transfixed cells, this phenomenon is more intense. The reasons for this are not clear, but expression of the construct cloned in these cells decreases over time. This change is considered normal, although it makes the storage of stably transfixed cell lines quite troublesome. Adhere to the following guidelines to ameliorate these problems:

- Maintain the cell line under constant selection pressure. Frequently or always cultivate in a selective medium to minimize the proportion of altered cells without a construct.
- Carry out clonal selection now and again. Pick out individual clones, evaluate their correct identity, and use them for further cultures.
- Refrain from cultivating the culture endlessly, because the cells become altered in any case. Instead, freeze the early cell passages from which you can later reactivate new, unaltered cells.

Or you can discover an answer to your problem dealing with transient, transfixed cells.

## 9.4.8 Reporter Genes

Often, you need to demonstrate whether an expression has occurred at all. You can find help using reporter gene constructs.

A reporter gene encodes a protein that is expressed when certain conditions are met. Reporter genes can signify the entire construct as well as the protein to be detected; in the case of fusion proteins, only a part of the protein demonstrates the activity to be detected. In principle, every DNA that can code for something and that can be demonstrated in some manner can serve as a reporter gene. In practice, expression constructs are selected that are simple and code for a protein that can be easily identified and is highly sensitive.

Reporter genes often are used in **promoter studies**. You clone the promoter ahead of a protein that can easily be demonstrated and, if possible, has only a slight or no injurious influence on the cell in which the construct will be expressed. The amount of protein or protein activation shows how active the promoter is, and in transgenic animals, you can examine in which tissues expression occurs or not.

If you are more interested in the location of a specific protein, you can forge a **fusion protein**: one half of your own protein and one half of a reporter gene. If you are lucky or have selected the proper system, such a fusion protein maintains the characteristics of the original protein and gains the activity of the reporter gene, by which the fusion protein can then be demonstrated. However, there is a risk that the fusion protein instead will develop its own way of life.

Reporter genes often are used for control, such as to ascertain whether a transfection has occurred at all, or as a standard, such as in promoter studies. The palette of customary reporter genes is relatively large, which leaves you with the problem of making a selection. In practice, the most important question is whether you want quantitative or qualitative evidence. Good quantitative reporters frequently require quite specific conditions to provide evidence, and good qualitative reporters are quantified quite poorly.

### Literature

Alam J, Cook JL. (1990) Reporter genes: Application to the study of mammalian gene transcription. Anal Biochem 188:245–254.

## Reporter Genes for Quanititative Evidence: Chloramphenicol Acetyltransferase

Chloramphenicol acetyltransferase (CAT) is the classic example among reporter genes. The enzyme catalyzes the transfer of an acetyl remnant of acetyl CoA to **chloramphenicol**. The cell lysate to be examined is incubated with acetyl CoA and [$^{14}$C]-labeled chloramphenicol. Acetylated and unacetylated chloramphenicol are separated from each other by thin-layer chromatography (TLC), and the TLC plate is autoradiographed. This provides a lovely picture and is quantifiable by scanning the x-ray film and evaluating it. A better approach is to scratch the radioactive spots from the plate and measure the activity in a scintillation counter. However, because there is usually more than a single preparation, analysis becomes somewhat troublesome.

As an alternative to $[^{14}C]$-chloramphenicol, you can insert $[^3H]$-labeled acetyl CoA. The application of $[^3H]$ instead of $[^{14}C]$ makes the procedure somewhat less expensive. Radioactive methods for providing evidence have been developed, such as anti-CAT antibody (Roche), which permits evaluation by means of enzyme-linked immunosorbent assay (ELISA). The quantity of available CAT is measured. Molecular Probes offers a kit (FAST CAT Chloramphenicol Acetyltransferase Assay Kit) with a fluorescing derivative of chloramphenicol, which offers greater sensitivity than the radioactive assay.

CAT is rather stabile in the cell and has a half-life of 50 hours. This can be advantageous if you want to have the largest possible yield or disadvantageous if you want to demonstrate the kinetics of the induction or inhibition of the CAT expression.

**Advantage:** Because this is a classic method, abundant data have been collected from past experiences. It is accepted readily for promoter studies, and it has a good signal-to-background ratio.

**Disadvantage:** The enzyme assay is simple and reliable, but it is very time-consuming. The entire procedure—manufacture of the cell lysate, acetylation, purification, chromatography, and autoradiography—takes at least 24 hours. The sensitivity is smaller than in later systems. Radioactivity is required, which is also quite expensive.

### Literature

Gorman CM, Moffat LF, Howard BH. (1982) Recombinant genomes which express chloramphenicol acetyltransferase in mammalian cells. Mol Cell Biol 2:1044–1051.

Neumann JR, Morency CA, Russian KO. (1987) A novel rapid assay for chloramphenicol acetyltransferase gene expression. Biotechniques 5:444.

## Luciferase

The enzyme, which causes love-stricken fireflies to glow green, is faster and more sensitive than CAT, and it functions without any radioactivity. Because of its high sensitivity, it is becoming more popular, especially in the investigation of weaker promoters. Because of its relatively short half-life of approximately 3 hours, it is well suited for induction studies.

The reaction is simple. Luciferin, ATP, and $Mg^{2+}$ are pipetted onto the cell lysate and immediately transformed into light; the light production is proportional to the quantity of luciferase. Because the reaction is also very fast (half-life <1 second), a suitable luminometer should be in a position to work quickly. The kinetics of the reaction can be slowed through CoA, pyrophosphate, and nucleotides (Ford et al., 1995). Promega, which has concentrated on the luciferase system and offers an entire assortment of luciferase vectors, has a complete test system that makes use of slower kinetics and a higher light intensity than the classic luciferase methods. The culmination is Promega's Dual-Luciferase Reporters Assay System, which is based on the use of two luciferases with different characteristics. With the luciferase from the firefly (*Photinus pyralis*), you can demonstrate the activity of your own promoter. The cotransfixed **jellyfish luciferase** (from *Renilla reniformis*) is under the control of a constitutive promoter and can therefore be used for standardization. The procedure is simple and fast. The firefly luciferase activity is measured, and the *Renilla* buffer is added. The buffer stops the firefly luciferase reaction, and starts the *Renilla* reaction.

**Advantage:** The sensitivity is 10 to 1,000 times higher than with CAT.

**Disadvantage:** A luminometer is needed to measure the enzyme activity. Because of the rapid kinetics of the reaction, some practice is needed to achieve uniformity, or the measurements will be somewhat random.

**Literature**

Ford SR, Buck LM, Leach FR. (1995) Does the sulfhydryl or the adenine moiety of CoA enhance firefly luciferase activity? Biochim Biophys Acta 1252:180–184.

# β-Galactosidase

β-Galactosidase is the same enzyme that is used for cloning to provide evidence of positive clones by means of blue-white detection. Providing the evidence is fast and sensitive, and no radioactivity is required.

An entire palette of substrates is at your disposal. For the colorimetric evidence, *o*-nitrophenyl-β-D-galactopyranoside (ONPG) is used, and fluorometric evidence is provided by 4-methylumbelliferyl-β-D-galactoside (MUG), which is not very sensitive. Most sensitive is the chemiluminescent evidence provided by **l,2-dioxetane substrates**, such as Galacton (Tropix) or Lumi-Gal 530 (Lumigen). Light production is measured with a luminometer or a scintillation counter. The sensitivity is then comparable with luciferase, and the linear area of the measurement is increased.

β-Galactosidase can be used as an internal standard for other identification (e.g., CAT, luciferase). A corresponding reporter construct is cotransfixed, because the cell lysate required for these tests can be used to obtain β-galactosidase evidence.

**Literature**

Bronstein I, Edwards B, Voyta JC. (1989) 1,2-Dioxetanes: Novel chemiluminescent enzyme substrates. Applications to immunoassays. J Biolumin Chemilumin 4:99–111.

Hall CV, Jacob PE, Ringold GM, Lee F. (1983) Expression and regulation of *Escherichia coli lacZ* gene fusion in mammalian cells. J Mol Appl Genet 2:101–109.

Jain VK, Magrath IT. (1991) A chemiluminescent assay for quantitation of beta-galactosidase in the femtogram range: Application to quantitation of beta-galactosidase in lacZ-transfected cells. Anal Biochem 199:119–124.

# Human Growth Hormone

Human growth hormone (hGH) is secreted from transfixed cells. The evidence for this comes from measurement of the hGH concentration in the medium. The expression of hGH can be followed over a long period without having to sacrifice any cells. The proof is offered immunologically by means of **radioimmunoassay** (RIA) and visibly by means of radioactively labeled antibodies, a method that is rapid and simple. The sensitivity is not good, although it is somewhat higher than with CAT evidence. The hGH-expression constructs with a constitutively expressed promoter are quite suitable as an internal standard for CAT, luciferase, and β-galactosidase.

**Literature**

Selden RF, Howie KB, Rowe ME, et al. (1986) Human growth hormone as a reporter gene in regulation studies employing transient gene expression. Mol Cell Biol 6:3173–3179.

# Secreted Alkaline Phosphatase

Secreted alkaline phosphatase (SEAP) is a shortened version of the human placental alkaline phosphatase (PLAP), which is not anchored in the membrane. Like hGH, it is secreted, which makes it possible to perform investigations in the living cell. Because the SEAP is very heat resistant and less sensitive to the phosphatase inhibitor L-homoarginine, there are few problems with the background caused by endogenous alkaline phosphatases. The evidence results colorimetrically with *p*-nitrophenyl phosphate, which is fast, simple, cheap, and insensitive. Alternatively, there is luminescent evidence from the transformation of luciferin by alkaline phosphatase-D-luciferin-*O*-phosphate, which can be demonstrated by means of luciferase. The most sensitive is the evidence gained using a chemiluminescent substrate, such as CSPD (Tropix), which is measured with a luminometer or a scintillation counter, and the linear range of measurement is the largest.

**Advantage:** SEAP is well-suited for kinetic studies, because only the medium is examined, and the cells can be used further. This technique can also be used as an internal standard like hGH, although it is cheaper and more sensitive.

**Literature**
Berger J, Hauber J, Hauber R, et al. (1988) Secreted placental alkaline phosphatase: A powerful new quantitative indicator of gene expression in eukaryotic cells. Genes 66:1–10.

## Reporter Genes for Qualitative Evidence

Typically, a qualitative reporter is used for evidence of expression in (living) cells, to see whether a certain cell has been transfixed or not, or to obtain some information concerning the tissues of the organism in which a promoter is active. A clear yes or no answer is important, despite the technical difficulties of dealing with cells on a plate or thin tissue layers on a slide.

**Luciferase** can be used to obtain qualitative evidence, and there are substrates that can help you overcome the plasma membrane and detect the enzyme in the cell. However, the sensitivity is low with this method.

**β-Galactosidase** is suited for quantitative applications and for obtaining qualitative evidence. It uses X-Gal (5-bromo-4-chloro-3-indolyl-β-D-galactoside) as a substrate. Cells, tissue sections, and whole transgenic embryos (e.g., flies, mice) can be stained. The blue-white evidence of tissue and specific, developmental regulation of the individual genes provides the most interesting images to be seen in scientific publications, although some sensitive souls may find that the hair on the nape of their neck stands on end when they view the stained images of mice embryos.

**Advantage:** The blue color is quite eye-catching. Any more detail will presumably be impossible.
**Disadvantage:** The material must be fixed before dyeing.

## Green Fluorescent Protein

*Was leuchtet so anmutig-schön?*

*What is it that gleams so prettily?*

The cloning of GFP (Prasher et al., 1992) marked the beginning of a small revolution. GFP is a fluorescing jellyfish protein that emits a green light when excited by blue or UV light. Because no substrate or other cofactors are required, results can be simply and directly demonstrated; a UV lamp is sufficient equipment. Because GFP is not cytotoxic, whole animals can be dyed with it. Examples are provided by the discretely green-fluorescing, healthy GFP mice from Okabe and coworkers (1997).

GFP has a broad spectrum of application. GFP functions in many varieties of species, including jellyfish, bacteria, yeasts, plants, and in animals such as *Drosophila*, zebra fish, or mice. Because the excitation and emission maxima are similar with regard to fluorescein, you can demonstrate GFP-positive cells using the same methods as when dyeing with fluorescein-coupled antibodies (i.e., with UV light under a fluorescence microscope or in the cell sorter [FACS]), with the difference being that the dye is unnecessary.

The whole 28-kDa protein is required to produce functional fluorescence. The amino acids most important for this characteristic have been described and mutated. The result is a palette of GFP variants that have had certain aspects optimized. Triplet-optimized variants, which should be translated especially fast, are available, as are intense luminescent variants and some with altered excitation and emission spectra. The latter types have an extremely broad range of applications because of double and triple labeling, as has been possible for many years in immunocytochemistry. However, a certain degree of caution is advised about the flood of variants offered by Clontech and Quantum (Table 9-4). The descriptions of these products can lead a novice down the garden path. A "red shift," for instance, is by no means an indication that the protein fluoresces red, but only that the excitation spectrum is shifted in the direction of red to improve the excitation. The greatest problem is that even the most colorful proteins do not help if the filters of the fluorescence microscope do not conform with them. Filters for fluorescence microscopy are a science in themselves and are quite expensive; expect to pay $1000 per set. These factors must be considered before you order the vectors.

GFP excites the ingenuity of the researcher like PCR does. Among the most fascinating developments are the destabilized GFP variants, with a half-life of 2 to 4 hours instead of the usual 12 to 24 hours, with which you can carry out real-time expression studies (e.g., promoter regulation), and the development of pH-sensitive GFP variants, with which you can demonstrate the transmission activity at individual neural synapses (Miesenböck et al., 1998). Just as exciting is a new variant from DsRed that fluoresces green (emission maximum at 500 nm) in the first hours after synthesis of the protein and, in the course of several hours, develops to form a red-fluorescing protein with an emission maximum at 580 nm. Through the relationship of green to red fluorescence, changes can be observed externally in the promoter activity (Terskikh et al., 2000). Even *fluorescence resonance emission transfer* (FRET; see Section 9.1.6) is possible with the fluorescing proteins, as long as the emission spectrum of the one overlaps with that of the other, as occurs in the case of SuperGloBFP and SuperGloGFP. Table 9-4 is based on the information from Clontech and Quantum. Values in parentheses indicate secondary maxima.

Three factors determine the characteristics of the GFP variants: the excitation spectrum, the emission spectrum, and the light intensity. To obtain an optimal result, you need a GFP with a high fluorescing intensity and an excitation and a rejection filter that are optimally adjusted to the variant being used (i.e., which permit excitation in the absorption maximum and observation of the emission maximum). The practice frequently appears to be quite different, because the ideal filter does not yet exist, the maximums are too close to one another to make a clean separation possible, or the excitation spectra of several fluorescing molecules unfortunately overlap. Occasionally, not enough money is available to purchase a fluorescing molecule for every filter set.

**Advantage:** The natural fluorescence enables fascinating investigations. With fusion proteins demonstrating the N- or C-terminal region of a GFP segment, you can determine the localization of proteins in the living cell. You can examine their migration in a real-time investigation, you can examine the activity of promoters in vivo by producing the corresponding transgenic animals.

**Table 9-4.** Excitation and Emission Maxima of Green Fluorescent Proteins

| Proteins | Excitation Maximum (nm) | Emission Spectrum (nm) | Color |
|---|---|---|---|
| EBFP | 380 | 440 | Blue |
| SuperGlo BFP | 387 | 450 | Blue |
| GFP | 395 (475) | 508 | Green |
| ECFP | 433 (453) | 475 (501) | Cyan |
| SuperGlo GFP | 474 | 509 | Green |
| EGFP | 488 | 507 | Green |
| EYFP | 513 | 527 | Yellowish green |
| DsRed | 558 | 583 | Red |

GFP-expressed cells can be sorted out in FACS devices. Thanks to specific antibodies, GFP and GFP-labeled proteins can also demonstrated immunologically, such as in a Western blot.

**Disadvantages:** Quantification of expression based on the fluorescence is difficult, because GFP fluoresces more weakly than the common fluorescing dyes. With the application of standard filters that are adjusted to suit the FITC dyes, the light intensity is only one-tenth that observed with FITC. Through the selection of a suitable UV lamp and filter (e.g., excitation at 395 nm, emission at 509 nm), the light intensity can be increased to one third of that with FITC, although no higher level is possible. More promising is the application of GFP variants with high light intensity, whose excitation maxima have been optimized for FITC filters. For the more colorful variants, analogous procedures are available.

GFP must mature; the fluorescence develops 1 to 3 hours after the protein can be demonstrated immunologically. The temporal resolution of kinetic investigations is therefore not impressive. Fusion proteins with a short half-life cannot be shown at all for the same reason. For the other variants, the situation is worse. DsRed, for example, requires 1 to 2 days until fluorescence can be demonstrated, and full fluorescence requires 3 to 5 days.

The gene for the wild-type GFP originates from a jellyfish, and the frequency of the codons differs from that in other organisms, causing a delay in the translation of the mRNA. EGFP (Clontech) has been optimized through passive mutation on the codon distribution in humans and is therefore (according to the manufacturers) expressed 5 to 10 times as intensely in humans.

**Suggestion:** There is an interesting web site for Green Fluorescent Protein Applications on the Internet (http://pantheon.cis.yale.edu/~wfm5/gfp_gateway.html).

**Literature**

Chalfie M, Tu Y, Euskirchen G, et al. (1994) Green fluorescent protein as a marker for gene expression. Science 263:802–805.

Miesenböck G, De Angelis DA, Rothman JE. (1998) Visualizing secretion and synaptic transmission with pH-sensitive green fluorescent proteins. Nature 394:192–195.

Okabe M, Ikawa M, Kominami K, et al. (1997) 'Green mice' as a source of ubiquitous green cells. FEBS Lett 407:313–319.

Prasher DC, Eckenrode VK, Ward WW, et al. (1992) Primary structure of the *Aequorea victoria* green-fluorescent protein. Genes 111:229–233.

Terskikh A, Fradkov A, Ermakova G, et al. (2000) "Fluorescent timer": Protein that changes color with time. Science 290:1585–1588.

# 9.5 Transgenic Mice

There are techniques available in molecular biology that are occasionally fashionable. The manufacture of transgenic mice is one of these. Today, it is rare to find a publication on mouse or human cDNA for which a *knock-out* mouse has not been manufactured to inhibit the function of the particular gene. However, the predictions made only rarely correspond to the phenotype of these animals. Usually, the animals are fully normal and differ at most in some detail from their wild-type relatives. The attempts to explain this situation appear to be rather helpless, and the researcher then makes another attempt with a *double knock-out* in the hope of seeing a phenotype that corresponds to the destruction of a sufficient number of genes. Occasionally, the experiment fails because the animals die in utero, and the researcher must recognize that the gene that should have been responsible for floppy ears proves to be important for embryonic survival. In some variants, the observed effect has nothing to do with the function known for the gene.

Even if the phenotype corresponds with the forecast, the problem is that those who manufacture such animals are primarily responsible for the support of molecular biology and do not even have the expertise to accomplish a neat, detailed analysis. Before you manufacture transgenic mice, consider whether you know somebody who is able to do something with these animals.

Nevertheless, the manufacture of transgenic animals has promise. In a time when the end of the discovery of new genes is looming and experimenters will be forced to explain life based on the few thousand of genes they already have, it is a good idea to concentrate on the functions of these genes. Examining the cooperation of genes in an organism is a task that will secure our jobs at least for the next 10 years. However, the production of a transgenic mouse still requires the work of many months, assuming that required supplies are available.

Not every method or gene in an animal can be used in the manufacture of transgenic animals. There are two forms of gene transfer, In the first case, **gene transfer occurs in somatic cells** of an organism. The DNA is not transmitted to the next generation and is usually only intermittently expressed in the cell concerned. It deals with the in vitro counterpart to the transfection of fewer cells in a cell culture. In the second case, **gene transfer** occurs **in the germline**, and the alteration is ultimately inherited by the next generation, which are truly transgenic animals. After the gene that was inserted can be found in the genome (i.e., **homologous gene transfer**), it is considered to be foreign from the original organism and consequently is a **heterologous gene transfer**. The DNA construct can be integrated at a random site in the genome (**random integration**) or inserted into a specific site through **homologous recombination**. In the latter case, there is the possibility of changing an individual gene in a targeted manner. If the mutation prevents expression of the gene, it is known as a **knock-out**. Usually, an exon is interrupted in the 5′ region of a coding region through the selection marker used, and no functional protein can be produced. Another possibility consists of deleting an important part of the gene. By means of homologous recombination, you can also introduce more subtle mutations, such as small deletions or point mutations, and only the function of the protein is changed. The result is called a **knock-in**, analogous to the *knock-out* form.

**Literature**
Schenkel J. (1995): Transgene Tiere. Heidelberg, Spektrum Akademischer Verlag.

# 9.5.1 Methods of Gene Transfer

Three techniques are known for gene transfer in mice embryos.

## Microinjection of DNA in the Fertilized Egg Cell

The largest possible yield of fertilized egg cells is obtained by stimulating a female to superovulation through the use of hormones. After successful pairing, the animal is killed, the egg cells are removed, and by means of a superfine capillary tube, micromanipulation, and much know-how, the transgenic DNA of the male pronucleus is injected into the fertilized egg cell (see Section 9.4.5). Because the prokaryotic sequences of the vector disturb the expression of the transgenic gene, only the purified insert is injected rather than the complete plasmid. The cells that survive the procedure are implanted into a pseudopregnant female and carried to term. After barely 3 weeks, the first transgenic mice are born.

The entire process is rather complex and should be left to individuals with experience. For that reason, it is important to have good relationships with an animal breeder with a transgenic unit. After the mouse progeny have developed, there is still much work for the molecular biologist to do.

The yield (i.e., probability of obtaining a mouse with an integrated transgene from a fertilized egg), if everything works properly, is about 10% to 20%. To obtain these few animals, a piece of the tip of the tail is removed from young mice at the age of 3 to 4 weeks, after they have been weaned from their mother. Sometimes, it is possible to get along with the bit of skin obtained during labeling of the animals' ears. From this material, genomic DNA is isolated and then tested for the presence of a transgene with the aid of Southern blot hybridization or PCR. The positive animals serve as a source for the initial or $F_0$ generation, also known as the *founder generation*, from which individual mice lines are bred. Because the integration site for each mouse is different, a single mouse represents the origin for each line; conversely, each mouse develops its own line. The larger the number of positive mice obtained, the more lines there are. It is important to have access to a functioning animal breeding laboratory with competent personnel, unless you would like to spend a large part of your time occupied with the breeding and care of mice.

The $F_0$ animal is paired with a normal mouse, and because the heredity of the integrated DNA occurs according to Mendel's laws, only one half of the developing $F_1$ generation demonstrates the transgene. Usually, there is somewhat less than one half, because the $F_0$ animals, depending on whether the integration of the transgene resulted during a single-cell stage or occurred later in the multiple-cell stage, form numerous genetically equivalent cells or develop into a so-called mosaic made up of genetically different cells. An $F_0$ mouse can be thoroughly positive in the tail-tip test and still deliver no transgenic progeny, because the transgene is found in the tail but not in the germline cells. A homozygous, transgenic animal is obtained in the $F_2$ generation, which results from the pairing of the $F_1$ animals. It is usually possible to determine whether a particular phenotype is available, so that you can begin with the investigations.

Because 9 weeks must pass from the fertilization of the egg cell up to the sexual maturity of the resulting animal, the manufacture of transgenic animals, even under favorable conditions, is a time-consuming affair. As with the manufacture of stabile, transfixed cell lines, several directly coupled copies of the transgene are integrated during microinjection in the egg cell. The number of copies may lie between a handful and several dozen. Because of the copy number of one transgenic mouse line compared with another, the expression level can be extremely different.

Another parallel with stabile transfection is that the DNA is integrated into the genome at an arbitrary site. The environment has a great influence on the expression level. If the transgene is integrated in a transcriptionally inactive heterochromatin region, the mouse will be unimpressed with even the most beautiful of DNA constructs. It is not sufficient to show the presence of the transgene in the genome; you must examine the quantity expressed.

The transgene may be integrated into a gene. Because considerable restructuring of the genomic DNA can occur at the site of integration, the effects are occasionally catastrophic. Occasionally, animals are obtained with a strange phenotype from such an experiment, which has nothing to do with the transgene but is instead a consequence of disturbance and destruction at the site of integration. Sometimes, these accidents prove to be more interesting than the actual transgenic mouse.

**Advantage:** The method permits the application of very long DNA chains, which may be several dozen kilobases (50 to 80 kb) long.

## Transfection of Early Embryonic Stages of Division with the Aid of Retroviruses

Retroviruses, the tiny transfection agents with guaranteed, installed integration, allow alteration of mouse embryos at a multiple-cell stage. As with microinjection, the integration results at an arbitrary site of the genome. Only a single copy is usually integrated, without any larger changes occurring at the site of integration. Because the integration results at another site in each cell, the $F_0$ animals are **mosaics** (i.e., composed of cells with different genomes). If the germline cells are infected, the $F_1$ generation also produces transgenic animals. Only then is an analysis sensible, because the cells from the tip of the tail in the $F_0$ generation and in the germ cells are different in all cases.

**Advantage:** The method is more simple, because the technical difficulties are to be found more in the area of production and in the reimplantation of embryos. The integration of the DNA in the genome is more exact, so that the characterization of the integration site is essentially easier. With this method, you can generate animals with a changed phenotype according to the random principle, and analyze which gene has been destroyed through integration.

**Disadvantages:** The size of the foreign DNA is limited by the margins of the viral system. The manufacture of retroviruses is somewhat more troublesome than that of an ordinary expression clone. The $F_1$ animals can be analyzed for the presence of the transgene about 15 weeks after the transfection.

## Transfer-Modified Embryonic Stem Cells in Blastocysts

The method of transfer-modified embryonic stem cells in blastocysts differs considerably from that of microinjection and viral transfection, because a large part of the work is carried out in the cell culture before approaching a mouse. Another advantage is that you can proceed in a goal-oriented manner, because your construct is introduced by means of *homologous recombination*, and a defined transgenic animal is obtained at the end. The embryonic stem cell technology enables production of a generation of *knock-out* mice (i.e., targeted inactivation of a specific gene). The *random integration* of DNA constructs into the genome is also possible.

**Embryonic stem cells** are obtained from the internal cells of the blastocyst. From these, permanent cell lines are manufactured that can be manipulated like other cell lines. This preparation makes it possible to select embryonic stem cells for transfixing until you have achieved the mutation desired.

Embryonic stem cell lines can be obtained from colleagues or acquired commercially. The real work begins with the manufacture of a DNA construct that contains the desired mutation (i.e., *targeting construct*). Two types of constructs are used. With **insertion constructs**, the entire construct is integrated into the desired gene; the normal gene structure is destroyed by insertion of the new sequences. **Replacement constructs** contain two flanking, homologous sequences between which a mutation is located (Figure 9-11). Through a double crossover in the two homologous regions, the original sequence is replaced by the mutated sequence. This procedure permits importation of more subtle mutations, such as of a point mutation, because the structure of the gene is not destroyed if one proceeds intelligently.

The second step consists of transfixing a construct into an embryonic stem cell. In a few cells, this calls for **homologous recombination** in which part of the target gene is exchanged with a mutated part of the construct (Figure 9-12). After identifying these cells, they are injected into blastocysts,

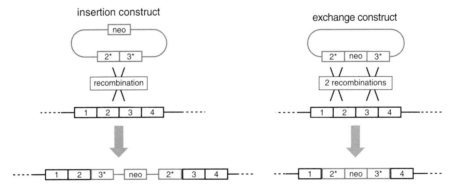

**Figure 9-11. Insertion and exchange constructs.** (Adapted from Ausubel, et al. Current Protocals in Molecular Biology.)

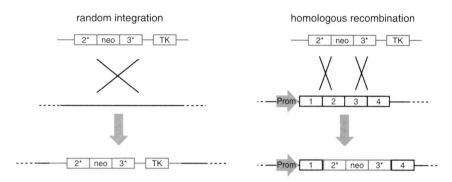

**Figure 9-12. Random integration and homologous recombination.** Integrated DNA fragments in the genome usually develop by chance, whereas a homologous recombination only rarely results. There are different techniques to select in the event of the latter. One possibility is to control the selection marker (here, the *neo* gene) with the aid of the gene's own promoter, so that it is expressed only in the event of a homologous recombination. Another possibility is the application of negative selection markers (here, the thymidine kinase gene), which is integrated randomly. (Adapted from Ausubel, et al. Current Protocols in Molecular Biology.)

which are then implanted into a **foster mouse**. The $F_0$ animals that develop are chimeras made up of normal and genetically altered cells, depending on how the embryonic stem cells nested in the blastocyst. If the germline cells are from the altered embryonic stem cells, the $F_1$ animals will also carry the transgene. In the $F_2$ generation, the first animals are obtained that are homozygous for the mutation.

Depending on how high the demands for the mutation are, you may wish to introduce different procedures or methods for planning your DNA construct and carrying out the selection. Random integration is the simplest method, for which only a normal DNA construct with a dominant **positive selection marker** is needed (see Section 9.4.7). The procedure is the same as that used in the manufacture of stably transfixed cell lines. The cells are exposed to an antibiotic over several days, and only the cells that have been transfixed stably are able to survive.

Homologous recombination is more difficult, because it requires a second round of selection until cells are obtained in which integration has occurred at the correct site. You can also install an additional, **negative marker** that lies outside of the region of the DNA construct that is exchanged in the course of homologous recombination. For this purpose, the gene for the herpes simplex virus thymidine kinase (HSV-TK) or for diphtheria toxin is commonly used. In random integration, the entire construct is integrated into the genome, and in addition to the resistance genes, the cells express the negative marker. Cells in a homologous recombination can be observed, although they contain only the resistance gene. Stably transfixed cells are selected as usual with the aid of the antibiotic.

If a substrate is added that is converted into a toxic product by the negative marker (e.g., ganciclovir in the case of thymidine kinase), the cells are killed, which leads to the expression of the negative marker. The cells that survive are those that underwent homologous recombination. The procedure is known as **positive-negative selection**.

If the gene that undergoes homologous recombination is normally expressed in the cell line, you can use the endogenous promoter for a positive selection. The resistance gene is cloned into a DNA construct in such a manner that a correct homologous recombination leads to fusion protein-resistant gene activity, whose expression is controlled through the endogenous promoter. In random integration, no expression occurs because of the lack of a promoter, unless the construct is integrated in the vicinity of an active promoter. In this way, the negative selection marker is spared. This procedure is practical for the manufacture of *knock-out* mice, because it prevents the development of a normal gene product.

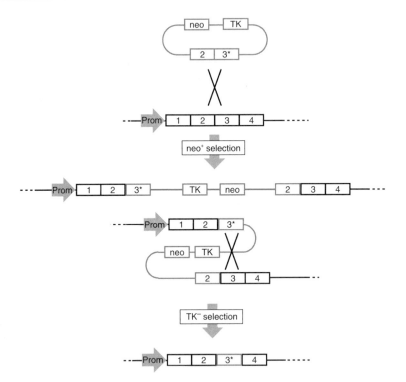

**Figure 9-13. Hit-and-run exchange is a two-stage process for screening for the presence of positive selection markers and for the absence of the negative selection marker.** At the end, a clone is found that contains no changes other than the desired mutation. (Adapted from Ausubel, et al. Current Protocols in Molecular Biology.)

By means of homologous recombination, mutations can be introduced without leaving behind any traces. Individual point mutations can be inserted without disturbing the gene structure by the insertion of a resistance gene. This procedure requires **hit-and-run** vector (i.e., in-and-out vector) that contains positive and negative selection markers. Both are far from the sequence segment that is exchanged (Figure 9-13). This construct is initially integrated into the desired site through homologous recombination, and clones are selected from which the entire vector portion has been eliminated with a second homologous recombination to retain the desired mutation. The process is associated with a substantial amount of screening work.

**Advantage:** Embryonic stem cell technology is the least expensive method for the production of transgenic animals, because it offers the most extensive spectrum of applications, from simple integration of a transgene to the targeted point mutation. Selection occurs in vitro and in living animals. It is ethically more pleasant and can reduce the confusion in the breeding unit.

More information about transgenic mice is available from a large number of books, such as that written by Joyner (1993).

**Literature**

Joyner AL (ed). (1993) Gene Targeting: A Practical Approach. New York, Oxford University Press.

Mansour SL, Thomas KR, Deng CX, Capecchi MR. (1990) Introduction of a lacZ reporter gene into the mouse int-2 locus by homologous recombination. Proc Natl Acad Sci U S A 87:7688–7692.

Robertson EJ. (1991) Using embryonic stem cells to introduce mutations into the mouse germ line. Biol Reprod 44:238–245.

# 9.6 Regulation of Transgenic Expression

## 9.6.1 The Tet System

Inserting a gene into a cell is one matter; controlling its expression is another. Frequently, the presence or lack of a protein can be fatal for the cell, and successfully transfixed cells die before they have even been found, and there is no possibility of controlling their expression. Even if the cell has survived, the absence of a protein can have a fatal consequence on the development of an entire organism. Homozygous *knock-out* mice often die in utero because development of the embryos has been confused. It would be much more practical to have a switch with which to turn the expression of a gene on or off and thereby regulate the expression accurately. This dream was fulfilled in 1992 by Gossen and Bujard with the development of a system that now haunts the catalogs of Clontech under the name of the Tet Expression System.

This Tet system is composed of two components: a **regulator** and a **response plasmid**. The regulator construct codes for the **tetracycline-responsive transcriptional activator (tTA)**, a fusion protein from the Tet repressor of the (Tn)10 transposon and the VP-l6 activation domains of the herpes simplex virus (everything clear so far?), which is under the control of a CMV promoter and is expressed constitutively. The response plasmid contains the cDNA to be expressed, which is under the control of the **Tet-responsive element** (TRE).

In addition to plasmids, a cell line can be stably transfixed using common methods. In the absence of tetracycline, tTA binds to the TRE and activates the transcription of the subsequent gene. If tetracycline is added instead, a tetracycline-tTA complex emerges that no longer binds to the TRE, and the amount of expression decreases. Clontech calls this variant **Tet-Off**, so that it can be distinguished from the **Tet-On system**, which is based on a mutated version of tTA that binds only to the TRE in the presence of doxycycline, a tetracycline derivative, and is therefore called a **reverse tTA (rtTA)** (Gossen et al., 1995).

Because inactivation or activation of transcription depends on the quantity of tetracycline or doxycycline used, the expression of the self-made construct can be upregulated or downregulated elegantly by a factor of about 1000 or more. With the Tet-On system, the response is first observed after a delay of 12 to 24 hours, whereas Tet-Off reacts much more slowly, especially in animals, because the retention time of doxycycline in the body of the animal is very long.

Clontech would not be a biotech company if it did not provide stably transfixed cell lines with Tet-Off or Tet-On. They also market different response plasmids and a retrovirus system (RetroTet), in addition to a monoclonal antibody against tTA to provide evidence of the tTA expression.

**Bidirectional Tet expression vectors** (pBI) permit the control of two genes through a single TRE. In this way, you can express the two subunits of a protein to the same extent or follow the expression of your protein indirectly by means of a reporter gene. Versions with built-in *enhanced green fluorescent protein* (EGFP), β-galactosidase, or luciferase are offered by Clontech.

The problems with the manufacture of the Tet-controlled cell lines are similar to others involving stable transfections. The expression of tTA can be too low because the integration of the regulator plasmid results in a region in which the transcription is inhibited, or the expression of the gene on the response plasmid can be too high or not be controllable because the integration of the response plasmid is in the vicinity of an endogenous promoter or enhancer. Cotransfection of both plasmids can lead to cointegration, so that the response plasmid falls under the control of the promoter on the regulator plasmid, and a high constitutive expression results.

A specific problem of this system involves the remnants of tetracycline in **fetal calf serum** (FCS), which is frequently used as an additive for media in cell cultures. In agriculture, antibiotics are applied almost everywhere, and the remnants are sufficient to influence the Tet system. However, Clontech also offers a tested serum.

Originally, the Tet system was developed for cell culturing, although it is also suitable for the manufacture of transgenic animals, plants, and yeast, and the possible applications are fascinating. You can produce a bank of transgenetic Tet-On or Tet-Off mice, whose tTA expression is under the control of tissue-specific promoters. If you pair one of these with mice whose transgene is regulated through tTA with another, you can accurately determine the quantity and the site of expression. Because it is much simpler to pair mice with one another than to manufacture transgenes, you can examine the role of individual genes in specific tissues at defined points in time. It is a truly wonderful perspective!

**Advantage:** The method offers very specific control of a gene, with a high level of expression in the "on" state.

### Literature

Baron U, Gossen M, Bujard H. (1997) Tetracycline-controlled transcription in eukaryotes: Novel transactivators with graded transactivation potential. Nucleic Acids Res 25:2723–2729.

Clontech. A detailed list of literature on the Tet expression system can be obtained on the Internet (http://www.clontech.com/clontech/TetRefs.html).

Gossen M, Freundlieb S, Bender G, et al. (1995) Transcriptional activation by tetracyclines in mammalian cells. Science 268:1766–1769.

Gossen M, Bujard H. (1992) Tight control of gene expression in mammalian cells by tetracycline-responsive promoters. Proc Natl Acad Sci U S A 89:5547–5551.

## 9.6.2 The Ecdysone System

Invitrogen offers an inducible mammalian expression system, which functions similar to the Tet system. In this case, the regulator plasmid contains the subunits of the *Drosophila* ecdysone receptor, and the response plasmid contains the induced-induced promoter. The addition of the ecdysone analogue, muristerone A, induces a 200-fold increase in expression. According to No and colleagues (1996) and Invitrogen, the system can be regulated quite finely and can be employed for transgenic mice. It demonstrates a greater difference in expression between the on and off states than the Tet system.

### Literature

No D, Yao TP, Evans RM. (1996) Ecdysone-inducible gene expression in mammalian cells and transgenic mice. Proc Natl Acad Sci U S A 93: 3346–3351.

# 9.7 Gene Therapy

*Was sagst du, Freund? das ist kein kleiner Raum.*
*Da sieh nur hin! du siehst das Ende kaum.*
*Ein Hundert Feuer brennen in der Reihe*
*Man tanzt, man schwatzt, man kocht, man trinkt, man liebt*
*Nun sage mir, wo es was Bessers gibt?*

*What say you? 'Tis no trifling space, my friend!*
*Regard it well. It never seems to end.*
*A hundred fires are burning in a row:*
*They dance, they talk, they cook, they drink, they court!*
*Now tell me where you'll meet with better sport?*

Gene therapy is fundamentally an attempt to grasp hereditary diseases at their origin. The underlying idea is to repair a mutation in a gene. Because this is a problem of medicine, it probably does not belong in a book on molecular biology, but it is an obvious topic to consider if you are discussing transgenic animals.

In gene therapy, you can distinguish between **somatic gene therapy**, an attempt to insert copies of a normal version of a gene that has been mutated into the (somatic) cells of a mature organism to replace its function, and **germline therapy**, in which you attempt to do the same thing in germ cells, from which only subsequent generations will profit. This dichotomy corresponds approximately with transient or stabile transfection and the manufacture of transgenic animals, and it should already be clear where the problems occur.

In practice, germline therapy is the manufacture of a transgenic individual. This form of therapy has been forbidden in Germany and should be forbidden elsewhere, if only because of the fear of breeding members of a master race in the test tube. The embryonic protection laws may continue to exist only until the Americans or Koreans have been successful in eradicating the first hereditary diseases, but this prohibition must be considered meaningful, particularly in view of the vast amount of embryonic waste that emerges from the production of transgenic mice, a situation that could also affect experiments on human embryos. It is also doubtful whether the expense is justified by the benefits, especially considering that the world's population grows by about 100 million individuals every year and that most of these new arrivals are healthy without the aid of high-tech medicine. It seems a poor bargain to pay tens of thousands of dollars for embryonic gene therapy while the money is lacking for the child's place in a kindergarten. Germline therapy would make sense only if embryos could be mutated precisely and the process could be free of waste, and not be subject to fads.

A variant of this problem could blossom despite all possible precautions. According to Wilmut and colleagues, who shocked the general public in February 1997 with their cloned sheep Dolly, which had been generated from a somatic cell, it is for the first time conceivable to accomplish a germline therapy on normal cells of the body, which can be obtained by taking a blood sample or perhaps by swabbing a mucous membrane. Some people are considering whether to manufacture replacement body parts, which could eventually allow us to be supplied with substitute kidneys or hearts. At the moment, this technology is still in its infancy. To complete the production of Dolly, a total of 277 experiments were required. Researchers are working eagerly on improving the methodology, although so far with only moderate success.

In addition to mice, oxen, and pets, cloning experiments are being carried out on humans. The line of approach is planned, and the rate of success will soon climb, and with it, the danger of abuse. In a world in which 60-year-old women are helped to have offspring; where children are born whose existence depends on donated eggs, donated sperm, and a surrogate mother; and where doctors are earnestly considering the possibilities of transplantation of entire heads, a lack of money appears to be the only serious obstacle for the realization of such global madness. Luckily, Mother Nature offers us a short breather. It looks as if cloning successfully is not the only problem; normal development of the individual may be just as challenging. Almost all cloned animals have demonstrated abnormalities in growth that have been caused by disturbances in transcription because of an unnatural imprinting pattern in the genome (Rideout et al., 2001). Dolly, for instance, had to be put to sleep at the youthful age of 5 years after arthritis and other abnormal geriatric complaints caused the experiment to degenerate into a form of cruelty to animals.

Somatic gene therapy is much more sensible and more realistic. The first experiments have begun. The experiments involving the treatment of cystic fibrosis (mucoviscidosis) in mice through lipo-fection were spectacular (Hyde et al., 1993). Because this approach involves transient transfections, whose benefits are limited to a specific period, researchers are working furiously to develop longer-term methods extending as far as to a stabile transfection. There is great hope for viral vectors, because they are wonderfully effective, even though they are also difficult to control, as was demonstrated in the death of a patient that occurred not too long ago (Teichler Zallen, 2000). Cell-type specificity is also a problem. A more targeted method consists of using endogenous cells, such as lymphocytes, and transfixing them in vitro to select and eliminate waste until treated cells are derived; these cells are then inserted into the body.

How successful these preparations will be is an exciting question, because no lasting cure is to be gained from transient transfections, and repeated treatments are associated with an increased risk for allergic reactions. Long-term improvement is possible only through the integration of DNA constructs in the genome of the cells, which is associated with a risk of cell degeneration and the development of cancer.

Nevertheless, the indications appear to be good. The men and women from the gene therapy front have had their own magazine since 1994 (originally called *Gene Therapy*), and there is also an American Society of Gene Therapy, which is sure to guarantee this field of research a long life.

For the time being, we should not expect to see any wonders from gene therapy. Perhaps it will have a contrary effect in the long run. In the same way as the availability and unrestrained application of antibiotics has led to drug-resistant infectious agents after 50 years, gene therapy may become an *effet pervers*. After the terrible hereditary diseases are cured, there will no longer be any reason to pass the genes on to the next generation. Because mutations develop on their own but do not disappear on their own, mild and severe hereditary diseases may someday become as common as defective vision is today. Whether humans will eventually benefit from this approach remains to be seen.

**Literature**

Hyde SC, Gill DR, HIggins CF, et al. (1993) Correction of the ion transport defect in cystic fibrosis transgenic mice by gene therapy. Nature 362:250–255.

Lavitrano M, Camaioni A, Fazio VM, et al. (1989) Sperm cells as vectors for introducing foreign DNA into eggs: Genetic transformation of mice. Cell 57:717–723.

Rideout WM, Eggan K, Jaenisch R. (2001) Nuclear cloning and epigenetic reprogramming of the genome. Science 293:1093–1098.

Teichler Zallen D. (2000) US gene therapy in crisis. Trends Genet 16:272–275.

Wilmut I, Schnieke AE, MvWhir J, et al. (1997) Viable offspring derived from fetal and adult mammalian cells. Nature 385:810–813.

# 9.8 Genomics

Since the human genome has been decoded, a new term has conquered the hearts of the populace and the media: **genomics**. It may be popular because only few know what it is. A good definition is hard to obtain, although it is used in the technical literature as if it were the most obvious matter in the world.

*Genomics* is derived from the term *gene*, which was coined long before the discovery of DNA as the carrier of hereditary information. Gene originally meant the "smallest unit in the world of hereditary information," and the field of research dealing with genes was called **genetics**. Genetics is concerned with the sense and the purpose of genes and their interactions. This research has produced important knowledge. It has provided explanations concerning the cause of many hereditary diseases, and it has increasingly revealed its limitations. Only a few diseases can be explained by mutations in a single gene, and in many cases, there was not even a logical connection between the affected gene or the proteins for which they coded and the clinical picture. It soon became clear that a breakthrough could not be achieved with this approach.

A remarkable technical development occurred in the 1990s. Paradoxically, however, no new techniques were offered at the time. There was instead another development: simplification and miniaturization of established methods. From these revisions, for example, nonradioactive sequencing (see Chapter 8, Section 8.1.2) and a small format of hybridization (*microarrays*; see Section 9.1.6) were developed. Simultaneously, substantial efforts were made to automate all of the techniques of the standard repertoire of molecular biologists. Carrying out plasmid minipreps and carrying out sequencing with DNA were improved in this way. Procedures that normally demand several days of work can now be accomplished automatically by a machine. The work can be accomplished faster, because machines are more reliable and not bound to 8-hour workdays, and the work can be done in a 96- or 384-well format. Even the cloning necessary for this procedure can be automated to a certain point (see Chapter 6, Section 6.1.7).

Combined miniaturization and automation have allowed the generation of truly large quantities of data, the basic requirement for previously unimaginable projects such as the sequencing of entire organisms. The focus has shifted from the individual gene to investigation of the genome. The gene is to genetics as the genome is to **genomics**. Genomics is not the end of this line of development. Because almost every gene codes for a protein, which finally completes the work, investigation involves all proteins of a cell (the **proteome**) and their interactions. The corresponding field of research is known as **proteomics**, and so far, its efforts have not produced the spectacular success demonstrated by genomics. The great technical breakthrough in proteomics will presumably take some time to occur, because proteins are much more different than DNAs. In the long term, proteomics no doubt will become as significant as genomics, and I can therefore advise the young researcher to keep up to date in proteomics. In principle, the same is true for glycomics, which deals with the analysis of the sugar chains of the extracellular protein components, a field of research that has been neglected for the most part.

Genomics is applied differently for different questions studied by various research organizations. The main difference is the goal of the work.

The term **structural genomics** unites various forms of preliminary work: preparation of gene libraries and gene maps, complete sequencing of entire genomes, and ultimately, processing of sequencing data.

**Comparative genomics** assesses relationships between the genome and the phenotype. For example, have you never asked yourself about skin color? Are the genes responsible for dark pigmentation the same for African and Asiatic blacks? Why do European whites not become black when exposed to UV radiation, but instead turn shades of brown? Useful (and absurd) questions of this kind fall in the field of comparative genomics. According to the particular interest, you can make comparisons between different species and analyze differences between and within different species of a population.

**Functional genomics** is the branch of genomics that tries to identify the biologic function of sequenced genes and the proteins they encode. Through the investigation of the entire genome, decisive information concerning the role of the individual parts of the puzzle is expected. Still more interesting, however, is the question of how the genes link the signal structures of the genome and the gene product. In this highly complex network, only certain basic characteristics of this relationship are known, a situation that limits the possibilities of having any influence on it. This point will lead to more intense debates in the near future and will likely result in massive restraints on researchers, approximately to the same degree as our knowledge of these relationships and the possibility of investigating them will increase.

**Clinical genomics** deals with the influence of differences in the genome on our health, and **pharmacogenomics** facilitates the development of new medications and seeks to make clearer statements concerning the possible efficacy of a certain medicine in a certain patient. **Single-nucleotide polymorphisms** (**SNPs**, pronounced snips), those small point mutations found on average every 1000 to 5000 bases in our genome and that are responsible for the differences between individuals,[1] are of great interest in clinical genomics and pharmacogenomics. They may help us understand why some disease affects only certain individuals or why a medicine functions in only 30% or 50% of the population and causes severe side effects in another 30%.

Common in all areas is the fact that large financial resources are required to pursue this kind of research. The most spectacular progress stems from financially strong firms or from global cooperations. The developmental demands are very large and the costs correspondingly high. Efforts are being made to collect microarray data in large databases and thereby create the quantity of data that is necessary for meaningful analyses. However, standards to make the data comparable and the necessary programs for such analyses are frequently lacking.

**Bioinformatics** is the backbone of genomics. Computers have made it possible to generate the current flood of data, and computers are the only possibility for dealing with it. With the help of gene banks, you can search for specific sequences, make comparisons between different sequences, establish genetic pedigrees through a comparison of homologous sequences, or combine the results of many individual sequences to form a united genome. This is only the beginning. Work is being done to find algorithms with which individual genes can be identified from the billions of bases in a genome. From individual mutations, attempts will be made to identify the structural changes of the affected proteins that lead to functional changes, or determinations can be made to discover which protein interacts with which site in a genome to have a strong or weak influence or no influence on the expression of different genes.

Bioinformatics is the gateway to theoretical biology. In the same way as theoretical physics provides new directions for practical physics, theoretical biologists will soon provide models that will provide the framework for the experiments of practical biologists. Golden times are awaiting us!

Because bioinformatics is still a new field, the possibilities for young biologists are quite good for getting a foot in the door. You presumably belong to those who believe that computers are the inherent terrain of scientists involved in informatics, but this is far from being true. The computing power of a personal computer is many times that which was available to the NASA computers at the time of the first moon landing, but computers are only as good as the information technologists who tell them what to do. Although these information technologists provide excellent work with computers, they do not have the vaguest idea of what a gene is. Because it is essentially quicker to learn the necessary basic knowledge in programming and to develop a relationship with computers than it is to acquire

---

[1] An international SNP consortium, composed of 11 large pharmaceutical companies, is working on the identification of genetic variations in the human genome. The data produced—so far more than 1.4 million SNPs—are accessible in the Internet (http://snp.cshl.org). The total number of variants may be several millions, only a small proportion of which are considered to be of functional interest, based on the fact that less than 5% of the base pairs code for proteins and regulatory areas. More than 95% of the sequences in the human genome are probably of no significant interest.

a deep understanding of molecular biology, the opportunities are quite good for biologists with an interest in computers. The need for such individuals is so great that some crash courses and targeted training in this field are occasionally sufficient for finding a place in the field of bioinformatics, because the remaining education chiefly takes place on the job. You do not automatically become a bioinformatic technologist, because you will have to deal with the consequences of this half-baked eduction for a number of years. However, if you are interested in the field of bioinformatics (or biostatistics, a similarly booming field), you can attain a specialty degree in time to establish a basis, and your prospects may then be quite good.

## Outlook

Despite the great interest in genomics, it deals only with a new form of data acquisition, which becomes truly interesting through cooperation with other fields of research, such as proteomics or epigenetics.[2] Despite tremendous quantities of money and work, the automation, and the extreme amount of data, genomics remains a kind of filter that can help us to find new genetic candidates, so-called targets, from the mass of the thousands of genes that are related to a particular disease or a specific phenotype. The explanation of the relationships and mechanisms requires classic methods and is associated with much manual labor. For this work, people are required who understand their craft. Some things do not change, even with progress.

---

[2] The Greek prefix *epi* can mean "on," "after," "above," or "in addition to," and *epigenetics* therefore refers to the influence of external (environmental) and internal (other genetic) factors on the genes that do not result in changes of the gene sequence (i.e., mutations). Key words in this relationship are DNA methylation, chromatin remodeling, and imprinting.

# 10 Using Computers

*Wenn ich sechs Hengste zahlen kann,*
*Sind ihre Kräfte nicht die meine?*

*If I've six horses, say, to show,*
*Are not their powers my own to make the most of?*

The times when you conquered empires for the good armed only with a pipette and centrifuge are over. In this trade, you must get used to working with the computer. The use of computers was still a matter for specialists 10 years ago, when the average girl or boy dared to touch them only when pressured to do so and when there was always a friendly freak to be found who would take over this job. Small personal computers (PCs) are now able to accomplish a sequence analysis, and they could easily search databases for sequences without any problem had not the dimensions of the databases also increased to an equal degree. The number of the computers and programs in the laboratories has multiplied, but the number of specialists has not, and each experimenter must carry out his or her own work.

I will not enter the debate about specific computer systems, because the differences are not particularly great, and the compatibility of one particular system or the other has more ideological rather than rational grounds. It is not the purpose of this chapter to explain the work with a particular computer. Because this field changes rapidly, such instruction would have little value. Instead, this chapter provides all of those who are suspicious of computers with an idea of their possibilities.

## 10.1 Something Quite Earnest

What can be done with a computer? Everything, according to the computer specialists. Quite a bit, say those who are in the know. A few things, says every person on the street.

In daily practice, the computer is used most frequently for **word processing**. Every day, workers must write letters, plan production schedules, and immortalize methods. Occasionally, papers must be written. Publishing is to the successful scientist as water is to a dehydrated soup. Nevertheless, the successful scientist often does not have a secretary, because he or she is reserved for the successful professor. As a consequence, every scientist is expected to type. Practice makes perfect, and proficiency can be gained over the course of a few years, even if only few scientists master the 10-finger system.

Regardless of what Microsoft says, word processing programs such as Word are conceived for writing letters and smaller papers, not for diplomas or for doctoral theses. Although these programs offer all kinds of comfort, such as automatic numbering and colorful graphics, longer texts that take full advantage of all of the possibilities of the program usually have increasing problems, and you may experience a total collapse, most likely when the work is practically perfect. You can chop up the text into smaller segments of approximately 50 pages as a precaution, or you can purchase a professional layout program if you have rich parents.

For many, the second most important use of the computer is *email* (*electronic mail*). With email, you can keep a collaboration alive, arrange to meet colleagues for a meal, assure your sweetheart that you love him or her, and send some communication to the boss whom you have not seen face to face for a long time.

There is also the **Internet**. Presumably, every child is aware of this concept. It is the world of colorful pictures and web sites that can be surfed with the aid of a browser. Connection to the Internet should be left to the responsible technical expert, who is paid to struggle with such problems. This expert also can install the browser and an email program, as well as all that is otherwise mandatory. The Internet earns the commendation "of utmost value to the researcher," and a few words concerning its practical aspect follow later.

The computer is indispensable for **literature searches**. The U.S. National Library of Medicine publishes PubMed for this purpose, which the more elderly researchers still know as MedLine. It contains a list of all articles over the past decades that have been found in the important and relatively important medical and scientific journals. Listings usually also contain abstracts of the articles, and entries are regularly updated. In the past, newer publications have appeared with a delay of 3 to 6 months, but publications now appear almost contemporaneously. Often, access to full articles found on MedLine must be purchased, but university libraries frequently have a collective license for members of the university faculty. Fortunately, access to this service is available over the Internet (http://www.ncbi.nlm.nih.gov/entrez/).

There is specific software for the **analysis of DNA sequences**, which is especially near to the heart of molecular biologists. It can also be considered a tool for the self-gratification of molecular biologists, with which they can kill time between individual cloning procedures.

Some programs can examine a polymerase chain reaction (PCR) primer for its usefulness. Some programs can be used for sequence analysis and predictions and used to search for functional areas in the DNA or the proteins coded by them. With other programs, the experimenter can look for sequences that demonstrate homologies to his or her own.

The Internet is particularly useful for searching for sequences. The multitude of sequences that are published every day quickly led to the idea of sequence data banks in which the researcher can rummage. The most important are the GenBank of the National Center for Biotechnology Information (NCBI) (http://www.ncb.nlm.nih.gov) and the EMBL bank of the European Molecular Biology Organisation (http://www.ebi.ac.uk), which compare one another's data daily and can therefore be considered as equivalent. Because the search for homologous sequences by hand is no longer possible, these institutions offer search programs for the databases, primarily the **BLAST** programs (Basic Local Alignment Search Tool), to which experimenters have access by way of email or over the Internet. Many additional databases with different emphases may occasionally be of interest to the molecular biologist, such as that for the Online Mendelian Inheritance in Man (OMIM), which deals with genetic illnesses (http://www.ncbi.nlm.nih.gov/Omim); SWISSPROT, the classic protein database (http://www.expasy.ch/sprot); or GeneCards, a data bank of human genes with information concerning the function of the coded proteins and their role in diseases (http://bioinformatics/weizmann.ac.il/cards).

The databases occasionally spare the experimenter the cloning process. A specific sequence can easily be searched for in the sequence banks, and if it is found, the clone can be ordered. In these mad times, there are institutions that isolate clones indiscriminately from cDNA banks, add sequences, and then deposit these sequences in data banks. These sequences are marked as **expressed sequence tags** (EST), and various EST banks and the respective clones are accessible for researchers (e.g., the dbEST data bank at http://www.ncbi.nlm.nih.gov/dbEST/index.html). In this way, a researcher can get hold of genes that have been previously unknown without having to do any cloning. From the several million human EST sequences available, an attempt has already been made to obtain an overview of the underlying genes (http://www.ncbi.nlm.nih.gov/UniGene/index.html). Some bacteria, such as *Saccharomyces cerevisiae* and *Caenorhabditis elegans* and larger organisms, such as *Drosophila melanogaster*, *Homo sapiens*, or *Mus musculus*, have practically been fully sequenced so that the end of any new discoveries is gradually reaching a tangible proximity.

More information is available from Harper (1994) and Bowtell (1999) and from regular contributions on this topic in journals such as *Trends in Biochemical Sciences* (TIBS) and *Nucleic Acids Research*. A good review of the current databases can be found in *Nucleic Acids Research* (Issue 1/9, Vol. 26, 1998).

**Literature**
Bowtell DDL. (1999) Options available—from start to finish—for obtaining expression data by microarray. Nat Genet 21:25–32
Harper R. (1994) Access to DNA and protein databases on the Internet. Curr Opin Biotechnol 5:4–18

# 10.2  A Matter of Practice

In practice, computer systems often fail to function because of a lack of care. Usually, someone with technical expertise can be found to provide instruction about how a computer functions. Often, a colleague has worked with the particular device for a longer period. You must first determine who is to be responsible for the care of the computers to obtain a user number for the computer or computers and for establishing an email address.

That individual is usually also the person who is responsible for installing new versions of a program, which appear without fail when you have finally gotten used to the older version. If you have some influence on the decision, strive to see that the old version survives. If possible, let the others try it out to see whether the new version offers any advantages or has problems involved in its use. A common problem, for instance, is that the new version consumes far more hard drive space and memory than the old version, which can overload the computer. Why should you struggle with such difficulties if the old program did everything you wanted?

A common error of noncomputer specialists is their trust in the computer and its components. Everyone thinks that computers are very reliable. That is the case so long as no one needs them. When you are in the final phase of your doctoral thesis—100 pages have been written, a mere 30 pages are still lacking, and the figures and layout are all correct—disaster is guaranteed to occur. This is the time when the hard disk crashes.

Always store back-up copies of the most important files. Memory sticks can be used for this purpose, although it is safer to store back-up copies on another hard disk or burn a CD.

Another popular sin is working directly on memory sticks. They are robust (more robust than floppy disks used to be), handy, and allow you to work on different computers without leaving you with a flood of copies of different stages of your work. Although being pretty reliable, they have still a higher tendency to fail than hard disks, they are slower (which can induce you to save your magnum opus less often than recommendable), and they can easily be unplugged, sometimes with unforeseeable consequences. To prevent problems, a back-up copy should always be saved on a well-preserved hard disk.

At some point, even the longest work is completed and must then be archived. A CD-ROM is the most suitable storage material. If you would like to take a look at your masterpiece in another 5 or 10 years, it is advisable to keep an old-fashioned, printed paper version somewhere. You can write what you like on paper, and even on the grayest of winter days, the sun supplies you with enough light to read it. Your electronic file, on the other hand, has become useless because you have moved at least once by then, purchased one or two new computers, and the program with which you originally wrote the text no longer exists or, after five upgrades, is no longer able to open the old file.

You should also be suspicious of other electronic blessings, such as email. The most important advantage of email is that it is fast. This is practical, although it frequently leads to neglect about what is being written. After clicking on the send button, the nonsense is received in a flash, which is between 10 seconds and 1 day. I have never been able to figure out how a communication wanders through the computer network for a matter of hours, but I know that it happens. Frequently, emails get lost. With luck, you receive a response explaining why your news could not be delivered, although frequently this is also not the case. To be certain that news has arrived, ask for a reply, if only the single line of "It has arrived."

Special characters, including ä, ö, ü, ß, often cannot be used, particularly when writing to English-speaking foreign countries, and the recipient must struggle through hieroglyphics such as /f{. or *{/ in the middle or at the end of some words.

If all prognoses about the Internet were correct, we would no longer have to leave our homes, because we could do all tasks through a simple click of the mouse. Somehow, everything worked out differently.

The **Internet** is a vast number of computers that communicate with one another over a network of lines, using a directory to assess who else is connected and how to reach them. If researcher A in Chicago wants to look at an Internet web site that has been prepared by researcher B in Harrisburg, he or she uses the Internet address to view the data. The first computer sends a corresponding message to computer Tom, which then transmits this to computer Dick and so forth, until it ultimately arrives at computer Harry. This computer can then send data back using the same principle, although not definitely according to the same route. The Internet is a network that functions according to the principle that "all roads lead to Rome," and it is surprisingly fast. As in real life, however, traffic jams occur on the data highway. Most are small and lead to delays of only seconds, although sometimes, nothing functions as it should.

The Internet offers inestimable advantages if you learn how to use it correctly. The largest problem is too much information. The network offers practical search services such as AltaVista (http://www.altavista.com), Excite (http://www.excite.com), Google (http:www.google.com), or MetaCrawler (http://www.metacrawler.com). Searches often produce dozens to tens of thousands of Internet addresses for each keyword typed in, and results may or may not be close to what you were searching for. Most of this information, however, is uninteresting and time consuming. It is best off to spend the free moments of the day relaxed in front of the computer and to go on a voyage of discovery. If you encounter something useful, the address can be stored with a bookmark. In the course of time, you can acquire a collection of useful addresses. With this healthy outlook, the Internet becomes a useful resource with which you can obtain information within seconds. However, if you do not heed this advice, surfing the network will turn out to be a great waste of time.

For researchers, the Internet is interesting because of the databases already mentioned; the news groups, which offer suggestions from like-minded individuals for solving problems (it is sometimes very motivational to see that many around the globe suffer from the same problems as you do); home pages from institutes, where you can learn what the people do there; sites where shareware or freeware programs can be obtained; online versions of scientific magazines; and much more.

Over time, using computers can provide more knowledge and help to develop a certain level of expertise. The more you know, the more frequently colleagues will come with their small problems, and before you know it, you have become the computer specialist that you had previously sought to find. However, you should not be officially condemned to caring for the computers in the laboratory. Computers generally cause problems, and the more you want from them, the more work they will require, until without realizing it, you have said goodbye to the ranks of pipette swingers. It is a bottomless pit, and it offers no rewards. Friendships and publishing will cede to computer problems— not the best career path for a molecular biologist.

# 11 Suggestions for Career Planning: The Machiavelli Short Course for Young Researchers

The life of the researcher is concerned with publications. Although you may think that the researcher's favorite pastime is research (offering the prospect of doing good and then talking about it), the focus on publishing has prevailed in research for a long time. The picture of the slightly fanatic but quite lovable researcher prevails in the general population, but it has nothing to do with reality. The true picture can be observed among physicians. The population of medical students at the beginning of their studies consists of a jumbled heap of average individuals, and at the end of their studies, there is a group of careerists who, after long years of arduous training, have discerned the rules of the system and now want to use the rules for their own purposes. These are the individuals who are responsible for the picture of a demigod in white, the omniscient genius who hovers through the passageways and through the entire world, who can at least explain every illness and pontificates from above to the others down below. After the completion of their medical studies, the careerists usually strive to make their career at a university clinic. The second group, the idealists, are students who have studied medicine, are able to help other people, and after many years, must recognize that they have failed with this concept. They are generally the more sympathetic of the two groups; they withdraw after completing their studies and practice in a town or village to be forgotten by the world and the system.

There also are careerists to be found in the natural sciences. Because biology classically plays no role in society, the researchers have unavoidably developed their own small world: the university, with its own rules, to which **publications** belong as a measurable dimension of success. The system is quite clever, because it permits the recording of quantity (the number of papers) and quality (the journals in which the experimenter has published). Even the position of the name in the long list of the authors involved is included in the calculations to allow a sort of *fine-tuning*. Only the conversion between these two groups seems to cause problems. How many run-of-the-mill papers does it take to counterbalance a paper for *Nature*? Is a third-author position in a *Cell* paper worth more or less than two first-author publications in the *Journal of Ophthalmic Turbulences*?

Because of the haziness of this evaluation system, the style of publishing has prevailed to allow researchers to play it safe. A doctoral thesis must be terminated with at least one publication, and dissertations are not immune to this mania, which—considering the short period in which a dissertation must be accomplished—is on the verge of perversion. *Publish or perish* is the battle cry. That is reasonable as long as you do not lose sight of reality. Publications in the circles not dealing with research have about as much significance as cowrie shells in a Swiss bank: quite decorative but worthless. Whoever engages in this game should not forget that it is mental masturbation at the highest level. The male peacock developed splendid plumage for this purpose.

Strange also is the manner in which this ritual, involving writing a paper, is carried out. The first draft is classically written by one individual, who has carried out the work. That is good, because he or she is the only person who truly knows what occurred. The draft is then presented to his or her boss, who therefore is awarded the last position in the list of authors. The boss then reaches for a red pencil and begins to make corrections. Generally, these are so comprehensive that nothing remains from the original draft. The corrected version is then returned to the first author, who performs revisions and generates a new version, which is then presented again to the boss and again comes back corrected. Depending on the particular boss, the number of cycles can easily be 20 or more, until a version has finally come into existence with which he or she is satisfied. It may also occur that earlier revisions are returned to the paper in the course of this process.

What qualifications does the boss have in regard writing papers to be published? With what right does he or she have the audacity to determine whether an article should appear? Usually, no special education is involved, and the knowledge has been acquired through a learning-by-doing process; the boss patiently submitted himself or herself to the corrections of his or her boss until he or she became the boss. Knowledge about writing papers has been transmitted from generation to generation. It is surprising to find the rites and behaviors of nomadic civilizations once again at the forefront of modern research.

This approach explains why corrections are not usually undertaken directly in the document, which would be the most economical method in the age of electronic word processing. The novice would learn how a paper is written through typing corrections in the paper.

Are there criteria for the objective quality of a paper? No. The style is not evaluated. Allegedly, the assessment depends on the content so that the question of style has no place in the brain of the scientist. Paradoxically, not even intelligibility appears to be an important criterion, although every researcher becomes angry when confronted by a poorly written article that does not get the point across, because it costs him or her time. If you get angry about the absurd corrections received from your boss, consider that he or she had no more training as a scientific writer than you did. Trust yourself occasionally to quietly insist on your formulations if you find that they are better. After all, it is your paper, and your name is found first in the list of authors.

In cultures whose knowledge is based on passing on information by word of mouth, the resulting tradition is highly dependent on the specific master. For that reason, careerists look at the list of publications from their future bosses before agreeing to take on a new position. How many papers has he or she written? In which journals have they been published? Important are primarily the publications of recent years, because they provide information on whether the writing style demanded by most important journals has been mastered. The right tone is decisive when striving to publish. Only those who are able to write a paper in the style of *Nature* will be published in *Nature*. The others will then have to publish their papers in third-class journals.

Please do not misunderstand me. The corrections made by the boss are necessary because the scientific journals have no editors who will transform your article into a readable product.

A readable product is an important aspect. You write so that other people can read your mental outpourings. If they cannot do so, you would have been able to spare yourself much work by not writing. In the universities, the idea frequently prevails, unfortunately, that researchers must demonstrate their brilliance, and that goal is achieved through writing complicated texts. This, however, is a fallacy. Explaining complicated things with simple words is clearly a greater art and considerably more difficult.

A completely different danger is to forget to publish because of all of the research. A person who enjoys performing research often finds that writing a paper is a horrible experience. Consider how many interesting experiments can be done during the time wasted on writing. If you are one of these people, look for a boss who sees that you bring your work to a conclusion and then publish. It is unpleasant to be under pressure, but it is even more unpleasant to have no publications to show. If the publications are lacking, your only prospect in the search for a new place of work is to have the support of your boss. Take this into consideration the next time you start procrastinating in writing a paper.

Establish a defined, final deadline. You will never be able to conduct all of the experiments that are necessary to produce an absolutely bulletproof paper. It is sufficient that the experiments were accomplished cleanly, that the results are reproducible, and that the story you want to tell is to some extent complete and coherent. The more your discovery corresponds to the *mainstream*, the fewer experiments you must provide to establish your evidence. Only those who want to sell a theory that goes against all of the currently valid doctrines must collect sufficient ammunition, and they will probably fail due to technical, publication-related reasons. Because such revolutionary ideas are rare, you normally need to pay attention only to the fact that your paper has been written well. Not every detail has to be demonstrated experimentally. Instead, plan a paper, in which you can argue that it has functioned for others and that it will therefore be the case for you.

Publications (and careers) can also be planned. Decisiveness is the right choice for a future boss or laboratory.

- Look for a laboratory with the proper critical mass. Three-man laboratories are comfortable but relatively unproductive. Successful laboratories have a specialist for every important technique, so that an exciting project can be carried out with a minimum delay in time until everything is finally ready for publication.

- Look for a laboratory that has recently published something that is current or even pioneering. In the coming years, the laboratory will be a source of well-placed publications, because the topic incites interest, and every detail of the paper will be exalted.

- The combination of a laboratory with a proper critical mass and with current topics is a guarantee for success. The work is performed in an assembly-line style, in which each member always performs the same thing—the cloner does the cloning, which the expresser carefully expresses—and the boss sees to it that each member of the group receives one first-author publication and a heap of second- and third-author listings in fairly well-known journals.

- It is best if the boss belongs to the right group. Usually, these people do not represent the embodiment of sympathy, but they know what they want—to increase their fame—and if an experimenter follows their rules and is consequently useful to them, he or she will be considered positively in exchange. The best positions in life are always distributed among friends and protégés, a procedure known as *networking*. Free yourself from the prejudice that such an exchange is morally reprehensible; otherwise, only others will profit from it.

If you find that the previous passage is not particularly to your taste, consider the fact that it also functions without a university! In the universities associated with this field, a certain amount of arrogance frequently develops, an attitude that is similarly popular among teachers. The individuals with the true ideals of the trade work at the universities while the others follow the orders of filthy lucre mongers. The company representatives who pay a visit to these laboratories frequently handle these simple candidates for a degree quite condescendingly.

You may therefore be happy to forget that the driving motivation at the university is rarely scientific curiosity. The research is left to the lower associates, especially the post-doctorates, and tactic, calculation, and professional relationships are more decisive for the progress than any pioneering knowledge. The one-sidedness of the work is usually at least as great as that in the industry, because a researcher frequently cannot perform only one specialty; an electrophysiologist is and remains primarily an electrophysiologist but must also commit himself or herself to a particular field of research if he or she wants to become known. It is therefore not unusual that a renowned or well-known researcher has spent 20 years studying the function of a small subunit in the ribosome of some specific fungus.

Between the 30th and the 40th year of life, the masters undergo a rapid crisis of conscience, because they finally understand the rules in this phase of the game and ultimately reject them. Meanwhile, their careers are in a rut, the scarcity of money causes increasing pressure, and the situation appears to be hopeless.

Do not allow it to go this far. Listen to your conscience sooner. Are you truly driven by scientific curiosity, by the delight in detail, by the feeling that you must find the answer to a problem? Are you convinced that your work is important for you and for the world? Are you certain that you will not stray from the path on your way to discovering this knowledge, but rather that your understanding of the world is continuously becoming larger? Is this feeling strong enough to compensate for the frustrations and the small salary? So strong that it will urge you onward even if you one day have to spend more time at your desk and in meetings than at the laboratory bench? Do you believe in what you are doing? If so, you are in the right place in university research. The world needs you, and it needs you there.

Have you instead experienced some doubt while reading the previous paragraph? Are you willing to work on every topic as long as the surroundings are suitable? Are you where you currently work because it is a mandatory step in your career, and you see no real alternative? Are you currently active in being conferred your doctor's degree because you will otherwise have no prospect of obtaining a job? Are you happy with your job because you can maintain that feeling of freedom you had enjoyed so much during your studies for a couple of years longer? If so, you have got the wrong idea. Enjoy your time but begin to prepare for your arrival in a new harbor.

The world outside of the university is large and that outside of research is even larger. Assume for the fun of it that you have no conception of how many possibilities there are for you. If you have come to favor these thoughts, progress with your search. Ask around, be a little inquisitive, and follow a new direction. Do not ask the people associated with you in your area of the university; if they knew what there was in the outside world, they would long since have disappeared. Make yourself a new circle of friends outside of the university, and talk with these people to find out what kind of wisdom they have to offer. Go through your city with your eyes wide open and make your thoughts about what you see. Run through a park. The plants were planted by gardeners, but who did the planning? Who worries about the questions dealing with nature in such a city? There is a fountain around the corner. Where does the water come from? Where does it go afterward? Who takes care of this? Does the water flow directly from the waterworks to the sewage treatment plant, or is the water recycled? How is the water quality maintained? Could it be performed microbiologically? Is not all of this microbiology? Do they need people with my expertise? On the street, there is a carton full of plastic chips, as used in the delivery of machines. These are now also available from natural materials, environmentally friendly and decomposing. Where are they produced? What kind of staff members does their manufacture require? Could I work for such a company?

With a little practice, you will come to the most original of questions, and some of them will stimulate your thoughts. You must invest some time, inform yourself, make inquiries of the responsible individuals, and initiate more actions. Do not do this with the goal of being offered a job at the first door where you knock, because that will certainly not be the case. However, in the course of time, you will get a clearer picture of what you find to be interesting and what you do not. The more precisely that you know what you want, the sooner you will get it.

In this way, you will come into contact with the most varied of individuals and learn how you can best "sell" yourself. An important difference between working in the university and industry is that you primarily appear at the university as a petitioner (Please, please may I perform my dissertation under your auspices?), whereas you become more interesting for industry when you have some additional value to offer. In your exploratory travels, you will recognize which of these are more or less attractive in regard to your abilities. You naturally have a considerable amount of specialist knowledge, but that will probably be of only moderate interest. During your studies, you have also acquired other qualities, such as analytic thinking and the ability to acquire knowledge on your own. Such capabilities are more interesting over the long term, because they can be made use of in varied situations. You also have inherent qualities, such as your perseverance at work and perhaps your flare for creativity, which makes you a good problem solver. Learn to recognize which of your capabilities will be interesting to others, and attempt to expand on these.

Do not forget that the goal is not to find a secure job that is as well paid as possible. This attitude is quite common, although it is more suitable for the born loser. A winner has a product—his or her good performance and impressive capabilities—and consequently expects to receive services in return, such as good pay, a pleasant working environment, an exciting task, or interesting possibilities for further developments. The job should not be an ejector seat, but a reliable position up to early retirement cannot be the goal of anyone who wants to develop himself or herself. Who wants to perform the same work for years on end? A perspective of 3 to 5 years is generally adequate, because it is impossible to plan beyond that.

Free yourself from the concept that the employees are the victims and the employer is the perpetrator. A good enterprise needs good staff members and must consider their needs and wishes to keep them, or the staff members will travel on, and the organization will go from a good enterprise to a bad one. This principle presupposes that the staff members (and therefore you) will travel onward if they consider that they have been handled poorly, instead of only lamenting and remaining in the position where they have been.

You will primarily have to lament and persevere if you have not provided yourself with an alternative in time. Do not concentrate only on your specialty, even if it appears to be very attractive temporarily. Create a broad, solid basis, and obtain an overview of other fields. In this way, opportunities will arise that you cannot currently envision. More knowledge is helpful; attain knowledge in medicine, in engineering, in geology, or in any other area that interests you. Knowledge in the field of economics is always helpful, because you will most likely be confronted with the world of money at some later time. The world of money is subject to its own laws, although it has an advantage in that it has been created by men and women and is consequently relatively simple. Economics is not particularly complex. In contrast to the way in which the economists view themselves, economics is not really a science, but rather an attempt to understand why people do what they do, with the futile hope of trying to extrapolate what people will do tomorrow, although they did something completely different yesterday. Common sense and some diligence are sufficient to get a grasp of this field, which is why there are quite a number of natural scientists found in the field of economy, but extremely few economists found in the field of the natural sciences.

In the field of economics, there are many interesting opportunities at all price levels to obtain further education, from the lectures offered by the faculties of universities involved with economics to programs leading to a master's degree in business administration (MBA) from a renowned private school. The most expensive path is not necessarily the best; if you do not want to become a chairman of the board for the Nestle company, you do not have to obtain a Harvard degree. Remember, a straight line is always the shortest path between two points, but not always the fastest!

I offer a final tip. If you find yourself in one of these low points, do not forget that the world is also full of individuals like you. You may be told that only the best have a chance, but that is not correct. Whatever clever people concoct on this planet, they must undertake its transformation with the aid of all individuals. You must find a place where associates are glad to see you and be able to live in peace with yourself and with others. And that is definitely not an impossible situation!

# 12 Concluding Thoughts

*Ihr beiden, die ihr mir so oft*
*in Not und Trübsal beigestanden,*
*sagt, was ihr wohl in deutschen Landen*
*von unsrer Unternehmung hofft!*

*You twain, who've oft befriended me*
*In hours of stress and fruitless toil,*
*Tell me what welcome you foresee*
*For our new venture upon German soil*

My problem with textbooks is that they make everything sound easy. Here and there, a little bit is cloned, and at the end, the result leaps to the eye. At lectures given for the general public, the situation is not different, especially in the course of lectures lasting longer than 20 minutes. The slim, slightly grayed lecturer is given 45 minutes to speak about a single topic. If he or she is a superstar in this field, the entire evening is given as a precaution, because he or she most likely will exceed the allotted time mercilessly. The message delivered in 45 minutes is likely to cause the listener to get gray hair. In fairness, pulling together all of the beautiful results in a meaningful way probably requires decades.

After some time in this field, you learn that this situation is not different from those in other disciplines. The elderly lecturer has given this lecture, in a slightly updated form, for many years, and in the worst of all cases, he even wrote his doctoral thesis on this subject. He carries it from one symposium to the next, can be invited between one or another lecture to give colleagues a brief lecture, and occasionally appears as a *Deus ex machina* in his laboratory, solving the problems that have collected there, gathering the last results, and proceeding again on tour, this time perhaps to provide an evaluation for a particular research society. He has not had a pipette in his hands for many years and may not even know where he can find one in the event that it is required. The results originate from 20 industrious helpers who, with "his" money and in his name, carry out the experiments that he has picked up on his journeys. The helpers work themselves to a frazzle, because they believe that they will be able to exist in this world only with a terrific dissertation, doctoral thesis, or list of publications. They continue to work in this way until they have their own money and their own helpers or until they are at the end of their tether somewhere along the way.

This appears to be a natural process that you cannot escape, but you should be careful to survive the individual stations along the way. We do not know what qualities the nice gentleman will demonstrate to maintain his position in the world of research and university politics or in the world of posts and claims. The helpers, for whom this book is written, primarily require the ability to handle failure. Science involves weeks and months of working in an attempt to clone a paltry 1-kb fragment into a run-of-the-mill vector, while each of the colleagues encountered in the corridors credibly maintains that it is technically no problem at all. At this point, the chaff is separated from the wheat, and it becomes evident who has the right mixture of obstinacy and of creativity. Only the beginners are able to perform experiments without any problem at the first go; everyone else should begin to think earnestly after the third failure about what to do differently. Eventually, the experiment will be successful, as long as you are strong enough to put up with 6 months of the deepest depression.

The system teaches the growing scientist the necessary modesty, which will be lost once again when he or she has become a full professor. If you need a sense of achievement, limit yourself to a smaller number of techniques. In this way, the probability is higher that the attempts will be successful, and the less you do, the less there is that can go wrong. "No experiments"—a slogan that is correct in politics—cannot be wrong in the sciences either, and considering the modest financial prospects that are offered by the universities, you are better off spending time planning a successful exit from this area than on unsuccessful attempts. True research is something for obsessed individuals, for those to whom no price is too high, and you should consider whether you are a member of this class of individuals before engaging fully in this adventure.

*Erfüll davon dein Herz, so großes ist,*
*und wenn du ganz in dem Gefühle selig bist,*
*nenn es dann, wie du willst:*
*Nenns Glück! Herz! Liebe! Gott!*
*Ich habe keinen Namen*
*dafür! Gefühl ist alles;*
*Name ist Schall und Rauch,*
*umnebelnd Himmelsglut.*

*Fill your heart, as big as it is, from that*
*And when you are completely blissful in the feeling,*
*Then call it what you like:*
*Call it happiness! Heart! Love! God!*
*I have no name for it!*
*Feeling is everything;*
*The name is sound and smoke,*
*Enshrouding heaven's glow.*

# Appendix 1

## Useful Figures and Tables

Figure 2-3. Relationship between the number of revolutions per minute and centrifugal force.
Figure 2-5. Determination of the concentration and the molarity of nucleic acid solutions.
Figure 2-9. Cesium chloride staging gradient.
Figure 7-1. Specific activity.
Table 2-2. Molecular Weight Cutoff of Amicon Concentrators.
Table 3-1. Examples of Isoschizomers and Neoschizomers.
Table 3-2. Activity of Restriction Enzymes in Taq Polymerase Buffer.
Table 3-3. Unusual and Unstable Restriction Enzymes.
Table 3-4. Agarose Concentrations and Respective Fragment or Separation Lengths.
Table 3-5. Range of Fragment Size Separation in Polyacrylamide Gels.
Table 4-1. Multiplication of DNA in the Polymerase Chain Reaction Tube.
Table 4-2. Characteristics of Thermostabile Polymerases.
Table 6-1. Enzymes with Compatible Ends.
Table 6-2. Modification at the 3′ End by Taq Polymerase
Table 6-3. Antibiotics.
Table 7-1. Labeling Methods for Particular Probes.
Table 9-1. Insertion of a Restriction Cleavage Site in a Protein-Coded DNA Sequence.
Table 9-2. Restriction Enzymes with a 3′ Overhang.
Table 9-3. Frequently Used RNA Polymerase Promoter Sequences.
Table 9-4. Excitation and Emission Maxima of Green Fluorescent Proteins.

## Standard Solutions and Bacterial Media

Some solutions are used repeatedly in molecular biology. Their compositions are given here.

### Solutions

| | |
|---|---|
| TE | Tris HCl, pH 7.4<br>1 mM EDTA, pH 8.0 |
| Phenol-chloroform solution | 25× phenol, pH 7.6 + 24× chloroform + 1× isoamyl alcohol (the isoamyl alcohol is not necessary) |
| RNase A solution, DNase-free (10 mg/mL) | RNase A is dissolved in $H_2O$ (10 mg/mL), and the vessel is then heated for 15 minutes in a beaker with boiling water. Aliquots are taken and stored at −20°C (−4°F). |

(*continued*)

| | |
|---|---|
| 20 ×SSC | 3 M NaCl<br>0.3 M sodium citrate<br>pH 7.0 |
| 5 ×TBE | 54 g Tris base<br>27.5 g boric acid<br>20 mL 0.5 M EDTA, pH 8.0<br>Fill to 1000 mL with $H_2O$ |
| 50 ×TAE | 242 g Tris base<br>57.1 mL acetic acid<br>100 mL of 0.5 M EDTA, pH 8.0<br>Fill to 1000 mL with $H_2O$ |
| SM solution | 10 mM Tris, HCl, pH 7.4<br>10 mM $MgCl_2$<br>10 mM $CaCl_2$<br>100 mM NaCl |
| 100 ×Denhardt's | 10 g Ficoll 400<br>10 g polyvinylpyrrolidone K-30<br>10 g bovine serum albumin (BSA)<br>Fill to 500 mL with $H_2O$, take aliquots, and store at $-20°C$ ($-4°F$). |
| TBS | 50 mM Tris, pH 8.0<br>150 mM NaCl |
| PBS | 0.2 g KCl<br>0.2 g $KH_2PO_4$<br>1.15 g $Na_2HPO_4$<br>8 g NaCl<br>Fill to 1000 mL with $H_2O$; eventually adjust to pH 7.4 with HCl or<br>    NaOH. Filter in a sterile manner. |

# Bacterial Media

| | |
|---|---|
| LB | 10 g trypton<br>5 g yeast extract<br>5 g NaCl<br>Fill to 1000 mL with $H_2O$, adjust to pH 7.5 with NaCl, and autoclave. |
| LB agar | LB medium + 1.5 % (w/v) agar<br>Autoclave and first add antibiotics when the solution has been cooled<br>    to 50°C (122°F); this is that temperature at which you can grasp the<br>    bottle in your naked hand without complaint. |
| TB | 12 g trypton<br>24 g yeast extract<br>4 mL glycerin<br>Fill to 900 mL with $H_2O$ and autoclave.<br>Before using, add 100 mL of a sterile $KHPO_4$ solution (0.17 M<br>    $KH_2PO_4$/0.72 M $K_2HPO_4$). |
| SOC | 20 g trypton<br>5 g yeast extract<br>0.5 g NaCl<br>10 mL 0.25 M KCl<br>5 mL of 2 M $MgCl_2$<br>20 mL of 1 M glucose<br>Fill to 1000 mL with $H_2O$, adjust to pH 7.0 with NaOH, and autoclave. |

# Glossary

**A:** adenine, one of the four purine bases of DNA

**aliquot:** something that is contained an exact number of times in something else; in laboratory jargon, it generally means a fraction, a portion, or part of a sample

**Amp:** ampicillin, an antibiotic

**AP:** alkaline phosphatase

**ATP:** adenosine triphosphate

**BAC:** bacterial artificial chromosome, a cloning vector

**BCIP:** 5-bromo-4-chloro-3-indolyl-phosphate

**biotin-11-dUTP:** biotin-labeled dUTP

**bp:** base pairs

**BSA:** bovine serum albumin

**C:** cytosine, one of the four purine bases of DNA

**CAT:** chloramphenicol acetyltransferase

**cDNA:** complementary DNA

**CHEF:** contour-clamped homogeneous electric field, a method of pulse-field gel electrophoresis

**CHO:** Chinese hamster ovary cells, a popular cell line

**CMV:** cytomegalovirus

**complexity:** an estimate for the number of different sequences in a DNA solution

**cpm:** counts per minute, a unit for measuring radioactivity

**CTAB:** cetyltrimethyl-ammonium bromide

**6-cutter:** restriction enzyme that recognizes a sequence of 6 base pairs

**D, Da:** dalton, 1 dalton is one-twelfth the mass of a $^{12}C$ atom

**DAB:** 3,3′-diaminobenzidine

**DAPI:** 4′,6-diamino-2-phenylindol

**dATP:** deoxyadenosine triphosphate

**dCTP:** deoxycytosine triphosphate

**DDBJ:** DNA Data Bank of Japan

**ddNTP:** dideoxynucleotide (e.g., ddATP)

**DEAE:** diethylaminoethanol

**DEPC:** diethylpyrocarbonate; used to make $H_2O$ RNase free

**dGTP:** deoxyguanidine triphosphate

**DMF:** dimethylformamide

**DMSO:** dimethylsulfoxide

**DNA:** deoxyribonucleic acid

**DNase:** deoxyribonuclease

**dNTP:** deoxynucleotide triphosphate; includes dATP, dCTP, dGTP, and dTTP

**dsDNA:** double-stranded DNA

**DTT:** dithiothreitol, a reducing agent

**dTTP:** deoxythymidine triphosphate

**dUTP:** deoxyuridine triphosphate

**email:** electronic mail

**EDTA:** ethylene diamine tetra-acetic acid, a chelator of such divalent cations as $Ca^{2+}$ or $Mg^{2+}$

**ELISA:** enzyme-linked immunosorbent assay

**EMBL:** European Molecular Biology Laboratory, a research institute with various branches throughout Europe
**EtOH:** ethanol, a beloved drug

**FACS:** fluorescence-activated cell sorting
**FITC:** conjugated fluorescein isothiocyanate, a fluorescent dye
**fluorochrome:** another word for a fluorescent dye
**FPLC:** fast-performance liquid chromatography

*g*: gravity
**G:** guanosine, one of the four purine bases of DNA
**gDNA:** genomic DNA
**GFP:** green fluorescent protein, a protein in the jellyfish, *Aequorea victoria*, which is fully nontoxic for mammalian cells and fluoresces under ultraviolet light.

**HEK293:** human embryonal kidney cells, a popular cell line
**HEPES:** *N*-(2-hydroxyethyl)piperazine-*N'*-2-ethane sulfonic acid, a buffer
**HPLC:** high-performance liquid chromatography
**HRP:** horseradish peroxidase
**hsDNA:** herring-sperm DNA

**IPTG:** isopropylthiogalactoside; induces the expression on the *lac* promoter

**kb:** kilobase, 1000 base pairs
**kDa:** kilodalton

**LB:** Luria broth, a bacterial medium

**MIPS:** Martinsreid Institute for Protein Sequences
**MMLV:** Moloney murine leukemia virus
**MOI:** multiplicity of infection
**MOPS:** 3-(*N*-morpholino) propane sulfonic acid, a buffer
**mRNA:** messenger RNA

**NBT:** nitroblue tetrazolium
**NCBI:** National Center for Biotechnology Information
*NEO*: neomycin resistance gene
**NP-40:** Nonidet P-40, a detergent
**nt:** nucleotide
**NTP:** nucleoside triphosphate

**OD$_{260}$:** optical density at a wavelength of 260 nm
**oligo-dT:** oligo-deoxythymidine
**ONPG:** *O*-nitrophenyl-β-D-galactopyranoside
**ORF:** open reading frame
**ori:** origin of replication
**PAC:** P1 artificial chromosome, an artificial P1 chromosome and a cloning vector
**PAGE:** polyacrylamide gel electrophoresis

**PCR:** polymerase chain reaction
**PEG:** polyethylene glycol
**PFGE:** pulse-field gel electrophoresis, a method for separating large DNA fragments
**pfu:** plaque-forming units
**PIPES:** piperazine-$N,N'$-bis(2-ethane sulfonic acid), a buffer
**poly A+:** polyadenyl-
**primer:** oligonucleotide that serves as a start fragment for DNA synthesis; the term is most frequently used in relation to the polymerase chain reaction

**RACE:** rapid amplification of cDNA ends, a polymerase chain reaction application
**RCF:** relative centrifugal force (measured in $g$, force of gravity)
**RFLP:** restriction fragment length polymorphism
**RNA:** ribonucleic acid
**RNase:** ribonuclease
**rpm:** revolutions per minute
**rRNA:** ribosomal RNA
**RT:** reverse transcriptase

**screening:** searching through large amounts of material (e.g., phage banks, patient DNA) for something that is expected to be found there
**SDS:** sodium dodecyl sulfate, a detergent
**SNP:** single nucleotide polymorphisms, naturally existing polymorphisms in a gene within a population
**specific activity:** provides information about how many markers are inserted per 1µg of DNA or RNA; the higher the specific activity, the more sensitive is the probe
**ssDNA:** single-stranded DNA
**supercoiled:** describes a condition involving circular DNA (e.g., plasmids), in which the DNA has so many additional rotations that it looks like a twisted rubber band
**SV40:** simian vacuolating virus 40

**T:** thymine, one of the four purine bases of DNA
**TAE:** Tris-acetate-EDTA buffer, an electrophoresis buffer
**Taq polymerase:** *Thermus aquaticus* DNA polymerase, primarily used for the polymerase chain reaction
**TB:** Terrific broth, a bacterial medium
**TBE:** Tris-borate-EDTA buffer; an electrophoresis buffer
**TE:** Tris-EDTA buffer
**TEMED:** $N,N,N',N'$-tetramethyl-ethylenediamine, a starter in the polymerization of acrylamide
**template:** draft, matrix; template DNA is the DNA employed as a "master copy" in the polymerase chain reaction or in labeling
**Tet:** tetracycline, an antibiotic
**$T_m$:** melting temperature; in nucleic acids, the temperature at which one half of the base pairs of a double-strand are separated
**Tris:** tris(hydroxymethyl)aminomethane, the most important buffer employed in molecular biology
**tRNA:** transfer RNA
**TTE:** Tris taurine EDTA buffer, an electrophoresis buffer

**u:** unit, quantity of measurement used for enzymes
**U:** uracil
**UTR:** untranslated region, of the mRNA
**UV:** ultraviolet

**v/v:** volume/volume; for example, 15% (v/v) = 150 mL per liter of a solution
**vortex:** mix well with the aid of a Vortex (reagent glass shaker) or some similar device

**w/v:** weight/volume; for example, 15% (w/v) = 150 g per liter of solution

**X-Gal:** 5-bromo-4-chloro-3-indolyl-β-D-galactoside, substrate of the β-galactosidase

**YAC:** yeast artificial chromosome, an artificial yeast chromosome and a cloning vector

# Appendix 2

## Suppliers

In the life sciences, things function as in life itself: It is chaos. The firms purchase one another, merge, and reorganize repeatedly. This is unfortunately true for the suppliers listed here. I have made an attempt to reflect the current state of affairs, although almost one-half of these addresses will no longer be correct within 2 years. *Panta rhei* (from the Greek, meaning "everything flows")!

**Affymetrix, Inc**.: 3380 Central Expressway, Santa Clara, CA 95051, USA (http://www. affymetrix.com)

**AGS** Angewandte Gentechnologie Systeme: see ThermoHybaid

**Aldrich**: see Sigma-Aldrich.

**Ambion, Inc.**: part of AMS Biotechnology GmbH. 2130 Woodward St., Austin, TX 78744-1832, USA (http://www.ambion.com)

**Amersham Pharmacia Biotech**: see GE Healthcare

**AMS Biotechnology**: GmbH: Henkelstr. 15, 65187 Wiesbaden, Germany (http://www.amsbio.com)

**Amicon**: see Millipore (http://www.millipore.comlamicon)

**Applied Biosystems** 850 Lincoln Centre Dr., Foster City, CA 94404, USA

**Appligene**: see QBiogene

**Bachofer** GmbH: Carl-Zeiss-Str. 35,72770 Reutlingen, Germany (Phone: 0-7121-54008; Fax 0-7121-54000)

**Becton Dickinson** Labware: Two Oak Park Drive, Bedford, MA USA 01730-990 (http://www.bdbiosciences.com)

**Bender & Hobein** GmbH: Fraunhoferstr. 7, 85737 Ismaning bei Munich, Germany (Phone: 0-89-996548-0; Fax: 0-89-996548-91)

**Biometra** GmbH: Rudolf-Wissell-Str. 30, 37079 Göttingen, Germany (http://www.biometra.de)

**BioRad** Laboratories: 1000 Alfred Nobel Drive, Hercules, CA 94547, USA

**BioWhittaker** Molecular Applications: 191 Thomaston Street, Rockland, ME 04841, USA (http://www.bmaproducts.com)

**Biozym** Diagnostik GmbH: Postfach, 31833 Hess, Oldendorf, Germany (http://www.biozym.com)

**Boehringer Mannheim**: see Roche Applied Science

**BTX**: part of QBiogene (http://www.BTXonline.com)

**Calbiochem**-Novabiochem: see EMD Biosciences.

**Cambrex** Corporation, One Meadowlands Plaza, East Rutherford, New Jersey 07073 (http://www.cambrex.com/bioproducts)

**Canberra Packard** GmbH: see Perkin Elmer

**Clontech**: 1290 Terra Bella Avenue, Mountain View, CA 94043, USA (http://www.clontech.com)

**Du Pont**: (http://www. dupont.com)
**DYNAL**: see Invitrogen

**Eastman Kodak**: part of Integra Biosciences (http://www.kodak.de)
**EMD Biosciences**, Inc., 10394 Pacific Center Court, San Diego, CA 92121, USA (http://www. merckbiosciences.com)
**Eurobio**: part of Fisher Scientific (http://www.eurobio.fr)
**Eppendorf** Vertrieb Deutschland GmbH: Peter-Henlein-Str. 2, 50389 Wesseling-Berzdort, Germany (http://www.eppendorf.com or http://www.eppendorf.de)

**Falcon**: see Becton Dickinson
**Fermentas** Inc., 7520 Connelley Drive, Hanover, MD 21076, USA (http://www.fermentas.com)
**Fisher Scientific** International Inc: Liberty Lane, Hampton, NH 03842, USA (http://www. fishersci.com)
**FMC** Bioproducts: see Cambrex
**Fuji**: part of Raytest (http://www.fujifilm.de)

**GE Healthcare**, 800 Centennial Avenue, P.O. Box 1327, Piscataway, NJ 08855-1327, USA
**GenomeSystems**: see Incyte (http://www.genomesystems.com)
**Gibco BRL**: see Invitrogen

**Hoefer** Scientific Instruments: part of Amersham Pharmacia Biotech
Thermo **Hybaid**: see Thermo Electron

**ICN Biomedicals**: PO Box 19536, Irvine CA 92713, USA (http://www.icnbiomed.com)
**Incyte** Corporation: 3160 Porter Drive, Palo Alto, CA 94304, USA (www.incyte.com)
**Integra Biosciences** GmbH: Ruhberg 4, 35463 Fernwald (http://www.integra-biosciences.com)
**Invitrogen** Corporation, 1600 Faraday Avenue, PO Box 6482, Carlsbad, California 92008, USA (http://www.invitrogen.com)

**Kendro**: see Thermo Electron (http://www.kendro.com)
**Kodak**: part of Integra Biosciences (http://www.kodak.de; www.kodak.com/go/scientific)

**Labtech** International GmbH: Obere Hauptstr. 67, 09235 Burkhardtsdorf, Germany (http://www. labtech-international.de)
**LI-COR**: part of MWG Biotech (http://www.licor.com; bio.licor.com)
**Life Technologies**: see Invitrogen
**LKB Instruments**: part of Pharmacia
**Lumigen**: (http://www.lumigen.com)

**Macherey-Nagel** GmbH: Postfach 101352, 52313 Düren, Germany (http://www.macherey-nagel.com)
**MBI Fermentas** GmbH: Opelstr. 9, 68789 St. Leon-Rot, Germany (http://www.fermentas.com)
**Merck** KGaA: Frankfurterstr. 250, Postfach 4119, 64293 Darmstadt, Germany (http://www.merck.de)
**Millipore**: 290 Concord Road, Billerica, MA 01821, USA (http://millipore.com)
**Molecular Probes**: see Invitrogen
**Molecular Dynamics**: see Amersham Pharmacia Biotech (http://www.mdyn.com)
**MWG-Biotech** Gesellschaft für angewandte Biotechnologie GmbH: Anzinger Str. 7, 85560 Ebersberg, Germany (http://www.mwgdna.com)

**Nalgene**: see Nalge
**Nalge Nune** International: 75 Panorama Creek Drive, Rochester, NY 14625, USA (http://www.nalgene.com)
**New Brunswick** Scientific Co., Inc.: Box 4005, 42 Talmadge Road, Edison, NJ 08818-4005, USA (http://www.nbsc.com)
**New England Biolabs**: 240 County Road, Ipswich, MA 01938-2723, USA (http://www.neb.com)
**Novagen**: part of Calbiochem (http://www.novagen.com)
**Novex** Electrophoresis: see Invitrogen
**Nunc:** see Nalge

**Oncor**: see QBiogene

**Pall Filtron** 2200 Northern Boulevard, East Hills, NY 11548, USA (http://www.pall.com)
**PE Biosystems**: see Applied Biosystems
**Peqlab** Biotechnologie GmbH: Carl-Thiersch-Str. 2b, 91052 Erlangen, Germany (http://www.peqlab.de)
**Perkin Elmer** Life and Analytical Sciences: 710 Bridgeprot Avenue, Shelton, CT 06484-4794, USA (http://las.perkinelmar.com)
**PerSeptive Biosystems**: see Applied Biosystems
**Pharmacia**: see Amersham Pharmacia Biotech
**Promega** Corporation: 2800 Woods Hollow Road, Madison, WI 53711, USA (http://www.promega.com)

**QBiogene**: Waldhofer Str. 102, 69123 Heidelberg, Germany (http://www.qbiogene.de)
**Qiagen** Inc.: 27220 Turnberry Lane, Valencia, CA 91355, USA (http://www.qiagen.com)
**Quantum Biotechnologies**: see QBiogene

**Raytest** Isotopenmessgeräte GmbH: Benzstrasse 4, 75334 Straubenhardt, Germany (http://www.raytest.de)
**Research Genetics**: see also Invitrogen; 2130 Memorial Pkwy, SW, Huntsville, AL 35801, USA (http://www.resgen.com)
**Roche** Applied Science, P.O. Box 50414, 9115 Hague Road, Indianapolis, IN 46250-0414
**Roche Diagnostics** GmbH: Sandhofer Str. 116, 68305 Mannheim, Germany (http://biochem.roche.com)
**Carl Roth** GmbH: Schoemperlenstr. 1-5, 76161 Karlsruhe, Germany (http://www.carl-roth.de)

**Sarstedt** AG: Postfach 1220, 51582 Nümbrecht, Germany (http://www.sarstedt.com)
**Sartorius** AG: Weender Landstrasse 94-108, 37075 Göttingen, Germany (http://www.sartorius.de or http://www.sartorius.com)
**Schleicher & Schüll**: Postfach 4, 37586 Dassel, Germany (http://www.s-und-s.de)
**Serva** Electrophoresis GmbH: Postfach 105 260, Carl-Benz-Str. 7, 69115 Heidelberg, Germany (http://www.serva.de)
**Sigma-Aldrich**, 3050 Spruce St., St Louis, MO 63103, USA
**Sigma** Chemie: Grünwalder Weg 30, Postfach, 82039 Deisenhofen, Germany (http://www.sigma-aldrich.com)
**Sorvall**: see Kendro (http://www.sorvall.com)
**Stratagene**, 11011 N. Torrey Pines Road, La Jolla, CA 92037, USA (http://www.stratagene.com)

**Techne**: part of Labtech International (http://www.techneuk.co.uk)
**Thermo-Dux** Gesellschaft für Laborgeräte mbH: Postfach 1622, 97866 Wertheim, Germany

**Thermo Electron** Corporation, 81 Wyman Street, Waltham, MA 02454, USA (http://www.thermohybaid.com)

**ThermoHybaid**: Sedanstr. 10. 89077 Ulm, Germany (Phone: 0-731-93579 290; Fax 0-731-93579 291) (http://www.thermohybaid.com)

**Tropix**: see Applied Biosystems (http://www.appliedbiosystems.com/tropix)

**USB**: part of Amersham Pharmacia Biotech (http://www.usbweb.com)

**Whatman**: see Biometra (http://www.whatman.co.uk; http://www.whatman.com)

A list of suppliers of biotechnologic products can be found in a special issue (volume 14, 2005) of *Nature Biotechnology*.

# Recommended Literature

Ausubel F et al. (eds). Current Protocols in Molecular Biology. New York, John Wiley & Sons. (This is definitely *the* laboratory handbook. It is three files that are updated quarterly.)

Ausubel F et al. (eds). (1995) Short Protocols in Molecular Biology. A Compendium of Methods from Current Protocols in Molecular Biology. New York, Wiley.

Beier M. (1996) Neue Strategien zum Aufbau von RNA- und DNA-Oligonucleotiden.

Bielka H, Bömer T. (1995) Molekulare Biologie der Zelle. Spektrum Akademischer Verlag. (This is no longer available.)

Brown TA. (1999) Moderne Genetik. Eine Einführung. Spektrum Akademischer Verlag.

Brown TA. (2002) Gentechnologie für Einsteiger. 3. Auflage. Spektrum Akademischer Verlag.

Caetano-Anolles G, Gresshoff PM (eds). (1997) DNA Markers. Protocols, Applications, and Overviews. New York, Wiley. (This is a good source for protocols.)

Campbell NA. (1997) Biologie. Spektrum Akademischer Verlag. (This is the way textbooks should be written.)

Clark M. (1996) In Situ Hybridisation. London, Chapman & Hall.

Drlica K. (1995) DNA und Genklonierung. Ein Leitfaden. Spektrum Akademischer Verlag. (This book is no longer available.)

Ganten D, Ruckpaul K (ed). (1997) Molekular und zellbiologische Grundlagen. In: Handb. d. molekularen Med. Gentherapie 1. Springer.

Glover DM, Harnes DB. (eds). (1996) DNA Cloning: The Practical Approach Series, New York, Oxford University Press. (This is a highly recommended series with various titles concerned with the techniques of biotechnology, such as polymerase chain reaction, gel electrophoresis, and cell cultures.)

Ibelgaufts H. (1993) Gentechnologie von A bis Z. Studienausgabe. New York, Wiley. (This book functions as the small version of the *Encyclopedia of Molecular Biology and Molecular Medicine*.)

Kessler C. (1990) Nachweis von DNA. Eine Einführung in das Digoxigenin-System. In: Driesel AJ (ed):. Hüthig/HVS.

Knippers R. (1995) Molekulare Genetik. Thieme.

Lee HH, Morse SA, Olsvik Ö. (1997) Nucleic Acid Amplification Technologies. Application to Disease Diagnosis. Birkhäuser.

Lindl T, Bauer J. (2002) Zell- und Gewebekultur, 5th ed. Spektrum Akademischer Verlag. (This is the standard work on cell culturing techniques.)

Lucotte G, Baueyx F. (1993) Introduction to molecular cloning techniques. New York, Wiley.

Martin BM. (1994) Tissue Culture Techniques: An Introduction. Birkhäuser.

Meyers RA. (ed), Encyclopedia of Molecular Biology and Molecular Medicine. New York, Wiley. (This is an excellent book, but it is somewhat too expensive.)

Nucleic Acids Research. Oxford University Press. (This is an interesting journal, in which interesting methods and variations of known methods are published. The editor obliges experimenters by including papers discussing methods, which are indicated separately in the publication's contents.)

Osterman LA. Methods of Protein and Nucleic Acid Research. Springer.

Pingoud A, Urbanke C. (1997) Arbeitsmethoden der Biochemie. de Gruyter.

Promega. Protocols and Applications Guide. (Promega has published their own handbook for many years, and it can be obtained free of charge on request.)

Pühler A (ed). (1993) Genetic Engineering of Animals. New York, Wiley.

Rehm H. (2002) Der Experimentator: Proteinbiochemie/Proteomics, 4th ed. Spektrum Akademischer Verlag. (This is the companion to this textbook. The English version should be available soon from Elsevier.)

Rickwood D, Harris JR (eds). (1996) Cell Biology. Essential Techniques. New York, Wiley.

Sambrook J, Fritsch EF, Maniatis T. (1989) Molecular Cloning: A Laboratory Manual. New York, Cold Spring Harbor Laboratory Press. (This is a classic among laboratory handbooks, although somewhat antiquated.)

Sambrook J, Russell DW. (2001) Molecular Cloning: A Laboratory Manual, 3rd ed. New York, Cold Spring Harbor Laboratory Press. (This is the new version of the classic *Maniatis*.)

Schleif RF, Wensink PC. (1981) Practical Methods in Molecular Biology. Springer.

Schmidt ER, Hankeln T. (eds). (1996) Transgenic Organisms and Biosafety. Horizontal Gene Transfer, Stability of DNA, and Expression of Transgenes. Springer.

Schrimpf G (ed). (2002) Gentechnische Methoden. Eine Sammlung von Arbeitsanleitungen für das molekularbiologische Labor. Spektrum Akademischer Verlag. (This is the German *Maniatis*; nice protocols zum Nachkochen.)

Scott TA, Mercer EI (eds). (1997) Concise Encyclopedia Biochemistry and Molecular Biology. de Gruyter.

Seyffert W et al. (eds). (1998): Lehrbuch der Genetik. Spektrum Akademischer Verlag. (This is the most complete textbook on genetics available on the German market.)

Spektrum Akademischer Verlag. (This is the companion to this text book. The English version should be available soon from Elsevier.)

Tschesche H (ed). (1990) Modern Methods in Protein and Nucleic Acid Research. Review Articles. de Gruyter.

Watson J. (1994) The polymerase chain reaction. In: Mullis KB, Ferre F, Gibbs RA (eds): Birkhäuser.

Westermeier R. (1997) Electrophoresis in Practice. A Guide to Methods and Applications of DNA and Protein Separations. New York, Wiley.

Winnacker EL. (1987) From Genes to Clones. Introduction to Gene Technology. New York, Wiley.

# Reading Material for Leisure Time

Bär S. (1993) Forschen auf Deutsch. Der Machiavelli für Forscher- und solche, die es noch werden wollen. Verlag Harri Deutsch.

Chargaff E. (1995) Ein zweites Leben. Klett-Cotta.

Dahl J. (1996) Die Verwegenheit der Ahnungslosen. Über Genetik, Chemie und andere schwarze Löcher des Fortschritts. Klett-Cottal/SVK.

Ridley M (ed). (1995) Darwin lesen. dtv.

Stümpke H. (2001) Bau und Leben der Rhinogradentia. 2. Auflage. Spektrum Akademischer Verlag.

Watson JD. (1997) Die Doppelhelix. Ein persönlicher Bericht über die Entdeckung der DNS-Struktur.

# Index